AF618674

Birkhäuser

Operator Theory: Advances and Applications

Volume 253

Founded in 1979 by Israel Gohberg

Volodymyr Koshmanenko • Mykola Dudkin

The Method of Rigged Spaces in Singular Perturbation Theory of Self-Adjoint Operators

Volodymyr Koshmanenko
Institute of Mathematics
National Academy of Sciences of Ukraine
Kyiv, Ukraine

Mykola Dudkin
Kyiv Polytechnic Institute
National Technical University of Ukraine
Kyiv, Ukraine

Translated by Nataliia Koshmanenko

Original Ukrainian edition published by Institute of Mathematics, NAS of Ukraine, Kyiv, 2013

ISSN 0255-0156 ISSN 2296-4878 (electronic)
Operator Theory: Advances and Applications
ISBN 978-3-319-29533-6 ISBN 978-3-319-29535-0 (eBook)
DOI 10.1007/978-3-319-29535-0

Library of Congress Control Number: 2016940830

Mathematics Subject Classification (2010): 47A10, 47A55, 28A80

Printed on acid-free paper

This book is published under the trade name Birkhäuser
The registered company is Springer International Publishing AG Switzerland (www.birkhauser-science.com)

To *Yu.M. Berezansky,*

the mathematician
with the inexhaustible
creating energy

Contents

Preface

This book deals with the singular perturbation theory for self-adjoint operators. Our approach is based on the method of rigged Hilbert spaces and employs the theory of singular quadratic forms. More precisely, the book is concerned with singular quadratic forms and singular perturbations of self-adjoint operators considered in rigged Hilbert spaces.

Among the mathematical objects and notions treated the three major ones are: Hilbert spaces, self-adjoint (or symmetric) operators, and quadratic forms.

These objects are closely related to each other. For instance, the inner product in a Hilbert space is always generated by a positive quadratic form. Each closed bounded from below quadratic form in a Hilbert space is naturally associated with a self-adjoint operator. A positive self-adjoint operator can be used, via the corresponding quadratic form, to change the inner product and construct a new Hilbert space. This ring of transformations can be extended by resorting to a wide arsenal of mathematical tools.

Standard courses on functional analysis, theory of linear operators, and mathematical physics treat the above-mentioned objects in connection with various problems and applications (see, e.g., the classical monographs by N.I. Akhiezer and I.M. Glazman [32], N. Dunford and J. Schwartz [75], and M. Reed and B. Simon [169]–[172]).

However, unexpected difficulties arise when one brings into consideration singular perturbations. The singularity phenomenon leads to essentially new features of standard mathematical objects and produces a series of problems, some unsolved until now.

In a wide sense, a singular perturbation means a very small change in the starting object. So small that an operator, quadratic form, or inner product in a Hilbert space stay the same almost everywhere (on a dense subset). Only naively one may think that such changes have no significant effects.

There is a rich literature devoted to singular perturbations of Schrödinger operators (see, e.g., the book by S. Albeverio, F. Gesztesy, R. Høegh-Krohn, and H. Holden [7] and references therein), where a series of explicitly solvable models are shown to exhibit a number new interesting features. The typical example is connected with one-point perturbations of the Laplace operator, $-\Delta + \lambda\delta$. Various generalizations, considered in the monograph by S. Albeverio and P. Kurasov

[23], involve differential operators of Schrödinger type. However, despite the wave of activity in this direction, so far no conventional universal approach for the satisfactory treatment of singular perturbations is available.

In this monograph, a new approach to the singular perturbation theory is developed. It essentially uses ordered triples of Hilbert spaces of the form $\mathcal{H}_- \supset \mathcal{H}_0 \supset \mathcal{H}_+$. These triples are treated as rigged Hilbert spaces, and provide a very fruitful tool of analysis. In particular, specific constructions of rigged Hilbert spaces have been already used in the study of the continuous spectrum of linear operators. The rigged Hilbert spaces method proved its efficiency in a wide class of models with explicitly given singular perturbations.

The material included in this book has been presented at several international conferences. Parts of the book were used in courses for advanced students in Kyiv State University, National Technical University of Ukraine (KPI), and National University of Kyiv-Mohyla Academy.

Acknowledgment

Most ideas concerning the topics presented in this book were discussed with S. Albeverio, W. Karwowski, P. Kurasov, and L. Nizhnik. We thoroughly enjoyed the collaboration with them and are grateful for many helpful suggestions and remarks. We were also strongly influenced by discussions with P. Cojuhari, P. Exner, F. Gesztesy, K. Makarov, H. Neidhardt, B. Pavlov, A. Posilicano, S. Kuzhel. We express them our warmest thanks.

We are grateful for support to the Institute of Mathematics in Kyiv, Institut für Angewandte Mathematik, Universität Bonn, Wroclaw University, and Banach Center in Warsaw.

Introduction

The singular perturbation theory for self-adjoint operators appeared as the newest part of perturbation theory for linear operators, which, together with the spectral theory, occupies a central place in functional analysis and mathematical physics.

There are a great number of publications devoted to perturbation theory problems and related applications. Here we cite only some of them, [23, 45, 75, 154, 166, 169–173, 178, 182, 183]. The best presentation of perturbation theory for linear operators have been done by T. Kato in his fundamental monograph [107].

The purpose of our monograph is to present a consistent exposition of the method of rigged Hilbert spaces in the theory of singular perturbation as a new approach to the construction of perturbed operators. This approach improves and develops the method of self-adjoint extensions of symmetric operators.

The key idea of our approach is to use a given singular perturbation in order to modify the starting rigged Hilbert space associated with the free Hamiltonian. In general, inner products and norms in Hilbert spaces are sensitive to changes on sets of zero Lebesgue measure, with more flexibility than for linear operators. So, instead of the standard procedure, where the perturbed operator is defined by the operator sum of the free Hamiltonian and a potential (often such a sum does not make sense), according to the method of rigged spaces, a singular perturbation is directly used to modify the inner products in the Hilbert spaces belonging to the rigging of the original state space of the free physical system. Then the perturbed Hamiltonian appears as the operator uniquely associated with the new rigged space.

We recall that the main problem of perturbation theory is to construct and investigate the spectral properties of the operator $H = H_0 + V$, where H_0 denotes a free Hamiltonian of some unperturbed physical system and V corresponds to a relevant perturbation. For example, in a single-particle model, $H_0 = -\Delta$ (the Laplace operator) and V describes the potential interaction with some external source. In the usual setting when V is a well-behaved object, the perturbation theory is completely developed. However, the problem of defining H becomes highly nontrivial in the case of singular perturbations.

Singularity means that V is concentrated on an extremely small set. More precisely, the singular potential is usually supported on a set of zero Lebesgue measure. The Dirac δ-potential provides a typical example of a singular perturbation.

In fact, the singularity phenomenon has a very deep origin and uncertain treating. So, for the so-called super-singular perturbations the problem of constructing the operator H looks in general unsolvable, since the support of the perturbation is so small that the free Hamiltonian H_0 is not influenced on its domain of essential self-adjointness.

This book is focused on a class of problems from mathematical physics which have been partly studied in numerous publications and several books, such as: "Solvable Models in Quantum Mechanics" by S. Albeverio, F. Gesztesy, R. Høegh-Krohn, and H. Holden (the first edition was published in 1988 and the second, with the appendix by P. Exner [7], in 2005), "Singular Bilinear Forms in Perturbation Theory of Self-adjoint Operators" (in Russian) by V. Koshmanenko [134], "Singular Perturbations of Differential Operators. Solvable Schrödinger Type Operators" by S. Albeverio and P. Kurasov [23], and "Singular Quadratic Forms in Perturbation Theory" by V. Koshmanenko [135].

All the aforementioned monographs actually discuss the same problem in various settings: how to construct the singularly perturbed operator and investigate its spectrum.

Let us recall that the singular perturbation theory started with the original problem connected with the motion of a one-dimensional particle in a periodic potential supported by isolated points. The corresponding model was very popular in solid state physics. The mathematical setting of this problem leads to the formal expression $-\Delta + \lambda\delta$. What is the self-adjoint operator corresponding to this sum in the space L_2? More specifically, this model is associated with the Schrödinger operator of the form

$$H = -\Delta + \sum_{i \in \mathbb{N}} \lambda_i \delta_{y_i}, \quad \lambda_i \in \mathbb{R}, \tag{1}$$

where $-\Delta$ denotes the Laplace operator in $L_2(\mathbb{R}^n)$, $n \geq 1$, and where the singular potential is presented as series in Dirac δ-functions δ_{y_i} concentrated at points $y_i \in \mathbb{R}^n$. This problem was known already in 1931 as the Kronig–Penney model describing the motion of an electron in a solid crystal lattice. This physically meaningful model dealing with potentials of zero range generated a significant mathematical problem.

The number of publications relating to Hamiltonians of the form (1) and more general ones, quickly increased. The fundamental difficulties caused by attempts to take into account singular perturbations, became a challenge for many physicists and mathematicians.

It should be clarified that "singular perturbation" is a mathematically meaningful term for an object having the following typical property. Its representation in the form of a linear functional, quadratic form, or operator in an appropriate Hilbert space, is always equal to zero on a dense linear subset. So, the non-trivial problem is, given a free Hamiltonian H_0 and some singular perturbation V, to determine a reasonable self-adjoint operator H in such a way that H and H_0 coincide almost everywhere.

The approximation approach, which is natural and highly fruitful in the standard perturbation theory, was also applied to treat some models with singular perturbations. However, this method has a rather limited success, especially in cases with singular potentials. Moreover, in a number of physically realistic models one needs additionally to apply the so-called renormalization procedure, otherwise the approximation method leads to ambiguous values. So, new ideas were required.

Such an idea appeared in 1961. In a short note [53], F. Berezin and L. Faddeev proposed to describe the singularly perturbed Hamiltonian $H = -\Delta + \lambda\delta(x)$, $x \in \mathbb{R}^3$ in terms of the theory of self-adjoint extensions of symmetric operators.

Surprisingly, the theory of self-adjoint extensions of symmetric operators, originally created by J. von Neumann, K. Friedrichs, and M.G. Kreĭn, was developed completely independently of the singular perturbation theory. Only in 1961 it becomes clear that for zero-range potentials one can naturally associate a continual family of perturbed operators which admit a parametrization analogous to the description of all self-adjoint extensions of a positive symmetric operator. The understanding of this fact generated a huge number of publications with new observations and deep results.

Nevertheless, a comprehensive theory of singular perturbations for self-adjoint operators is still missing. One of the reasons is that different approaches to the construction of singularly perturbed operators are yet inconsistent and there exist distinct ways for parametrization of these operators. The most significant achievements of the theory are connected with a large number of explicitly solvable models and original methods for solving specific problems. However, there are few general abstract results. Among the wide collection of tools and ways for determining the perturbed operator, the method of self-adjoint extensions of symmetric operators is the most productive. Moreover, there are a series of unsolved problems in the most difficult situations related, for example, to a class of so-called super-singular perturbations. In the latter case, perturbations have supports so small that the functions, defined beyond these supports, form a domain of essential self-adjointness for the free Hamiltonian. How to take into account the non-trivial effect of such perturbations? This question remains without conventional answer until now.

Thus, it should be again emphasized that in a contrast to the standard perturbation theory (systematically presented by T. Kato), the singular perturbation theory still lacks a consistent mathematical formulation.

This book is intended to fill this gap to some extent. To this end we use and develop a rather powerful tool, the method of rigged Hilbert spaces. This method have already proved its efficiency in the spectral theory of self-adjoint operators. And it is no doubt about its prospects in the theory of singular perturbation. However, the time for a consistent and complete presentation of the theory, perhaps, has not come yet.

Thus, our book is devoted to the setting out of a new approach to the problem of constructing singularly perturbed operators. This approach is based on the consideration of a chain of Hilbert spaces, which yields a so-called rigged Hilbert space. More precisely, resorting to specific triplets of Hilbert spaces we combine two important tools, the theory of self-adjoint extensions of symmetric operators and the theory of singular quadratic forms. Constructions with linear operators and quadratic forms are more productive in rigged spaces than in a single Hilbert space.

Let us briefly explain the essence of our approach.

It is well known that each bounded from below closed quadratic form γ on a Hilbert space $\mathcal{H}$ is uniquely associated with a self-adjoint operator $A = A^*$. This fact is usually expressed as $\gamma(\varphi,\psi) = (A\varphi,\psi)$, where $(\cdot,\cdot)$ denotes the inner product in $\mathcal{H}$. We assume for simplicity that $A \geq 1$. Then the operator domain $\operatorname{Dom} A$ constitutes a Hilbert space $\mathcal{H}_+$ with the positive inner product $(\varphi,\psi)_+ = (A\varphi, A\psi)$. $\mathcal{H}_+$ is densely and continuously embedded in $\mathcal{H}_0$. The completion of $\operatorname{Ran} A$ with respect to the negative inner product $(f,g)_- := (A^{-1}\varphi, A^{-1}\psi)$ gives a new Hilbert space $\mathcal{H}_-$ which includes $\mathcal{H}$ densely and continuously. Thus, there arises the triple

$$\mathcal{H}_- \supset \mathcal{H} \supset \mathcal{H}_+ = \operatorname{Dom} A,$$

which is called the *rigged Hilbert space* associated with the operator A. That is, $\mathcal{H}_-$ coincides with the dual space to $\mathcal{H}_+$. This fact ensures the existence of an isometric mapping $D_{-,+} : \mathcal{H}_+ \to \mathcal{H}_-$, called the Berezansky canonical isomorphism. In fact, $D_{-,+} = A^{\mathrm{cl}}A$, where A^{cl} denotes the closure of the map $A : \mathcal{H}_0 \to \mathcal{H}_-$. It is important that A can be restored from $D_{-,+}$. We generalize the latter fact in regard to the singularly perturbed operator denoted by $\tilde{A}$. Namely, in the method of rigged spaces, $\tilde{A}$ arises as an operator associated with a new equipped space

$$\tilde{\mathcal{H}}_- \supset \mathcal{H} \supset \tilde{\mathcal{H}}_+ = \operatorname{Dom} \tilde{A}$$

and defined by the canonical Berezansky isomorphism $\tilde{D}_{-,+} : \tilde{\mathcal{H}}_+ \to \tilde{\mathcal{H}}_-$. In turn, the spaces $\tilde{\mathcal{H}}_-$, $\tilde{\mathcal{H}}_+$ appear after changing the inner products in $\mathcal{H}_-$ and $\mathcal{H}_+$ by means of the singular perturbation.

The crucial step is to use the mathematical expression that defines a singular perturbation in order to introduce new inner products and corresponding positive or negative spaces $\tilde{\mathcal{H}}_+$ and $\tilde{\mathcal{H}}_-$. So, the perturbed operator $\tilde{A}$ appears directly through the Berezansky canonical isomorphism in a new rigged space, but not as a usual perturbation of the initial operator A.

More specifically, it is convenient to first construct the inverse operator $\tilde{A}^{-1}$ as a restriction of the Berezansky canonical isomorphism $\tilde{I}_{1,-1} : \tilde{\mathcal{H}}_{-1} \to \tilde{\mathcal{H}}_1$, and then to define the perturbed operator $\tilde{A}$ associated with the rigged space $\tilde{\mathcal{H}}_- \supset \mathcal{H} \supset \tilde{\mathcal{H}}_+$. Here $\tilde{\mathcal{H}}_{-1}$ and $\tilde{\mathcal{H}}_1$ are produced as perturbations of the inner products $(\cdot,\cdot)_1 = (A\cdot,\cdot)$ and $(\cdot,\cdot)_{-1} = (A^{-1}\cdot,\cdot)$ in $\mathcal{H}_1$ and $\mathcal{H}_{-1}$, respectively. Then for $\tilde{A}^{-1}$ one has the representation by the M. Kreĭn formula

$$\tilde{A}^{-1} = A^{-1} + \tilde{B}, \tag{2}$$

where the self-adjoint operator $\tilde{B}$ on $\mathcal{H}$ is uniquely related to a singular perturbation.

It is worth explaining the meaning of formula (2) from another point of view. We assume that a perturbation of the operator $A = A^* \geq 1$ on $\mathcal{H}$ is given by a quadratic form γ such that the set $\mathfrak{D} = \operatorname{Ker} \gamma$ is dense in $\mathcal{H}$ and contained in $\operatorname{Dom} A$. Then the restriction

$$\mathbf{A} := A \restriction \mathfrak{D} \tag{3}$$

is a symmetric operator in $\mathcal{H}$. In accordance with the Berezin–Faddeev approach, the family $\tilde{A} \in \mathcal{A}_+(\mathbf{A})$ of all self-adjoint extensions of the operator $\mathbf{A}$ can be understand as the family of singularly perturbed operators corresponding to the heuristic sum $\tilde{A} = A \tilde{+} \gamma$. In the abstract situation the formula (2) has the particular meaning of the resolvent $\tilde{R}_z = (\tilde{A} - z)^{-1}$ under the assumption that the point 0 belongs to the resolvent set. Thus, to construct $\tilde{A}$ we need to establish an evident connection between the starting quadratic form γ, as a singular perturbation, and the operator $\tilde{B}$. It is for this aim that we use the rigged Hilbert spaces.

Let us explain this. Assume that the positive and singular on $\mathcal{H}$ form γ is closable in $\mathcal{H}_+$. Assume also that the set $\mathfrak{D} := \operatorname{Ker} \gamma$ is dense in $\mathcal{H}$. It follows that $\mathcal{H}_+$ can be decomposed into the orthogonal sum of subspaces $\mathcal{H}_+ = \mathcal{M}_+ \oplus \mathcal{N}_+$, where $\mathcal{M}_+ = \mathfrak{D}^{\mathrm{cl},+}$ ("cl, +" denotes a closure in $\mathcal{H}_+$). Note that $\mathcal{M}_+$ is also dense in $\mathcal{H}$. Further, since γ is closable in $\mathcal{H}_+$, there exists the positive operator S associated with γ after its closure. Precisely, $(S\varphi, \psi)_+ = \gamma^{\mathrm{cl},+}(\varphi, \psi)$. That is, $\operatorname{Ker} S = \mathcal{M}_+$. The formula

$$\tilde{B} := ASA^{-1}$$

plays an important role in our constructions. Since $\mathfrak{D} = \operatorname{Ker} \gamma$ is dense in $\mathcal{H}$, the subspace $\mathcal{N}_0 = A\mathcal{N}_+$ ($\mathcal{N}_+ = \mathcal{M}_+^{\perp}$ in $\mathcal{H}_+$) has a zero intersection with $\mathcal{H}_+$. This shows that the operator $\tilde{A}$ defined in (2) coincides with A on $\mathfrak{D}$. Therefore, $\tilde{A}$ is a self-adjoint extension of the symmetric operator $\mathbf{A}$ from (3).

Let us briefly describe the contents of the book by chapters.

In the first chapter we recall the definitions of basic concepts and notions concerning spaces, linear operators, and quadratic forms, give a series of preliminary assertions needed for understanding of this monograph, and formulate several well-known but important theorems.

The second chapter is devoted to a brief presentation of the theory of unbounded symmetric and self-adjoint operators, including only some selected facts. In particular, we briefly examine the necessary for what follows information from the theory of extensions of symmetric operators, recall the basic self-adjointness criterion, and introduce the notion of sesquilinear quadratic form. We state the theorem about the operator representation for bounded quadratic forms. Further treatment of these topics can be found in the fundamental monographs [32, 56, 75, 159, 169–172].

In the third chapter we briefly present more specific facts and details from the theory of self-adjoint extensions of symmetric operators. There is a large literature on the theory of extensions for linear operators. Here we refer only to some sources which influenced our work in this field [4, 24, 28, 30, 32, 34, 41, 54, 56, 62–65, 68, 70, 74, 86, 93, 109, 110, 147, 148, 151, 152, 156, 174, 179, 181].

Let us note that importance of the self-adjoint extensions of symmetric operators derives from their wide physical applications. For this reason, we consider only semi-bounded from below symmetric and self-adjoint operators. We use the description of self-adjoint extensions in terms of auxiliary operators and quadratic forms. In addition to the purely theoretical problem of description of all self-adjoint extensions of a symmetric operator, there is the problem of choosing the parameters describing extensions so that they fit various applications. From this point of view, the theory of extensions developed by J. von Neumann is less suitable for applications and spectral analysis. It soon became clear that a more practical theory of extensions is that associated with the names of K. Friedrichs, M. Kreĭn, M. Vishik, and M. Birman. In their version, the parameter of extensions plays the role of abstract boundary conditions. They are usually written in terms of linear functionals, named boundary forms. It is precisely this version of the extensions theory of symmetric operators that we use in our constructions. But it is worth noting that the most popular approach to the extensions theory is based on the so-called abstract boundary triplets (see the papers by V.M. Bruk [64], A.N. Kochubei [109, 110], M.L. Gorbachuk [91–93], V.A. Derkach and M.M. Malamud [73, 74], S.O. Kuzhel [152]).

The first three chapters of the book can be considered as preliminary material. The main content of the monograph begins in Chapter 4.

Starting with Chapter 4 we consistently describe the classical construction of rigged Hilbert spaces, which have been completely developed by P.D. Lax, I. Gel'fand, and Yu. Berezansky (a comprehensive treatment is given in the fundamental monograph by Yu. Berezansky [42]). In our description we slightly simplify and modernize the construction of rigged Hilbert spaces from [42].

A rigged Hilbert space is a triple of spaces

$$\mathcal{H}_- \supset \mathcal{H}_0 \supset \mathcal{H}_+, \tag{4}$$

densely and continuously embedded into each other and satisfying a series of additional requirements. This notion appeared in [153] as a specific version of the topological rigged (equipped) space, $\Phi^* \supset \mathcal{H}_0 \supset \Phi$, which is well known as a Gel'fand triple [87]. In applications, when the rigged Hilbert space is fixed, the space $\mathcal{H}_-$ corresponds to a certain class of generalized functions and $\mathcal{H}_+$ is a completion of the set of test functions in a norm induced by the considered operator on $\mathcal{H}_0$. The notion of rigged Hilbert space is an important tool in the spectral theory of linear operators. Thus, according to the general spectral theorem, for each self-adjoint operator A on $\mathcal{H}_0$ there is a rigged Hilbert space of the form (4), such that all generalized eigenfunctions corresponding to the continuous spectrum of A belong to $\mathcal{H}_-$.

It is important that there is a one-to-one correspondence between positive self-adjoint operators on $\mathcal{H}_0$ and the rigged spaces of the form (4). We intensively explore this correspondence in the method developed in this book related to the theory of singular perturbations of self-adjoint operators. Moreover, the expansion of the triples of type (4) into an infinite scale of Hilbert spaces (the so-called A-scale), opens vast prospects for investigating singular perturbations of higher orders (super-singular perturbations) (see Section 8).

When one passes to the theory of operators in rigged Hilbert spaces one needs to dwell into the corresponding theory for quadratic forms. In fact, the concept of singular quadratic form inevitably leads to constructions in rigged spaces.

Accordingly, Chapter 5 is devoted to the theory of singular quadratic forms in rigged Hilbert spaces. We note that the general concept of singular quadratic form was introduced by V. Koshmanenko (see [135] and references therein). We also note that the term "singular quadratic form" was used by B. Simon in [177] in the decomposition of a positive quadratic form γ into two components, $\gamma = \gamma_{\mathrm{r}} + \gamma_{\mathrm{s}}$. There, the singular part γ_{s} has a meaning similar to μ_{sing} in the well-known Lebesgue decomposition of a Borel measure $\mu = \mu_{\mathrm{ac}} + \mu_{\mathrm{sing}}$.

A typical property of a singular quadratic form γ in the Hilbert space $\mathcal{H}$ is that its kernel subspace is dense in $\mathcal{H}$. We write this as $\operatorname{Ker}\gamma \sqsubset \mathcal{H}$. Note that in the most modern applications, the condition $\operatorname{Ker}\gamma \sqsubset \mathcal{H}$ is satisfied, although, in general, it is not necessary, but only sufficient for the singularity. Furthermore, in the treatment of a quadratic form γ in the rigged Hilbert space, the question about singularity of γ is often reduced to the fulfillment of the above-mentioned property.

Chapter 6 is entirely devoted to the study of orthogonal decompositions $\mathcal{H}_+ = \mathcal{M}_+ \oplus \mathcal{N}_+$, such that $\mathcal{M}_+$ is dense in $\mathcal{H}_0$. This problem has an additional aspect concerning singular perturbations of self-adjoint operators and a possibility to introduce the symmetric operator $\mathbf{A} = A \restriction \operatorname{Ker}\gamma$. In particular, we have to determine whether $\mathbf{A}$ is essentially self-adjoint.

So, in Chapter 6 we present all known abstract results on the denseness of the linear subset $\mathcal{M}_+$ from the positive space $\mathcal{H}_+$ in $\mathcal{H}_0$. In other words, we are interested, when $\mathcal{H}_0 \sqsupset \mathcal{M}_+$. This problem appears in different constructions. Indeed, the theory of self-adjoint extensions always begins with the question whether a Hermitian operator is densely defined. Of course, it is assumed that the operator is given on a linear set, which, in turn, is determined by some formally symmetric expression. Similarly, in the theory of singular perturbations of self-adjoint operators, the kernel subspace of the form giving the perturbation must be dense. So, we have to prove that this property holds or find a way for it to be satisfied. Therefore, the problem of finding conditions for the dense embedding of linear subspaces $\mathcal{M}_+$ from the positive space $\mathcal{H}_+$ into the central Hilbert space $\mathcal{H}_0$ is really of current importance.

Chapter 7 is the main part of this monograph. It contains the basic constructions and results obtained by the method of rigged Hilbert spaces in the theory of singular perturbations. Here we investigate various ways of defining the singularly perturbed operators and exhibit connections between these operators and new chains of rigged spaces. Special attention is devoted to the problem of uniqueness for singularly perturbed operator (see the introduction to Subsection 7.4). A detailed discussion is given regarding the singularity phenomenon, which implies the possibility of coexistence of non-equivalent Hilbert metrics on one and the same linear set. We show that this phenomenon is closely related to the right interpretation of the extensions theory for symmetric operators in terms of rigged spaces. Finally, we describe in detail the construction of the self-adjoint operator $\breve{A}$ associated with a fixed dense subspace $\mathcal{M}_+ \subset \mathcal{H}_+ = \mathrm{Dom}\, A$ in $\mathcal{H}_0$ and discuss its relation with A. These constructions add yet another way of accounting for singular perturbations.

In Chapter 8 we apply the method of rigged spaces to treat super-singular perturbations. Briefly, the idea of this method (it is better to call it the method of scales of Hilbert spaces) is a generalization of the usual form-sum method to the case of singular perturbations for the operator A^k for $k > 1$. In fact, we consider the super-singular perturbations of the operator A^k for $k \geq 2$. After constructing the perturbed operator $\tilde{A}^k$ by the form-sum method in the rigged space $\tilde{\mathcal{H}}_{-k} \sqsupset \mathcal{H} \sqsupset \tilde{\mathcal{H}}_k$, we define $\tilde{A}$ as a positive root of the last operator.

Finally, in Chapter 9 we demonstrate some applications of the method of rigged spaces. We investigate the inverse eigenvalue problem for finite-rank singularly perturbed operators. In particular, we give an explicit construction of the operator $\tilde{A} = A \tilde{+} \alpha\langle\cdot,\omega\rangle\omega$ which solves the eigenvalue problem $\tilde{A}\psi = \lambda\psi$ for an arbitrarily prescribed $\lambda \in \mathbb{R}^1$ and $\psi \in \mathcal{H} \setminus \mathrm{Dom}\, A$. Then we generalize this construction for a finite sequence of non-positive numbers $\lambda_1, \ldots, \lambda_N$.

Chapter 1

Preliminaries

We hope that our reader is already familiar with the contents of standard courses of mathematical analysis, functional analysis, and mathematical physics (see, e.g., [29, 32, 35, 40, 41, 48, 56, 75, 84, 154, 159, 166, 168–172, 178]). Therefore, in this chapter we recall only briefly basic concepts that are used it the sequel, and formulate some well-known theorems needed for understanding the theories of this monograph. The definitions and notations that we use are mostly standard, although there are several differences. In particular, we introduce some new notations that are more convenient, in our view, to use.

1.1 Notations

1.1.1 Sets, subsets, points

The set (synonyms – family, collection) is a primary mathematical term abstracted from experience and usually consisting of similar elements. We denote sets by capital letters of different alphabets. Elements of sets are called points and are denoted by lower case letters. The symbol $\varnothing$ denotes the empty set.

The usual relations between points, sets, and subsets we will be denoted by the standard, generally accepted symbols. In particular, we use the notations

$$x \in A,\ x \notin A,\ A = B,\ A \subset B,\ A \cap B,\ A \setminus B, \ldots$$

For known sets we use standard notations: $\mathbb{N} = \{1, 2, \ldots\}$ is the set of natural numbers; $\mathbb{N}_0 = \mathbb{N} \cup \{0\}$; $\mathbb{Z} = \{0, 1, -1, 2, -2, \ldots\}$ is the set of integers; $\mathbb{Q}$, $\mathbb{R}$, and $\mathbb{C}$ denote the sets of rational, real, and complex numbers, respectively. $\mathbb{R}^d$ denotes d-dimensional Euclidean space. The term *space* refers to sets with some additional properties.

The symbol $\{x \in A \mid P(x)\}$ denotes the subset of elements A that satisfy the condition $P(x)$.

1.1.2 Functions and mappings

The term *function* (synonym – mapping) is understood as the rule: an element $x \in X$ (not necessarily all elements from X) there corresponds an $y \in Y$. We use the notations: $y = f(x)$, or $X \xrightarrow{f} Y$, or $f : X \to Y$. The domain of the function (we denote it by $\mathrm{Dom}\, f$) is the set of all elements from X, on which the function f is defined. The set

$$\mathrm{Ran}\, f = \{y \in Y \mid y = f(x),\ x \in \mathrm{Dom}\, f\}$$

is called the range of the function f.

In our book we will usually encounter only single-valued mappings, i.e., for each element $x \in \mathrm{Dom} f$ we have one and only one element $y = f(x)$ from $\mathrm{Ran} f$. The inverse mapping f^{-1} is the rule which to each element $y \in \mathrm{Ran}\, f$ associates the set $\{x \in \mathrm{Dom}\, f \mid y = f(x)\}$.

A mapping f is called *injective* (or *one-to-one*), $\mathrm{Dom}\, f = X$ and if for each $y \in \mathrm{Ran}\, f$ there exists only one $x \in \mathrm{Dom}\, f$ such that $y = f(x)$; f is called *surjective* (or *onto*), if $\mathrm{Ran}\, f = Y$. If f has both properties, i.e., it is injective and surjective, then we will say that it is *bijective* . The restriction of f to a subset $A \subset \mathrm{Dom}\, f$ is denoted by $f \restriction A$.

A function $f = f(x)$ defined on a closed interval $[a, b]$ is called *absolutely continuous* if for each $\varepsilon > 0$ there exists $\delta > 0$ such that

$$\sum_{i=1}^{n} |x_i - x_i'| < \delta \Longrightarrow \sum_{i=1}^{n} |f(x_i) - f(x_i')| < \varepsilon,$$

for an arbitrary finite set of pairwise disjoint intervals $[x_i, x_i'] \subset [a, b]$. Absolutely continuous means that the function $f(x)$ is not only continuous, but also differentiable almost everywhere and its derivative is integrable: $\int_a^b f'(x)dx < \infty$.

1.2 Spaces

1.2.1 Topological spaces

A *topological space* is a set S endowed with a family $\mathcal{T}$ of subsets of S that are called open sets and satisfy the following properties:

i) $\mathcal{T}$ is closed under finite intersections, i.e., if $A_1, \dots, A_n \in \mathcal{T}$, then $A_1 \cap \dots \cap A_n \in \mathcal{T}$;
ii) $\mathcal{T}$ is closed under arbitrary unions, i.e., if $A_\alpha \in \mathcal{T}$ for all α in some index set I, then $\bigcup_{\alpha \in I} A_\alpha \in \mathcal{T}$;
iii) the whole set S and the empty set $\varnothing$ belong to $\mathcal{T}$, i.e., $\varnothing \in \mathcal{T}$ and $S \in \mathcal{T}$.

A family $\mathcal{T}$ with the above properties is called a topology on S. We occasionally write $(S, \mathcal{T})$ for a topological space.

Examples of topological spaces are the metric spaces (as defined below).

A set O is called a neighborhood of a point $x \in S$ in a topological space if there exists an open set U such that $x \in U \subset O$.

A set $N \subset S$ is called closed if it is the complement of an open set $M \in \mathcal{T}$, $N = M^c$.

Let $A \subset S$ be a subset of a topological space. The *closure* of A, denoted A^{cl}, is the smallest closed set containing A.

The *interior* of A is the largest open set contained in A and is denoted by $\operatorname{int} A$. The *boundary* of A is the set $A^{\partial} := A^{\mathrm{cl}} \setminus \operatorname{int} A$.

Let $(S, \mathcal{T})$ and $(T, \mathcal{U})$ be two topological spaces. A mapping $f : S \to T$ is called *continuous* if $f^{-1}(A) \in \mathcal{T}$ for every $A \in \mathcal{U}$.

A topological space $(S, \mathcal{T})$ is called *Hausdorff*, if its topology separates all points of the space, i.e., if for each pair $x, y \in S$, $x \neq y$, there exists open sets $O_1, O_2 \in \mathcal{T}$ such that $x \in O_1$, $y \in O_2$, and $O_1 \cap O_2 = \varnothing$.

1.2.2 Metric spaces

A *metric space* is a set M endowed with a real-valued function $d(x, y)$, which satisfies the following conditions:

i) $d(x, y) \geq 0$;
ii) $d(x, y) = 0$ if and only if $x = y$;
iii) $d(x, y) = d(y, x)$;
iv) $d(x, z) \leq d(x, y) + d(y, z)$ (*the triangle inequality*) for all $x, y, z \in M$.

The function d is called a *distance* or a *metric*.

Example 1.2.1. The Euclidean n-dimensional space

$$\mathbb{R}^n = \{x = (x_1, \dots, x_n),\ x_i \in \mathbb{R}\},$$

is a metric space with the distance function

$$d(x, y) = \sqrt{\sum_{i=1}^{n} (x_i - y_i)^2}.$$

A sequence of points $x_n \in M$, $n \in \mathbb{N}$, is said to *converge* to a point $x \in M$, if for each $\varepsilon > 0$, there exists a number N such that $d(x, x_n) < \varepsilon$ for all $n > N$. A sequence $x_n \in M$, $n \in \mathbb{N}$, is called *fundamental* (or *Cauchy*), if for each $\varepsilon > 0$ there exists N such that $d(x_m, x_n) < \varepsilon$ for all $m, n > N$.

1.2.3 Linear topological spaces

We denote by $\mathbb{K}$ a field that will be either the field of real numbers $\mathbb{R}$, or the field of complex numbers $\mathbb{C}$. A set L is called a *linear* (*vector*) *space* over the field $\mathbb{K}$ if it is equipped with an operation of addition and an operation of multiplication by number from $\mathbb{K}$ such that the following axioms hold:

i) $x + y = y + x, \quad \forall x, y \in L$;
ii) $x + (y + z) = (x + y) + z, \quad \forall x, y, z \in L$;
iii) there exists an element $0 \in L$ such that $x + 0 = x, \quad \forall x \in L$;
iv) $\forall x \in L$ there exists an element $(-x) \in L$ such that $x + (-x) = 0$;
v) $\forall \alpha, \beta \in \mathbb{K}, \quad \forall x \in L, \ \alpha(\beta x) = (\alpha\beta)x$;
vi) $\forall x \in L$, $1 \cdot x = x$, where 1 stands for the identity element of the field $\mathbb{K}$;
vii) $\forall \alpha, \beta \in \mathbb{K}, \ \forall x \in L, \quad (\alpha + \beta)x = \alpha x + \beta x$;
viii) $\forall \alpha \in \mathbb{K}, \ \forall x, y \in L, \quad \alpha(x + y) = \alpha x + \alpha y$.

The elements of L are called *vectors*. The zero symbol 0 denotes the zero vector of the space as well as the number zero of the field $\mathbb{K}$.

A set E is called a *linear topological space* if it is simultaneously topological and linear and the operations addition and the multiplication by a scalar are continuous.

Let M be an arbitrary subset of a linear topological space E. The linear space

$$\operatorname{span} M := \left\{ \sum_{i=1}^{p} \alpha_i x_i \mid \alpha \in \mathbb{K}, \ x_i \in M, \ p \in \mathbb{N} \right\}.$$

is called the *linear span* of M. The set M is *total* in E if $\operatorname{span} M$ is dense in E, namely, $(\operatorname{span} M)^{\text{cl}} = E$.

1.2.4 Linear normed spaces

Let L be a linear space over a field $\mathbb{K}$. The scalar function

$$L \ni x \longmapsto \|x\| \in \mathbb{R}$$

is called a *norm*, if it satisfies following conditions:

1) $\forall x \in L, \ \|x\| \geq 0$, and $\|x\| = 0$ if and only if $x = 0$;
2) $\forall \alpha \in \mathbb{K}, \ \forall x \in L, \ \|\alpha x\| = |\alpha|\|x\|$;
3) $\forall x, y \in L, \ \|x + y\| \leq \|x\| + \|y\|$.

A linear space equipped with a norm is called a *normed* linear space over the field $\mathbb{K}$. Let us emphasize that a linear space can have several norms. In this case, we distinguish norms by appending indices.

Two different norms $\|\cdot\|_1$ and $\|\cdot\|_2$ on a linear space L are *equivalent* if

$$c_1\|x\|_1 \leq \|x\|_2 \leq c_2\|x\|_1, \ x \in L,$$

where the constants $c_1, c_2 \geq 0$ are independent of x.

Each normed space is also a metric one with the metric defined by $\rho(x, y) = \|x - y\|$.

In a normed space one can define convergence for a sequence of vectors. We will write $x_k \to x_0$, if $\lim\limits_{k\to\infty} \|x_k - x_0\| = 0$.

A sequence $\{x_n\} \subset L$ is called *fundamental* (*Cauchy*) *sequence* if

$$\lim_{m,n\to\infty} \|x_n - x_m)\| = 0.$$

Every convergent sequence is fundamental. The converse is not always true.

The space L is called *complete* if every fundamental sequence $\{x_n\}$ converges to some vector x_0 in L.

1.2.5 Banach and Hilbert spaces

A complete normed space is called a *Banach* space.

Let L be a linear vector space over the field $\mathbb{C}$. The complex-valued function

$$L \ni x, y \longmapsto (x, y) \in \mathbb{C}$$

is called an *inner* (*scalar*) *product* if it satisfies the following conditions:

1) $\forall x \in L,\ \ (x,x) \geq 0$, and $(x,x) = 0$ if and only if $x = 0$;
2) $\forall \alpha, \beta \in \mathbb{C},\ \ \forall x, y, z \in L,\ \ (\alpha x + \beta y, z) = \alpha(x,z) + \beta(y,z)$;
3) $\forall x, y \in L,\ \ (x,y) = \overline{(y,x)}$ (the bar denotes complex conjugation).

We emphasize that a linear space can have various inner products. In this case, we distinguish between the inner products by appending indices.

Let us recall some properties of an inner product.

The inner product is conjugate linear in the second variable:

$$(x, \alpha y) = \bar{\alpha}(x,y), \quad x, y \in L,\ \alpha \in \mathbb{C}.$$

The Cauchy–Schwarz inequality:

$$|(x,y)|^2 \leq (x,x)(y,y), \quad x, y \in L.$$

In the case of function spaces, this inequality is often called the Cauchy–Bunyakovsky inequality.

Definition 1.2.2. A linear space over the field of complex numbers, with an inner product and complete with respect to the norm $\|x\| = \sqrt{(x,x)}$ is called a complex *Hilbert space.*

We well usually denote Hilbert spaces by $\mathcal{H}$, possibly with additional indices. We consider only the separable spaces, i.e., we assume that there exist a countable total subset.

So, a Hilbert space is a Banach space in which the norm is generated by an inner product.

The norm of each Hilbert space satisfies the so-called polarization identity

$$(x,y)=\frac{1}{4}(\|x+y\|^2-\|x-y\|^2+i\|x+iy\|^2-i\|x-y\|^2),\ x,y\in\mathcal{H}.$$

If the vectors of a Banach space satisfy the parallelogram identity:

$$\|x+y\|^2+\|x-y\|^2=2(\|x\|^2+\|y\|^2),\quad x,y\in B,$$

then, using the polarization identity, we can introduce an inner product and obtain a Hilbert space.

1.2.6 Function spaces

Now we recall, briefly, definitions of some function spaces used further in the text. For more information about their properties and needed basic concepts see, e.g., [29, 32, 35, 40, 48, 87, 88, 157, 169, 170, 182, 183]. For an open set $\Omega\subseteq\mathbb{R}^n$, the space $C(\Omega)$ is defined as a set of complex-valued continuous functions $f(x)$, $x\in\Omega$, with the topology of uniform convergence on compact subsets. If K is a compact subset of $\mathbb{R}^n$, then the space $C(K)$ with the norm

$$\|f\|_\infty:=\operatorname*{ess\,sup}_{x\in K}|f(x)|$$

is usually denoted by $L_\infty(K)$. We denote by $C_0(\Omega)$ the subset of $C(\Omega)$ consisting of the functions with compact support on Ω.

The space of functions that together with all partial derivatives $D^\alpha f$, $|\alpha|\le l\in\mathbb{N}$, belong to $C(\Omega)$ is denoted by $C^l(\Omega)$. Here

$$D^\alpha=\partial_1^{\alpha_1}\cdots\partial_n^{\alpha_n},\quad |\alpha|=\alpha_1+\cdots+\alpha_n,$$

and $\partial_i^{\alpha_i}$ denotes a partial derivative of order α_i.

The subset of $C^l(\Omega)$ consisting of functions with compact support is denoted by $C_0^l(\Omega)$.

For $1\le p<\infty$ and any measurable set $\Omega\subseteq\mathbb{R}^n$, the space $L_p(\Omega)$ consists of the equivalence classes of measurable functions $f(x)$ for which

$$\|f\|_{L_p}\equiv\|f\|_p=\left(\int\limits_\Omega|f(x)|^pdx\right)^{1/p}<\infty$$

and is equipped with the norm $\|f\|_{L_p}$. Analogously, we define the Banach space $L_p(\Omega,d\mu)$, where μ is a Borel measure on Ω. If $p=2$, then $L_2(\Omega,d\mu)$ is a Hilbert space with the inner product

$$(f,g)=\int\limits_\Omega f(x)\overline{g(x)}d\mu(x).$$

The set of infinitely differentiable functions on Ω is denoted by $C^\infty(\Omega)$, and its subset of functions with compact support in Ω is denoted by $C_0^\infty(\Omega)$.

The Schwarz space (the space of test functions) is denoted by $\mathcal{S}(\mathbb{R}^n) \equiv \mathcal{S}$. It consists of infinitely differentiable functions rapidly decaying at infinity. The convergence in $\mathcal{S}(\mathbb{R}^n)$ is defined by a system of semi-norm (see, e.g., [87, 88]).

The space of linear continuous functionals on $\mathcal{S}(\mathbb{R}^n)$ consists of the Schwarz generalized functions (distributions) and is denoted by $\mathcal{S}'(\mathbb{R}^n)$. The expression $\langle \omega, \varphi \rangle$ denotes the duality pairing of $\omega \in \mathcal{S}'$ and $\varphi \in \mathcal{S}$.

A generalized function ω is positive if $\langle \omega, \varphi \rangle \geq 0$ for all non-negative $\varphi \in \mathcal{S}$.

Theorem 1.2.3. *Each positive generalized function $\omega \in \mathcal{S}'$ defines a positive measure μ by $\langle \omega, \varphi \rangle = \int\limits_{\mathbb{R}^n} \varphi(x) d\mu(x)$.*

Proof. See, e.g., in [182]. □

1.2.7 Sobolev spaces

For natural numbers k and $1 \leq p < \infty$, $W_p^k(\mathbb{R}^n) = W_p^k$ denotes the *Sobolev spaces*. These spaces consist of all functions belonging to $L_p(\mathbb{R}^n)$, and such that all theirs derivatives up to order k, also belong to this space:

$$W_p^k = \{f \in L_p \mid D^\alpha f \in L_p, \ |\alpha| \leq k\}.$$

The norm in W_p^k is defined as

$$\|f\|_{W_p^k} = \left(\sum_{|\alpha| \leq k} \|D^\alpha f\|_{L_p}^p \right)^{1/p}. \tag{1.1}$$

All W_p^k are Banach spaces. Moreover, for $p = 2$, the space W_2^k is a Hilbert space with the norm

$$(f, g)_{W_2^k} = \sum_{|\alpha| \leq k} (D^\alpha f, D^\alpha g)_{L_2}, \quad k = 0, 1, \ldots.$$

Often, the norm and the inner product in W_2^k are defined in some what simpler, but equivalent way:

$$\begin{aligned} \|f\|_{W_2^k} &= \left(\|f\|_{L_2}^2 + \sum_{|\alpha| = k} \|D^\alpha f\|_{L_2}^2 \right)^{1/2}, \\ (f, g)_{W_2^k} &= (f, g)_{L_2} + \sum_{|\alpha| = k} (D^\alpha f, D^\alpha g)_{L_2}. \end{aligned} \tag{1.2}$$

It is clear that the space W_p^k can be obtained as the completion of $C_0^\infty(\mathbb{R}^n)$ or $\mathcal{S}(\mathbb{R}^n)$ with respect to the norm

$$\|f\|_{W_p^k} = \left(\sum_{|\alpha|\leq k} \int_{\mathbb{R}^n} |D^\alpha f(x)|^p dx \right)^{1/p}. \tag{1.3}$$

The space W_p^k can also be defined using the Fourier transform

$$(\mathcal{F}f)(\xi) = \frac{1}{(2\pi)^{n/2}} \int_{\mathbb{R}^n} e^{-ix\xi} f(x)dx.$$

It is known that

$$\mathcal{F}(D^\alpha f)(\xi) = (i\xi)^\alpha (\mathcal{F}f)(\xi), \quad f \in \mathcal{S}(\mathbb{R}^n).$$

Hence

$$W_2^k = \{f \in L_2 \mid \|f\|_k^2 = \int_{\mathbb{R}^n} (1+|\xi|^2)^k |(\mathcal{F}f)(\xi)|^2 d\xi < \infty\},$$

since, due to the Plancherel theorem (see, e.g., [32]), $D^\alpha f \in L_2$ if and only if $(i\xi)^\alpha(\mathcal{F}f)(\xi) \in L_2$, $|\alpha| \leq k$.

In general, the Sobolev space W_p^k, $p > 1$, can be identified with the space of Bessel potentials,

$$W_p^k = \{f \in L_p \mid f = G_k * g,\ g \in L_p\},$$

where $*$ stands for convolution and G_k is the integral operator generated by the Bessel function:

$$\begin{aligned} G_k(x) &= (2\pi)^{-n/2} \mathcal{F}^{-1}((1+|\xi|^2)^{-k/2})(x) \\ &= \frac{1}{(2\pi)^n} \int_{\mathbb{R}^n} \frac{e^{ix\xi}}{(1+|\xi|^2)^{k/2}} d\xi. \end{aligned}$$

The equivalence of these definitions can be proved based on a fundamental theorem of Calderón (see, e.g., [32]), which establishes of inequalities

$$c_1\|f\|_{W_p^k} \leq \|(1-\Delta)^{k/2} f\|_{L_p} \leq c_2 \|f\|_{W_p^k}, \quad c_1, c_2 > 0,$$

where Δ is the Laplacian. In particular, in the case $p = 2$, using the Fourier transform, we have:

$$f = G_k * g = (1-\Delta)^{-k/2} g, \quad g \in \mathcal{S}(\mathbb{R}^n).$$

Hence, one of the equivalent norms in W_2^k can be defined by the formula

$$\|f\|_{W_p^k} = \|(1-\Delta)^{k/2} f\|_{L_2} = \|g\|_{L_2}. \tag{1.4}$$

Thus, the mapping

$$(1-\Delta)^{-k/2} : L_2 \longrightarrow W_p^k$$

originally defined only on functions $g \in \mathcal{S}(\mathbb{R}^n)$, can be extended to a unitary operator on the closure.

The negative Sobolev space W_2^{-k}, the space with the index $-k < 0$, is dual to W_2^k, $k > 0$, and is defined as the subset of generalized functions $\mathcal{S}'(\mathbb{R}^n)$ with a finite norm

$$\|\omega\|_{W_2^{-k}} := \|\mathcal{F}^{-1}((1+|\xi|^2))^{-k/2}(\mathcal{F}\omega)(\xi)\|_{L_2} < \infty, \quad \omega \in \mathcal{S}'(\mathbb{R}^n).$$

In this case, the closure of the mapping

$$(1-\Delta)^{-k/2} : W_p^{-k} \longrightarrow L_2$$

is also a unitary operator.

In different applications, the local Sobolev spaces $W_p^k(\Omega)$ are used, where Ω is a compact or open set in $\mathbb{R}^n$. The space $W_p^k(\Omega)$ is defined as the completion of $C^k(\Omega)$ with respect to one of the equivalent norms given in (1.1), (1.2), (1.3), or (1.4), where $\mathbb{R}^n$ is replaced by Ω.

It should be noted that the closure of $C_0^\infty(\Omega)$ in $W_2^k(\Omega)$, in general, is only a proper subspace, denoted by $\overset{\circ}{W}{}_2^k(\Omega)$. So, in general,

$$\dim(W_2^k(\Omega) \ominus \overset{\circ}{W}{}_2^k(\Omega)) > 0.$$

Let us recall here, the Sobolev theorem, which is very important in applications. This theorem regards the embedding of the Sobolev space into the space of continuous or continuously differentiable functions (for the proof see, e.g., [29, 42]).

Theorem 1.2.4 (Sobolev). *If $k > n/2$, then the Sobolev space $W_2^k(\Omega)$, $\Omega \subseteq \mathbb{R}^n$, consists of continuous functions:*

$$W_2^k(\Omega) \subset C(\Omega), \quad k > n/2. \tag{1.5}$$

Moreover, one has the embedding

$$W_2^k(\Omega) \subset C^l(\Omega), \quad k > n/2 + l, \tag{1.6}$$

where $l \in \mathbb{N}_0$.

1.3 Linear transformations

1.3.1 Continuous linear functionals

Let L be a linear normed space over the field $\mathbb{K}$ of either real or complex numbers. A mapping

$$L \supseteq \operatorname{Dom} l \ni x \longmapsto l(x) \in \mathbb{K}$$

is called a *functional*; here $\operatorname{Dom} l$ denotes the domain of l.

A functional l is called *continuous at the point* x_0, if any vector sequence $x_n \in \operatorname{Dom} l$ that converges to $x_0 \in \operatorname{Dom} l$ the number sequence $l(x_n)$ converges to $l(x_0)$. A functional l is *continuous* on a set $\mathfrak{D}$ if it is continuous at each point of this set $\mathfrak{D}$.

A functional l is called *linear*, if $\operatorname{Dom} l$ is a linear set and

$$l(\alpha x + \beta y) = \alpha l(x) + \beta l(y), \quad \alpha, \beta \in \mathbb{C}, \ x, y \in \operatorname{Dom} l.$$

A functional l is called *bounded*, if for some constant $c > 0$,

$$|l(x)| \leq c\|x\|, \quad x \in \operatorname{Dom} l.$$

The next two facts are important for applications.

Proposition 1.3.1. *If a linear functional l is continuous at some point of* $\operatorname{Dom} l$, *then it is continuous everywhere on* $\operatorname{Dom} l$.

Proposition 1.3.2. *A linear functional l is continuous on a linear subset* $\operatorname{Dom} l \subseteq L$ *if and only if it is bounded.*

The norm of a linear functional l is defined as a real number

$$\|l\| = \sup \{|l(x)| \mid x \in \operatorname{Dom} l \subseteq L, \ \|x\| = 1\}. \tag{1.7}$$

It is obvious that each continuous linear functional satisfies the inequality

$$|l(x)| \leq \|l\| \cdot \|x\|, \quad x \in \operatorname{Dom} l.$$

Let L be a linear normed (not necessarily complete) space over the field $\mathbb{K}$ of real or complex numbers. Let us denote all linear continuous functionals on L by L^*.

Proposition 1.3.3. *The space L^* with the operations of addition and multiplication by numbers is a Banach space with the norm* (1.7).

The space L^* is called the *conjugate* or (*dual*) of L.

Let L be linear normed space and $G \subset L$ is a linear subset. A functional $\hat{l}$ on L is called an extension of the functional l defined on G, if $l(f) = \hat{l}(f)$ for all $f \in G$. Conversely, a functional l defined on $G \subset L$, is called the restriction of the functional $\hat{l}$ on L, if $l(f) = \hat{l}(f)$ for all $f \in G$, and one notes $\hat{l} \upharpoonright G = l$.

Let us notice some important properties of linear functionals.

Proposition 1.3.4. *Let L be a linear normed space and $G \subset L$ be a linear dense subset. Then for any continuous linear functional l defined on G, there exists a unique continuous linear functional $\hat{l}$ on L as an extension by continuity such that $\hat{l} \upharpoonright G = l$. In addition, $\|\hat{l}\| = \|l\|$.*

Theorem 1.3.5 (Hahn–Banach). *Let L be a linear normed space and $G \subset L$ be a linear subset of L. Then, for any continuous linear functional l defined on G, there exists a functional $\hat{l} \in L^*$ such that $\hat{l} \restriction G = l$ and $\|\hat{l}\| = \|l\|$.*

Theorem 1.3.6 (Riesz). *Let $\mathcal{H}$ be a Hilbert space with the inner product $(\cdot, \cdot)$ and the norm $\|\cdot\|$. Then for any continuous linear functional l on $\mathcal{H}$, there exists a unique vector $h \in \mathcal{H}$ such that*

$$l(x) = (x, h), \quad \forall x \in \mathcal{H}.$$

In addition, $\|l\|_{\mathcal{H}} = \|h\|_{\mathcal{H}}$. Conversely, for each fixed $h \in \mathcal{H}$, the expression $l(x) := (x, h)$, $x \in \mathcal{H}$, defines a continuous linear functional l on $\mathcal{H}$ with the norm $\|l\|_{\mathcal{H}} = \|h\|$.

1.3.2 Linear operators

Let L_1 and L_2 be two complete normed linear spaces (Banach spaces). A mapping

$$L_1 \supseteq \operatorname{Dom} A \ni x \longmapsto Ax \in \operatorname{Ran} A \subseteq L_2$$

is called a *linear operator* if $\operatorname{Dom} A$ is a linear set and

$$A(\alpha x + \beta y) = \alpha Ax + \beta Ay, \quad x, y \in \operatorname{Dom} A, \ \forall \alpha, \beta \in \mathbb{K}.$$

The set Dom$A$ is called the domain of the operator A, and Ran$A$ is the range of A.

An operator A is called *bounded*, if

$$\|Ax\|_{L_2} \leq C\|x\|_{L_1}, \quad \forall x \in \operatorname{Dom} A.$$

for some fixed $C > 0$.

The norm of a linear operator A is defined as

$$\|A\| := \sup\left\{\|Ax\|_2 \mid x \in \operatorname{Dom} A, \ \|x\|_1 = 1\right\}.$$

A linear operator is *continuous at a point* $x_0 \in \operatorname{Dom} A$, if the convergence of a sequences $x_n \in \operatorname{Dom} A$ to x_0 in L_1 implies the convergence of the sequence Ax_n to Ax_0 in the norm of L_2.

A linear operator A is called *continuous*, if for any sequence $x_n \in \operatorname{Dom} A$, its convergence in L_1 implies the convergence of the sequence $Ax_n \in \operatorname{Dom} A$ in the norm of L_2:

$$\|x_n - x_m\|_{L_1} \longrightarrow 0 \implies \|Ax_n - Ax_m\|_{L_2} \longrightarrow 0, \quad n, m \longrightarrow \infty.$$

The proofs of the next two theorems are left to the reader.

Theorem 1.3.7. *A linear operator acting from L_1 into L_2 is continuous if it is continuous at a least one point.*

Theorem 1.3.8. *A linear operator acting from L_1 into L_2 is continuous if and only if it is bounded.*

Theorem 1.3.9 (About bounded linear mappings). *Let a linear operator A acting from a Banach space L_1 into a Banach space L_2, be continuous and densely defined, $(\mathrm{Dom}\, A)^{\mathrm{cl}} = L_1$. Then A has a unique continuous linear extension on the whole space L_1 preserving the norm.*

Proof. Since $\mathrm{Dom}\, A$ is dense in L_1, for each vector $h \in L_1 \setminus \mathrm{Dom}\, A$ there exist a sequence $\varphi_n \in \mathrm{Dom}\, A$ such that $\varphi_n \to h$, $n \to \infty$. Since A is a bounded operator, the sequence $\psi_n = A\varphi_n$ converges in L_2: $\psi_n \to \psi \in L_2$. For the same reason the vector ψ is independent of the choice of a converging to h sequence φ_n. Hence, the mapping defined by $A^{\mathrm{cl}}h = \psi$ is an extension of A to the whole space L_1. This extension is linear, continuous, and unique. In addition, automatically $\|A^{\mathrm{cl}}\| = \|A\|$. □

If the linear operator A from L_1 onto L_2 has a zero kernel,

$$\mathrm{Ker}\, A := \{x \in L_1 \mid Ax = 0\} = \{0\},$$

then there exists an operator A^{-1}, called the *inverse* of A one. It acts from $\mathrm{Ran}\, A \subseteq L_2$ onto $\mathrm{Dom}\, A \subseteq L_1$. In addition, the composition $A^{-1}A$ is then identity mapping from $\mathrm{Dom}\, A$ onto L_1.

It is clear that the inverse operator A^{-1} from L_2 onto L_1 is bounded (continuous) if and only if there exists a constant $C > 0$ such that

$$\|Ax\|_2 \geq C\|x\|_1, \quad \forall x \in \mathrm{Dom}\, A.$$

Theorem 1.3.10 (Banach; on the inverse operator). *If a linear operator A from a Banach space L_1 onto a Banach space L_2 is continuous, then there exists the inverse operator A^{-1} and it is also continuous.*

Let A be a densely defined linear operator which acts from L_1 onto L_2. The *adjoint operator* A^* acts from the space L_2^* onto L_1^* and has the following descriptive definition. A continuous linear functional $l^* \in L_2^*$ belongs to the domain $\mathrm{Dom}\, A^*$, if the functional $l(x) := l^*(Ax)$, $\quad x \in \mathrm{Dom}\, A$, is continuous on L_1 (has an extension by continuity to the whole space L_1). Hence $l \in L_1^*$. Then, by definition $A^*l^* := l$.

It is easy to show that if A is bounded, then A^* is also bounded, and moreover, $\|A\| = \|A^*\|$.

1.3.3 Bounded sesquilinear forms

Let L be a Banach space over the field of complex numbers $\mathbb{C}$. A mapping

$$L \supseteq Q(\gamma) \ni x, y \longmapsto \gamma(x, y) \in \mathbb{C}$$

is called a *sesquilinear form*, if it is linear in the first variable and antilinear for the second; $Q(\gamma)$ denotes the domain of the form.

Any sesquilinear form γ in a complex Banach space satisfies the equality

$$4\gamma(x,y) = \gamma[x+y] - \gamma[x-y] + i\gamma[x+iy] - i\gamma[x-iy], \ \forall x,y \in Q(\gamma),$$

where $\gamma[x] \equiv \gamma(x,x)$. It is called the *polarization identity.*

A form $\gamma(x,y)$ is called *bounded* if there exist a constant $C>0$ such that

$$|\gamma(x,y)| \leq C \cdot \|x\|_L \cdot \|y\|_L, \quad \forall x,y \in Q(\gamma).$$

Next we consider sesquilinear forms on a complex Hilbert space. $\mathcal{H}$.

Theorem 1.3.11. *Between the set of bounded sesquilinear forms γ with $Q(\gamma) = \mathcal{H}$ and the set of bounded linear operators A with* $\operatorname{Dom} A = \mathcal{H}$*, there exists a bijective correspondence given by the relation*

$$\gamma(x,y) = (Ax,y), \quad x,y \in \mathcal{H}. \tag{1.8}$$

Moreover,

$$\|A\| = \sup_{\|x\|,\|y\|=1} |\gamma(x,y)|.$$

A sesquilinear form γ on $\mathcal{H}$ is called *Hermitian* (*symmetric*), if $\gamma(x,y) = \overline{\gamma(y,x)}$, $\forall x,y \in Q(\gamma)$, where the bar denotes complex conjugation.

If the form γ bounded in $\mathcal{H}$ is Hermitian, then by (1.8), the operator A corresponding to γ is self-adjoint $A = A^*$ in the sense that $(Ax,y) = (x,Ay)$, $\forall x,y \in \mathcal{H}$.

According to the definition, the bounded on $\mathcal{H}$ operator A is nonnegative if a sesquilinear form corresponding to it via (1.8) is also nonnegative:

$$A \geq 0 \Longleftrightarrow \gamma(x,x) = (Ax,x) \geq 0, \quad \forall x,y \in \mathcal{H}.$$

Let G be a subspace of the Hilbert space $\mathcal{H}$. We denote its orthogonal complement by $G^\perp := \mathcal{H} \ominus G$. Then each vector $x \in \mathcal{H}$ has a unique representation as a sum $x = y + z$, where $y \in G$, $z \in G^\perp$, since $\mathcal{H} = G \oplus G^\perp$. Here $y = P_G x$ is called the *projection* of the vector x onto the subspace G, and $z = P_{G^\perp} x$ is the projection of x onto the subspace $G^\perp$. The operator P_G, acting as $\mathcal{H} \ni x \mapsto P_G x$ is called the *projection* (*orthogonal projection*) in $\mathcal{H}$ onto G.

A continuous linear operator V from a Hilbert space $\mathcal{H}_1$ to a Hilbert space $\mathcal{H}_2$ is called isometric if

$$(Vx,Vy)_2 = (x,y)_1, \quad \forall x,y \in \operatorname{Dom} V \subseteq \mathcal{H}_1.$$

An isometric operator U with domain $\operatorname{Dom} U = \mathcal{H}_1$ and range $\operatorname{Ran} U = \mathcal{H}_1$ is called *unitary.*

Every unitary operator U satisfy the equality $U^{-1} = U^*$. The latter is often used as the definition of a unitary operator.

1.4 Unbounded operators, closability

A linear densely defined operator A on a Hilbert space $\mathcal{H}$ is called *unbounded* if $\sup_{\|x\|=1} \|Ax\| = \infty$.

Two linear operators A and B on $\mathcal{H}$ are said to be equal if $\operatorname{Dom} A = \operatorname{Dom} B$ and $Af = Bf$, $\forall f \in \operatorname{Dom} A$.

If $\operatorname{Dom} A \subseteq \operatorname{Dom} B$ and $Af = Bf$, $\forall f \in \operatorname{Dom} A$, then the operator B is called an extension of the operator A, and in turn A is called a restriction of B.

Let A and B be unbounded operators on a Hilbert space $\mathcal{H}$. The following expressions can be interpreted as definitions:

$$\begin{aligned}
(\alpha A)f &:= \alpha(Af), && f \in \operatorname{Dom} A = \operatorname{Dom}(\alpha A),\ \alpha \in \mathbb{C};\\
(A+B)f &:= Af + Bf, && f \in \operatorname{Dom}(A+B) := \operatorname{Dom} A \cap \operatorname{Dom} B;\\
(AB)f &:= A(Bf), && f \in \operatorname{Dom}(AB) := \{f \in \operatorname{Dom} B \mid Bf \in \operatorname{Dom} A\}.
\end{aligned}$$

For an unbounded operator A, we define its inverse A^{-1} by the equality $A^{-1}(Af) = f$, where $f \in \operatorname{Dom} A$. It acts from $\operatorname{Ran} A$ onto $\operatorname{Dom} A$. Of course, it is correctly defined only under the condition $\operatorname{Ker} A = \{0\}$.

Definition 1.4.1. The operator A on the Hilbert space $\mathcal{H}$ is called *closed*, if for any sequence $(f_n)_{n=1}^{\infty}$ from $\operatorname{Dom} A$, the convergence $f_n \to f$ and $Af_n \to g$, $n \to \infty$ implies that $f \in \operatorname{Dom} A$ and $Af = g$.

There are several criterias for closability of a linear operator.

We define the *graph norm* of an operator A by

$$\|f\|_+ := (\|f\|^2_{\mathcal{H}} + \|Af\|^2_{\mathcal{H}})^{1/2}, \quad f, g \in \operatorname{Dom} A.$$

Then A is closed if and only if its domain $\operatorname{Dom} A$ is a complete space with respect to the graph norm.

Let us consider the orthogonal sum of two copies of a Hilbert space, $\mathcal{H} \oplus \mathcal{H}$, which consists of pairs $\{f, g\}$, $f, g \in \mathcal{H}$. The linear operations for the orthogonal sum are defined component wise, and the inner product in $\mathcal{H} \oplus \mathcal{H}$ is defined as

$$(\{f_1, g_1\}, \{f_2, g_2\})_{\mathcal{H}\oplus\mathcal{H}} = (f_1, f_2) + (g_1, g_2), \quad f_1, g_1, f_2, g_2 \in \mathcal{H}.$$

The set

$$\Gamma_A := \{\{f, Af\} \in \mathcal{H} \oplus \mathcal{H} \mid f \in \operatorname{Dom} A\}$$

is called the *graph* of the operator A.

It is obvious (see, e.g., [41]) that a linear operator A on the Hilbert space $\mathcal{H}$ is closed if and only if its graph Γ_A is closed subspace in $\mathcal{H} \oplus \mathcal{H}$.

Theorem 1.4.2 (Banach). *If a linear operator is closed and defined everywhere in a Hilbert space, then it is continuous.*

Of course, not every closed subspace in $\mathcal{H}\oplus\mathcal{H}$ is the graph of a linear operator on $\mathcal{H}$.

A linear operator A on a Hilbert space $\mathcal{H}$ is called *closable*, if it has a closed extension. We write $A \subseteq A^{\mathrm{cl}}$, where cl stand for the *closure*. This means that one extends the domain $\operatorname{Dom} A$ by adding vectors $f \in \mathcal{H}$, for which there exist sequences $(f_n)_{n=1}^{\infty}$ from $\operatorname{Dom} A$, converges to f and such that $\lim_{n\to\infty} Af_n = g \in \mathcal{H}$, where the vector g does not depend on a choice of the sequence f_n, and then one defines the closure A^{cl} as the extension of A to vectors f by setting $A^{\mathrm{cl}} f = g$.

Theorem 1.4.3. *A linear operator A on a Hilbert space is closable, if and only if the closure of its graph is also the graph of some linear operator.*

Let A be a linear operator with a dense domain $\operatorname{Dom} A$ in $\mathcal{H}$. Consider a vector $g \in \mathcal{H}$, for which there exists $g^* \in \mathcal{H}$ such that $(Af, g) = (f, g^*)$, $\forall f \in \operatorname{Dom} A$. The set of all such vectors is denoted by $\mathfrak{D}^*$. The operator A^* with the domain $\operatorname{Dom} A^* = \mathfrak{D}^*$, acting by the rule $A^* g = g^*$, is called the *adjoint* of the operator A on $\mathcal{H}$. If $A^* = A$, then the operator A is called *self-adjoint.*

We note several properties of the operator A^*.

1) A^* is automatically closed, $A^* = (A^*)^{\mathrm{cl}}$.
2) If A is closed, then $(A^{\mathrm{cl}})^* = A^*$.
3) If $\operatorname{Ran} A = \mathcal{H}$ and there exist the inverse operator A^{-1}, then $(A^*)^{-1}$ exists and $(A^{-1})^* = (A^*)^{-1}$.
4) The inclusion $B \supseteq A$ implies $A^* \supseteq B^*$.
5) If for operators A and B, the set $\operatorname{Dom}(A+B)$ is dense in $\mathcal{H}$, then $(A+B)^* \supseteq A^* + B^*$. In particular, if B is a bounded operator, then $(A+B)^* = A^* + B^*$.
6) If for operators A and B, the set $\operatorname{Dom}(AB)$ is dense in $\mathcal{H}$, then $(BA)^* \supseteq A^* B^*$. In particular, if B is a bounded operator, then $(BA)^* = A^* B^*$.
7) If the operator A has a closure $A \subseteq A^{\mathrm{cl}}$, then $(A^*)^* = A^{\mathrm{cl}}$.

Finally, we recall the simple, but widely used Hellinger–Toeplitz theorem.

Theorem 1.4.4. *Each symmetric operator defined on the whole Hilbert space is self-adjoint and bounded.*

As a corollary, every densely defined symmetric operator A, for which

$$\operatorname{Ran} A = \operatorname{Dom}(A^{-1}) = \mathcal{H},$$

is self-adjoint.

Chapter 2

Symmetric Operators and Closable Quadratic Forms

In this chapter we recall, with more details, properties of unbounded symmetric and self-adjoint operators and their connection with quadratic forms. We give a brief review of some standard facts on the extension theory of symmetric operators and the general theory of quadratic forms. In particular, we formulate and prove the theorem that gives the basic criterion for self-adjointness. These facts will be used in the sequel. We recall also the well-known theorems on operator representations of quadratic forms. A more extensive treatment of the theory can be found in the well-known books by N.I. Akhiezer and I.M. Glazman [32], by T. Kato [107], and by M. Reed and B. Simon [169, 170].

2.1 Resolvent, spectrum, deficiency indices

Let A be a linear operator with dense domain $\operatorname{Dom} A$ in the separable Hilbert space $\mathcal{H}$. The operator A is called *symmetric* if $A \subseteq A^*$, i.e.,

$$(Af, g) = (f, Ag), \quad f, g \in \operatorname{Dom} A.$$

A symmetric operator A is called *self-adjoint*, if it is maximal, i.e., if it does not have any symmetric extensions in $\mathcal{H}$. A symmetric operator A is called *essentially self-adjoint* if its closure (see Definition 1.4.1) is self-adjoint.

Usually, the terms "symmetric operator" and "Hermitian operator" are used as identical. However, we follow [148] and distinguish the Hermitian operators from the symmetric ones. Namely, the domain of a Hermitian operator A is not necessarily dense in $\mathcal{H}$. On the other hand a symmetric operator is Hermitian and has dense domain $(\operatorname{Dom} A)^{\mathrm{cl}} = \mathcal{H}$.

A point $\lambda \in \mathbb{C}$ is called *regular* if the operator $A - \lambda I \equiv A - \lambda$ defines a bijective mapping from $\operatorname{Dom} A$ onto the whole space $\mathcal{H}$. In this case we say that λ belongs to the *resolvent set* of the operator A and write $\lambda \in \rho(A)$. It is clear that for all $\lambda \in \rho(A)$ the inverse operator $(A-\lambda)^{-1}$ is bounded and defined everywhere on $\mathcal{H}$. So, in accordance with this definition, the resolvent set is

$$\rho(A) := \{\lambda \in \mathbb{C} \mid \operatorname{Ran}(A-\lambda) = \mathcal{H}\}.$$

The operator-valued function

$$R_\lambda(A) = (A-\lambda)^{-1}, \quad \lambda \in \rho(A),$$

is called the *resolvent* of the operator A. The complement to the set $\rho(A)$ in $\mathbb{C}$ of all regular points of the operator A is called the *spectrum* of A, denoted $\sigma(A) = \mathbb{C} \setminus \rho(A)$. If the operator A has eigenvalues, then they form a subset of the purely point spectrum $\sigma_{\mathrm{pp}}(A) \subseteq \sigma(A)$. We recall that λ is an *eigenvalue* of the operator A if $A\varphi = \lambda\varphi$, where $\varphi \neq 0$; then the vector φ is called the *eigenvector* of A.

In the next considerations we are interested in unbounded symmetric or self-adjoint operators.

It is easy to see (Theorem 2.2.1) that points $\lambda \in \mathbb{C}$ with $\operatorname{Im}(\lambda) \neq 0$ belong to the resolvent set of any self-adjoint operator. As a corollary, the spectrum of a self-adjoint operator belongs to the real axis.

It is well known (see, e.g., [32, 107, 169]), and easy to show, that the resolvent set $\rho(A)$ of each self-adjoint operator A is open in $\mathbb{C}$ and the resolvent $R_\lambda(A)$ is an analytic operator-valued function on $\rho(A)$. Moreover, the operators $R_\lambda(A)$, $R_\mu(A)$, $\lambda, \mu \in \rho(A)$ commute as bounded operators and satisfy the equality

$$R_\lambda(A) - R_\mu(A) = (\lambda-\mu)R_\lambda(A)R_\mu(A), \quad \lambda, \mu \in \rho(A),$$

which is called the *Hilbert identity.*

Let A be a symmetric operator. A point $\lambda \in \mathbb{C}$ is called a point of *regular type,* if there exist a bounded operator $(A-\lambda)^{-1}$, although its domain not necessarily coincides with $\mathcal{H}$ and can be a proper subspace of $\mathcal{H}$. The set of all regular type points of a symmetric operator A is called the *domain of regularity* of A.

It is easy to see that $\lambda \in \mathbb{C}$ is a point of regular type of A if there exists $k > 0$ such that

$$\|(A-\lambda)f\| \geq k\|f\|, \quad \forall f \in \operatorname{Dom} A.$$

As a consequence we have

Proposition 2.1.1. *Let A be a symmetric operator on $\mathcal{H}$. Then all points*

$$z \in \mathbb{C}, \quad z = \mu + i\lambda, \quad \operatorname{Im}(z) = \lambda \neq 0$$

are of regular type.

Proof. One has the obvious estimate:

$$\|(A-z)f\|^2 = \|(A-\mu)f\|^2 + |\lambda|^2\|f\|^2 \geq k\|f\|, \quad k = |\lambda|^2. \qquad \square$$

Thus (for details see, e.g., [32]), the domain of regularity of any symmetric operator is an open set in $\mathbb{C}$ that contains all points $z = \mu + i\lambda$, $\lambda \neq 0$ and consists of either one or two connected components.

If A is a symmetric, but not self-adjoint operator, $A \neq A^*$, then due to its closability for each point z of regular type (in particular for all $z \in \mathbb{C}$, $\mathrm{Im}(z) \neq 0$), the range $\mathrm{Ran}\,(A - \bar{z})$ is a closed subspace:

$$\mathcal{M}_z = \mathcal{M}_z^{\mathrm{cl}} = \mathrm{Ran}(A - \bar{z}) \neq \mathcal{H}.$$

This follows from the fact that $(A - \bar{z})^{-1}$ is bounded and closed. The orthogonal complement

$$\mathcal{N}_z := \mathcal{H} \ominus \mathcal{M}_z \neq \{0\}$$

is the *deficiency* (or *defect*) *subspace* of the operator A at the point $z \in \mathbb{C}$. It turns out (see [32]) that the dimension of deficiency subspace $\mathcal{N}_z \equiv \mathcal{M}_z^{\perp}$ is a constant for all z in each connected component of the domain of regularity. The numbers

$$\mathbf{n}_+ = \dim \mathcal{N}_z, \quad \mathbf{n}_- = \dim \mathcal{N}_{\bar{z}}, \quad \mathrm{Im}(z) > 0$$

are called the *deficiency indices* of the operator A.

2.2 Adjoints and self-adjoint operators

2.2.1 The construction of the adjoint operator

Let us consider in detail the construction of the adjoint operator. Let A be linear unbounded operator with dense domain $(\mathrm{Dom}\,A)^{\mathrm{cl}} = \mathcal{H}$ in the separable Hilbert space. For each such operator (regardless of whether it is symmetric and closed) there exists an adjoint operator A^* which is constructed as follows.

Let us fix an arbitrary vector $g \in \mathcal{H}$. We suppose that the linear functional

$$l_g(f) := (Af, g), \quad f \in \mathrm{Dom}\,A,$$

is continuous on $\mathcal{H}$. Since it is densely defined, it can be extended continuously to the whole space. Then, by the Riesz theorem, it has the representation

$$l_g(f) = (Af, g) = (f, g^*)$$

with some $g^* \in \mathcal{H}$. According to the definition, all such vectors g form the domain of the adjoint operator A^* to A, the corresponding vectors g^* form the range of A^*:

$$g \in \mathrm{Dom}\,A^*, \quad A^*g = g^* \in \mathrm{Ran}\,A^*.$$

In general, it may happen that $\mathrm{Dom}(A^*) = \{0\}$. However, in what follows we consider only operators A for which the adjoint operator also has domain dense in $\mathcal{H}$. If, in addition, the conditions

$$\mathrm{Dom}\,A \subseteq \mathrm{Dom}\,A^*, \quad Af = A^*f, \quad f \in \mathrm{Dom}\,A,$$

are fulfilled, then A is symmetric, $A \subseteq A^*$. If $A = A^*$, then the operator A is *self-adjoint.* We recall that a symmetric operator A is called *essentially self-adjoint* if its closure is self-adjoint, $A^{\mathrm{cl}} = A^*$.

It is worth noting that if A is a densely defined operator, then its adjoint operator is automatically closed $(A^*)^{\mathrm{cl}} = A^*$. If A is closable, then $A^* = (A^{\mathrm{cl}})^*$.

We emphasize that if A is a symmetric, but not self-adjoint, then its adjoint A^* is necessarily an extension of A, i.e., $A \subset A^*$.

2.2.2 Self-adjointness

In applications, we often meet formally symmetric operators, for which it is initially not clear whether they are essentially self-adjoint.

What additional conditions on symmetric operators guarantee their self-adjointness? In the literature one can find many sufficient conditions for (essential) self-adjointness, mainly for symmetric differential operators in different settings and specific problems (see, e.g., [42] and the bibliography listed there).

The following theorem contains simple necessary and sufficient conditions for self-adjointness.

Theorem 2.2.1 (Basic self-adjointness criterion, see [169]). *Let $A \subseteq A^*$ be a symmetric operator on a Hilbert space $\mathcal{H}$. Then the following three statements are equivalent:*

(a) $A = A^*$.

(b) *A is closed and* $\operatorname{Ker}(A^* \pm i) = \{0\}$.

(c) $\operatorname{Ran}(A \mp i) = \mathcal{H}$.

Proof. Let us show that (a) $\Longrightarrow$ (b). Indeed, the operator A is closed, since A^* is closed by construction and $A = A^*$. Suppose that

$$A^*\varphi = i\varphi \quad (\text{or } A^*\varphi = -i\varphi)$$

for some vector $\varphi \in \operatorname{Dom} A^*$.

Then from the equality $A = A^*$ it follows that $A\varphi = \pm i\varphi$ and

$$\pm i(\varphi, \varphi) = (A\varphi, \varphi) = (\varphi, A^*\varphi) = (\varphi, A\varphi) = \mp i(\varphi, \varphi).$$

It means that $\varphi = 0$. So, (b) holds.

Let us prove that (b) $\Longrightarrow$ (c). We claim that if the equation $A^*\varphi = \pm i\varphi$ has only the trivial solution, then the range $\operatorname{Ran}(A \mp i)$ is dense in $\mathcal{H}$. Indeed, otherwise for $\varphi \perp \operatorname{Ran}(A \pm i)$ we would have

$$((A \pm i)\psi, \varphi) = 0, \quad \forall \psi \in \operatorname{Dom} A.$$

But then $\varphi \in \operatorname{Dom} A^*$ and $(A^* \mp i)\varphi = 0$. However, for $\varphi \neq 0$, this contradicts (b). Hence $(\operatorname{Ran}(A \mp i))^{\mathrm{cl}} = \mathcal{H}$. Conversely, it is clear that if $(\operatorname{Ran}(A \mp i))$ is dense in $\mathcal{H}$ then $\operatorname{Ker}(A^* \pm i) = \{0\}$. Further, from the equality

$$\|(A \mp i)\varphi\|^2 = \|A\varphi\|^2 + \|\varphi\|^2,$$

we obtain $\operatorname{Ran}(A \mp i) = \mathcal{H}$, due to the closability of the operator A.

Let us finally verify the implication (c) $\Longrightarrow$ (a). Let $\varphi \in \operatorname{Dom} A^*$. Since $A \subseteq A^*$ and $\operatorname{Ran}(A \mp i) = \mathcal{H}$, for each $\varphi \in \operatorname{Dom} A^*$ there exists a vector $\eta \in \operatorname{Dom} A$ such that

$$(A \pm i)\eta = (A^* \pm i)\varphi.$$

Hence $(A^* \pm i)(\varphi - \eta) = 0$. Using the already established fact $\operatorname{Ker}(A^* \pm i) = \{0\}$, we conclude that $\varphi - \eta = 0$. Hence, $\varphi = \eta$, and so $\operatorname{Dom} A = \operatorname{Dom} A^*$ and $A = A^*$. $\square$

Let us note that for a symmetric operator which is not closed, the equality $\operatorname{Ker}(A^* \pm i) = \{0\}$ follows from the fact that $\operatorname{Ran}(A \mp i)$ is dense in $\mathcal{H}$. In this case the condition (c) can be rewritten as (c$'$) $\operatorname{Ran}(A \mp i)^{\mathrm{cl}} = \mathcal{H}$. This means that for an essentially self-adjoint operator A, the set $\operatorname{Ran}(A \mp i)$ is dense in $\mathcal{H}$. The equality $\operatorname{Ran}(A \mp i) = \mathcal{H}$ holds true only for a self-adjoint operator. Indeed, due to the closability of A, if the sequence $\varphi_n \in \operatorname{Dom} A$ converges to φ and the sequence $A\varphi_n$ is convergent, then $\varphi \in \operatorname{Dom} A$ and the vector $\psi = (A \mp i)\varphi \in \operatorname{Ran}(A \mp i)$. But the convergence of $\psi_n = (A \mp i)\varphi_n$ is equivalent to the convergence of the two sequences φ_n and $A\varphi_n$. Hence $(\operatorname{Ran}(A \mp i))^{\mathrm{cl}} = \operatorname{Ran}(A \mp i) = \mathcal{H}$.

2.3 Basic concepts of extension theory

We describe briefly one of the variants of extension theory of symmetric operators founded by J. von Neumann [161] (for details see [32, 65, 75, 82, 92, 147, 148, 168, 179]).

Let $\mathbf{A} \subset \mathbf{A}^*$ be a closed symmetric (but not self-adjoint) operator on the Hilbert space $\mathcal{H}$. Since $\mathbf{A}$ is symmetric, for each $z \in \mathbb{C}$, $\operatorname{Im}(z) \neq 0$, it holds that

$$\|(\mathbf{A} - z)\varphi\|^2 \geq b^2\|\varphi\|^2, \quad \varphi \in \operatorname{Dom}(\mathbf{A}),\ z = a + ib,\ b \neq 0.$$

This inequality shows that the mapping

$$(\mathbf{A} - \bar{z})^{-1} : \mathcal{M}_z \longrightarrow \operatorname{Dom} \mathbf{A}$$

is a bounded operator defined on the subspace

$$\mathcal{M}_z := \operatorname{Ran}(\mathbf{A} - \bar{z}) \neq \mathcal{H}.$$

It is clear that $\mathcal{M}_z$ is closed for $z \in \mathbb{C}$, $\operatorname{Im}(z) \neq 0$. Hence, the operator $(\mathbf{A} - \bar{z})^{-1}$ is bounded, but not densely defined in $\mathcal{H}$. Next, since $\mathbf{A} \neq \mathbf{A}^*$, the subspace

$$\mathcal{N}_z = \mathcal{M}_z^{\perp} = \mathcal{H} \ominus \mathcal{M}_z \neq \{0\}.$$

It is easy to see that this subspace consists of eigenvectors of the adjoint of $\mathbf{A}$:

$$\mathbf{A}^*\eta_z = z\eta_z, \quad \eta_z \in \mathcal{N}_z.$$

Thus,

$$\mathcal{N}_z = \operatorname{Ker}(\mathbf{A}^* - z).$$

Recall that, $\mathcal{N}_z$ is called the *deficiency* (or *defect*) *subspace* of the operator $\mathbf{A}$ the point z. The dimension $\dim \mathcal{N}_z =: \mathbf{n}_z$ is a numerical characteristic measuring the non-self-adjointness of the operator $\mathbf{A}$. It is known that the value $\mathbf{n}_z$ remains constant when z runs in the open upper half-plane $\mathbb{C}_+$ and in the open lower half-plane $\mathbb{C}_-$ of the complex plane $\mathbb{C}$. Hence, two numbers (deficiency indices of the operator $\mathbf{A}$) measure the degree of non-self-adjointness:

$$\mathbf{n}_+ = \dim \mathcal{N}_z, \quad \mathbf{n}_- = \dim \mathcal{N}_{\bar{z}}, \quad \operatorname{Im}(z) > 0.$$

In general $\mathbf{n}_+, \mathbf{n}_-$ can be arbitrary natural numbers, and can be equal to $+\infty$.

It is easy to see that the symmetric operator $\mathbf{A}$ has a self-adjoint extension in $\mathcal{H}$ only if

$$\mathbf{n}_+ = \mathbf{n}_-. \tag{2.1}$$

This equality holds automatically for the operators bounded from below, namely, if

$$(\mathbf{A}\varphi, \varphi) \geq m\|\varphi\|^2.$$

In such a case, $\dim \mathcal{N}_z$ is constant for all $z \in \mathbb{C} \setminus [m, \infty)$.

It is clear that every self-adjoint operator has null deficiency indices: $\mathbf{n}_- = \mathbf{n}_+ = 0$.

Let us consider a domain $\operatorname{Dom} \mathbf{A}^*$ of the adjoint operator. Due to the closability of $\mathbf{A}^*$, its domain is a complete Hilbert space with respect to the graph-norm of $\mathbf{A}^*$. We denote this space by $\mathcal{H}_{+,*}$. The inner product in $\mathcal{H}_{+,*}$ is given by

$$(\varphi, \psi)_{+,*} = (\varphi, \psi) + (\mathbf{A}^*\varphi, \mathbf{A}^*\psi), \quad \varphi, \psi \in \operatorname{Dom} \mathbf{A}^*.$$

Clearly, $\mathcal{H}_{+,*}$ is positive space with respect to $\mathcal{H}$ in the sense of [42], which means that

$$\|\cdot\| \leq \|\cdot\|_{+,*}, \quad \mathcal{H} \sqsupset \mathcal{H}_{+,*},$$

where $\sqsupset$ denotes the dense and continuous embedding of $\mathcal{H}_{+,*}$ into $\mathcal{H}$. It turns out that the domain $\operatorname{Dom} \mathbf{A}$ of the operator $\mathbf{A}$ is a proper closed subspace in $\mathcal{H}_{+,*}$, though this set is dense in $\mathcal{H}$. So, we can write

$$\mathcal{H}_{+,*} = \operatorname{Dom} \mathbf{A}^* = \operatorname{Dom} \mathbf{A} \oplus \mathcal{K}, \tag{2.2}$$

where $\dim \mathcal{K} > 0$. If a nontrivial subspace $\mathcal{K}$ exists, then $\mathbf{A}$ is a non-self-adjoint operator. The dimension of $\mathcal{K}$ characterizes the non-self-adjointness defect of the operator $\mathbf{A}$. In other words, the value $\dim \mathcal{K}$ shows, how far from a self-adjoint operator is our operator $\mathbf{A}$.

Next, we suppose that the condition (2.1) holds. We denote

$$\mathcal{K}_+ \equiv \mathcal{N}_i = \operatorname{Ker}(\mathbf{A}^* - i), \quad \mathcal{K}_- \equiv \mathcal{N}_{-i} = \operatorname{Ker}(\mathbf{A}^* + i).$$

Using the introduced subspaces $\mathcal{K}_\pm$, one can give a more precise description of the domain $\operatorname{Dom}\mathbf{A}^*$ of the adjoint operator $\mathbf{A}^*$.

Proposition 2.3.1. *The domain* $\operatorname{Dom}\mathbf{A}^*$ *of the adjoint operator* $\mathbf{A}^*$ *has the orthogonal decomposition:*

$$\operatorname{Dom}\mathbf{A}^* = \mathcal{H}_{+,*} = \operatorname{Dom}\mathbf{A} \oplus \mathcal{K}_+ \oplus \mathcal{K}_-. \tag{2.3}$$

Proof. The subspaces $\operatorname{Dom}\mathbf{A}$, $\mathcal{K}_+$, and $\mathcal{K}_-$ are mutually orthogonal in the space $\mathcal{H}_{+,*}$. This fact follows directly from the relations $\mathbf{A}^*\mathcal{K}_\pm = (\pm i)\mathcal{K}_\pm$. In order to prove (2.3), namely that each $h \in \operatorname{Dom}\mathbf{A}^*$ as a vector of the space $\mathcal{H}_{+,*}$ is decomposed into an orthogonal sum of three components, $h = \varphi \oplus k_+ \oplus k_-$, where $\varphi \in \operatorname{Dom}\mathbf{A}$, $k_\pm \in \mathcal{K}_\pm$, we assume that there exists a vector ψ in $\operatorname{Dom}\mathbf{A}^*$ that is orthogonal to

$$\operatorname{Dom}\mathbf{A} \oplus \mathcal{K}_+ \oplus \mathcal{K}_-$$

in $\mathcal{H}_{+,*}$. Then, in particular, for all $\varphi \in \operatorname{Dom}\mathbf{A}$ we have

$$(\varphi, \psi)_{+.*} = 0 = (\varphi, \psi) + (\mathbf{A}\varphi, \mathbf{A}^*\psi),$$

or

$$(\varphi, \psi) = -(\mathbf{A}\varphi, \mathbf{A}^*\psi).$$

Then the linear functional

$$l_g(\varphi) = (\varphi, \psi) = (\mathbf{A}\varphi, g), \quad g = -\mathbf{A}^*\psi, \quad \varphi \in \operatorname{Dom}\mathbf{A},$$

is obviously continuous in φ in $\mathcal{H}$. Then the vector $g \in \operatorname{Dom}\mathbf{A}^*$, and

$$\mathbf{A}^*g = -\mathbf{A}^*\mathbf{A}^*\psi = \psi.$$

Consequently,

$$(\mathbf{A}^*\mathbf{A}^* + 1)\psi = 0 = (\mathbf{A}^* + i)(\mathbf{A}^* - i)\psi,$$

and so $(\mathbf{A}^* - i)\psi \in \mathcal{K}_-$. But for all vectors $k_- \in \mathcal{K}_-$ $(A^*k_- = -ik_-)$ we have

$$i((\mathbf{A}^* - i)\psi, k_-) = (\psi, k_-) + (\mathbf{A}^*\psi, \mathbf{A}^*k_-) = (\psi, k_-)_{+,*} = 0,$$

since, by the assumption, $\psi \perp \mathcal{K}_-$ in $\mathcal{H}_{+,*}$. It means that $(\mathbf{A}^* - i)\psi = 0$, namely $\psi \in \mathcal{K}_+$. But this contradicts our assumption on ψ. Hence $\psi = 0$, and the equality (2.3) is proved. □

Of course, the domain $\operatorname{Dom}\mathbf{A}^*$ can also be represented as a sum of three subspaces in the space $\mathcal{H}$. To make it we use the direct sum instead of the orthogonal sum by:

$$\operatorname{Dom}\mathbf{A}^* = \operatorname{Dom}\mathbf{A} \dotplus \mathcal{K}_+ \dotplus \mathcal{K}_-. \tag{2.4}$$

But we need to be sure that $\operatorname{Dom}\mathbf{A}$ has a trivial intersection with each of the subspaces $\mathcal{K}_\pm$.

Proposition 2.3.2. *The domain* Dom $\mathbf{A}$ *of each symmetric operator has trivial intersection with each of the subspaces* $\mathcal{K}_\pm$*:*

$$\operatorname{Dom}\mathbf{A}\cap\mathcal{K}_+ = 0 = \operatorname{Dom}\mathbf{A}\cap\mathcal{K}_-$$

Proof. If $\varphi \in \operatorname{Dom}\mathbf{A}\cap\mathcal{K}_+$, then

$$\mathbf{A}\varphi = \mathbf{A}^*\varphi = i\varphi.$$

However, for a symmetric operator, this equality holds only for $\varphi = 0$. □

The next observation was crucial in the von Neumann extension theory of symmetric operators.

Consider the mapping

$$U := (\mathbf{A}-i)(\mathbf{A}+i)^{-1} : \mathcal{K}_+^\perp \longrightarrow \mathcal{K}_-^\perp. \tag{2.5}$$

In the literature (see, e.g., monograph [32]), the operator U is called the Cayley transformation. It is isometric, since

$$\|(\mathbf{A}+i)\varphi\|^2 = \|(\mathbf{A}-i)\varphi\|^2, \quad \varphi\in\operatorname{Dom}\mathbf{A},$$

and maps the whole subspace $\mathcal{M}_i = \mathcal{K}_+^\perp$ onto the whole subspace $\mathcal{M}_{-i} = \mathcal{K}_-^\perp$.

It is easy to see that for any closed symmetric extension $\tilde{A}$ of the operator $\mathbf{A}$,

$$\mathbf{A}\subset\tilde{A}\subseteq\tilde{A}^*\subset\mathbf{A}^*,$$

the corresponding Cayley transformation $\tilde{U}$ is an extension of U. Therefore, the operator $\tilde{A}$ is self-adjoint only if the mapping $\tilde{U}$ is unitary, or equivalently, only if the corresponding deficiency subspaces is trivial, i.e., if $\tilde{\mathcal{K}}_+ = 0 = \tilde{\mathcal{K}}_-$, and then $\tilde{\mathcal{K}}_+^\perp = \mathcal{H} = \tilde{\mathcal{K}}_-^\perp$. Thus, the symmetric (self-adjoint) extensions of the operator $\mathbf{A}$ correspond to isometric (unitary) extensions of the Cayley transformation U.

The following theorem is the main result in this direction.

Theorem 2.3.3. *Let* $\mathbf{A}\subset\mathbf{A}^*$ *be a closed symmetric operator on* $\mathcal{H}$ *and*

$$\mathcal{K}_+ = \mathcal{N}_i = \operatorname{Ker}(\mathbf{A}^*-i), \quad \mathcal{K}_- = \mathcal{N}_{-i} = \operatorname{Ker}(\mathbf{A}^*+i)$$

be its defect subspaces. Let $V:\mathcal{K}_+\to\mathcal{K}_-$ *be an arbitrary (partially) isometric transformation between the deficiency subspaces. Then the operator* $\tilde{A}$ *defined by the von Neumann formula*

$$\tilde{A}\tilde{\varphi} = \mathbf{A}\varphi + i\varphi_+ - iV\varphi_+, \tag{2.6}$$

$$\operatorname{Dom}\tilde{A} := \{\tilde{\varphi}\in\mathcal{H} \mid \tilde{\varphi} = \varphi+\varphi_+ + V\varphi_+,\ \varphi\in\operatorname{Dom}\mathbf{A}\}, \tag{2.7}$$

where $\varphi_+\in\operatorname{Dom}V\subseteq\mathcal{K}_+$*, is a closed symmetric extension of the operator* $\mathbf{A}$*. Conversely, every closed symmetric extension* $\mathbf{A}\subset\tilde{A}$ *uniquely corresponds to some extension of the Cayley transformation* (2.5)*,* $U\subset\tilde{U}$*, which is defined by a fixed (partially) isometric operator* $V:\mathcal{K}_+\to\mathcal{K}_-$*.*

To prove this theorem we need some preparation.

Note that each symmetric extension $\mathbf{A} \subset \tilde{A}$ yields a restriction of the adjoint operator, $\tilde{A}^* \subset \mathbf{A}^*$. So, always

$$\mathbf{A} \subset \tilde{A} \subseteq \tilde{A}^* \subset \mathbf{A}^*. \tag{2.8}$$

Indeed, let $\tilde{A}$ be some symmetric extension of the symmetric operator $\mathbf{A}$. Then, for all $\varphi \in \operatorname{Dom} \mathbf{A}$ and $\psi \in \operatorname{Dom} \mathbf{A}^*$ we have

$$(A\varphi, \psi) = (\tilde{A}\varphi, \psi) = (\varphi, \tilde{A}^*\psi).$$

This means that $\psi \in \operatorname{Dom} \mathbf{A}^*$ and, moreover, $\mathbf{A}^*\psi = \tilde{A}^*\psi$. Thus, $\tilde{A}^*$ is a restriction of $\mathbf{A}^*$. Therefore, every asymmetric extension $\tilde{A}$ of the operator $\mathbf{A} \subset \mathbf{A}^*$ is contained in the adjoint operator: $\tilde{A} \subset \mathbf{A}^*$.

From (2.2), (2.3), and (2.5) it follows that the domain of an arbitrary closed symmetric extension $\tilde{A}$ of the operator A in the space $\mathcal{H}$ has a form

$$\operatorname{Dom} \tilde{A} = \operatorname{Dom} \mathbf{A} \oplus \tilde{\mathcal{K}}, \tag{2.9}$$

where $\tilde{\mathcal{K}}$ is a subspace in $\mathcal{K} = \mathcal{K}_+ \oplus \mathcal{K}_-$. It is clear that $\tilde{\mathcal{K}}$ is not an arbitrary subspace in $\mathcal{K}$, but it has a specific property, due to the fact that the operator $\tilde{A}$ is symmetric. To explain this in more detail, let us consider the sesquilinear form on $\mathcal{H}_{+,*}$ defined by

$$B(\varphi, \psi) := (\varphi, \mathbf{A}^*\psi) - (\mathbf{A}^*\varphi, \psi), \quad \varphi, \psi \in \operatorname{Dom} \mathbf{A}^*, \tag{2.10}$$

which is called the *boundary form* of the operator $\mathbf{A}$. The following statement is obvious.

Proposition 2.3.4. *The domain* $\operatorname{Dom} \mathbf{A}$ *of a symmetric operator* $\mathbf{A}$ *consists of isotropic vectors of the boundary form:*

$$B(\varphi, \varphi) \equiv B[\varphi] = 0, \quad \forall \varphi \in \operatorname{Dom} \mathbf{A}. \tag{2.11}$$

Therefore, the boundary form $B(\varphi, \psi)$ of a symmetric operator, as a quadratic form on the space $\mathcal{H}$, is singular in the sense that it equals to zero on a dense subset (the precise definition of a singular quadratic form is given in Chapter 5.

A subspace $\tilde{\mathfrak{D}}$ in $\mathcal{H}_{+,*}$ is called $\mathbf{A}$-symmetric (see [170]), if

$$B(\varphi, \psi) = 0, \quad \varphi, \psi \in \tilde{\mathfrak{D}}. \tag{2.12}$$

Now, it is clear that the symmetric extensions of an operator $\mathbf{A}$ are determined by the $\mathbf{A}$-symmetric extensions $\tilde{\mathfrak{D}}$ of the domain $\operatorname{Dom} \mathbf{A}$. The following statement serves in describing all symmetric extensions.

Lemma 2.3.5. *A subspace* $\tilde{\mathfrak{D}} = \operatorname{Dom} \mathbf{A} \oplus \tilde{\mathcal{K}}$ *of* $\mathcal{H}_{+,*}$ *is* $\mathbf{A}$*-symmetric if and only if* $\tilde{\mathcal{K}}$ *is also* $\mathbf{A}$*-symmetric.*

Proof. Exercise for the reader. □

The following fact is deeper.

Proposition 2.3.6. *A subspace* $\tilde{\mathcal{K}} \subset \mathcal{K}_+ \oplus \mathcal{K}_-$ *is* **A**-*symmetric if and only if it is the graph of the (partially) isometric mapping*

$$V : \mathcal{K}_+ \longmapsto \mathcal{K}_-.$$

Proof. Let $\tilde{\mathcal{K}}$ be **A**-symmetric. Then for each vector $\tilde{\varphi} \in \tilde{\mathcal{K}}$, expressed as a sum $\tilde{\varphi} = \varphi_+ \oplus \varphi_-$, we have

$$B(\tilde{\varphi}, \tilde{\varphi}) = 0 = 2i(\varphi_+, \varphi_+) - 2i(\varphi_-, \varphi_-).$$

Hence $\|\varphi_+\| = \|\varphi_-\|$. Then, since $\tilde{\mathcal{K}}$ is a linear space, there exists a partially isometric operator V acting from $\mathcal{K}_+$ into $\mathcal{K}_-$, such that

$$V\varphi_+ = \varphi_-.$$

Conversely, if $\tilde{K}$ is the graph of some (partially) isometric operator $V : \mathcal{K}_+ \mapsto \mathcal{K}_-$, then we check directly that $B(\tilde{\varphi}, \tilde{\psi}) = 0$ for arbitrary vectors $\tilde{\varphi} = \varphi_+ \oplus V\varphi_+$ and $\tilde{\psi} = \psi_+ \oplus V\psi_+$, where

$$\varphi_+, \psi_+ \in \operatorname{Dom} V \subseteq \mathcal{K}_+, \quad V\varphi_+, V\psi_+ \in \mathcal{K}_-.$$ □

Proof of Theorem 2.3.3. Let there be given a transformation $V : \mathcal{K}_+ \mapsto \mathcal{K}_-$. Then, by Proposition 2.3.4 and Lemma 2.3.5, the set $\operatorname{Dom} \tilde{A}$, defined in (2.7), is **A**-symmetric in $\mathcal{H}_{+,*}$. Indeed, this set, as a closed subspace in $\mathcal{H}_{+,*}$, has the representation:

$$\operatorname{Dom} \tilde{A} = \operatorname{Dom} \mathbf{A} \oplus \tilde{\mathcal{K}},$$

with

$$\tilde{\mathcal{K}} = \{\varphi_+ \oplus V\varphi_+ \mid \varphi_+ \in \operatorname{Dom} V \subseteq \mathcal{K}_+, \; V\varphi_+ \in \mathcal{K}_-\}.$$

Therefore, one immediately has that

$$B(\tilde{\varphi}, \tilde{\psi}) = 0, \tilde{\varphi}, \quad \tilde{\psi} \in \operatorname{Dom} \tilde{A}. \tag{2.13}$$

It follows from (2.13) that the operator $\tilde{A}$, defined by the equality (2.6), is symmetric and closed. Indeed, since $\tilde{A}$ is a restriction of the operator $\mathbf{A}^*$, we have

$$(\tilde{A}\tilde{\varphi}, \tilde{\psi}) = (\mathbf{A}^*\tilde{\varphi}, \tilde{\psi}) = (\tilde{\varphi}, \mathbf{A}^*\tilde{\psi}) = (\tilde{\varphi}, \tilde{A}\tilde{\psi}), \quad \tilde{\varphi}, \tilde{\psi} \in \operatorname{Dom} \tilde{A}.$$

Conversely, if $\tilde{A}$ is a closed symmetric extension of an operator **A** (and it is also a restriction of the adjoint operator $\mathbf{A}^*$), then its domain, according to Proposition 2.3.4, consists of isotropic vectors, i.e., $B(\tilde{\varphi}, \tilde{\psi}) = 0$, $\tilde{\varphi}, \tilde{\psi} \in \operatorname{Dom} \tilde{A}$. Since the set $\operatorname{Dom} \tilde{A}$ in the space $\mathcal{H}_{+,*}$ has a representation (2.9), due to Lemma 2.3.5, we conclude that $\tilde{\mathcal{K}}$ is **A**-symmetric. Now, Proposition 2.3.6 ensures the existence of the (partially) isometric operator $V : \mathcal{K}_+ \mapsto \mathcal{K}_-$, which gives an extension of the Cayley transformation: $\tilde{U} = U \oplus V$. □

Let us recall some definitions (see [75, 82]). Any linear continuous on $\mathcal{H}_{+,*}$ functional $l(h)$, $h \in \mathcal{H}_{+,*} = \mathrm{Dom}(\mathbf{A}^*)$, which is equal to zero on $\mathrm{Dom}\,\mathbf{A}$, $l(\varphi) = 0$, $\varphi \in \mathrm{Dom}\,\mathbf{A}$, is called a *boundary value*, for the symmetric operator $\mathbf{A}$.

Since $\mathcal{H}_{+,*}$ is a Hilbert space, then by the Riesz theorem, each boundary value for the operator $\mathbf{A}$ is determined by some vector $\psi \in \mathcal{H}_{+,*} \ominus \mathrm{Dom}\,\mathbf{A}$. In fact, (2.3) shows that the vector $\psi \in \mathcal{K} = \mathcal{K}_+ \oplus \mathcal{K}_-$.

Next, if the operator $\mathbf{A}$ has finite deficiency indices, i.e., $\mathbf{n}^+ = \mathbf{n}^- = \mathbf{n} < \infty$, then there exist only $2\mathbf{n}$ linearly independent boundary values, since $\dim \mathcal{K} = 2\mathbf{n}$. An arbitrary set of linearly independent boundary values $l_i(\cdot)$, $i = 1, 2, \ldots, 2\mathbf{n}$ is called a complete system. If a complete system of boundary values is fixed, then the boundary form $B(\varphi, \psi)$ has a unique representation

$$B(\varphi, \psi) = \sum_{i,j=1}^{2\mathbf{n}} b_{ij} l_i(\varphi) \overline{l_j(\psi)}.$$

In addition, the matrix $\{b_{ij}\}$ is symmetric (see more details [75]).

If $l(\cdot)$ is some boundary value for $\mathbf{A}$, then the equality

$$l(h) = 0, \quad h \in \mathrm{Dom}\,\mathbf{A}^*$$

is called a *boundary condition*. Of course, an extension of a symmetric operator $\mathbf{A}$ is defined by a certain set of boundary conditions $l_i(h) = 0$, $i = 1, \ldots, k$. The set of boundary conditions

$$l_i(h) = 0, \quad i = 1, \ldots, k \tag{2.14}$$

is called symmetric, if the boundary form is equal zero for all $h, g \in \mathrm{Dom}\,\mathbf{A}^*$ which satisfy (2.14): $B(h, g) = 0$.

The following theorem is important for the sequel.

Theorem 2.3.7 ([75]). *Let* $\mathbf{A}$ *be a symmetric operator with finite and equal deficiency indices* $\mathbf{n}^+ = \mathbf{n}^- = \mathbf{n} < \infty$. *The restriction of* $\mathbf{A}^*$ *to some subset of vectors from* $\tilde{\mathfrak{D}} \subset \mathrm{Dom}\,\mathbf{A}^*$, *is a self-adjoint operator if and only if all vectors* $h \in \tilde{\mathfrak{D}}$ *satisfy the symmetric boundary conditions* (2.14) *given by a system of* $k = \mathbf{n}$ *linearly independent boundary values.*

2.3.1 Examples

Here are a number of examples regarding the foregoing.

Example 2.3.8. Take $\mathcal{H} = L_2(\mathbb{R})$ and consider the operator of multiplication by the independent variable, $Tf(x) = xf(x)$, with a domain

$$\mathrm{Dom}\,T = \{f \in L_2 \mid xf(x) \in L_2\}.$$

It is easy to see that this operator is self-adjoint, $T = T^*$.

Example 2.3.9. Take $\mathcal{H} = L_2(\mathbb{R})$ and $Tf(x) = f'(x)$, with the domain

$$\operatorname{Dom} T = \{f \in L_2 \mid f \in C_0^1(\mathbb{R}^1)\}.$$

It is obvious that the operator T is not symmetric. Also it is easy to see that the operator $(Sf)(x) = if'(x)$ with the same domain $\operatorname{Dom} S = \operatorname{Dom} T$ is symmetric, but not self-adjoint.

Example 2.3.10. Let $T = i\frac{d}{dx}$ be given in the space $L_2(0,1)$ with the domain

$$\operatorname{Dom} T = \{\varphi \in L_2 \mid \varphi \in AC[0,1],\ \varphi(0) = 0 = \varphi(1)\},$$

where $AC[0,1]$ is the set of absolutely continuous functions, which belong to L_2 together with their derivatives. It is easy to see that the adjoint operator acts by the same rule, but has a larger domain:

$$T^* = i\frac{d}{dx}, \quad \operatorname{Dom} T^* = AC[0,1].$$

The functions $\psi \in \operatorname{Dom} T^*$ need not satisfy any boundary conditions at the points 0 and 1. Indeed, for $\varphi \in \operatorname{Dom} T$, $\psi \in \operatorname{Dom} T^*$, since

$$(\varphi', \psi) - (\varphi, \psi') = \varphi(1)\bar{\psi}(1) - \varphi(0)\bar{\psi}(0) = 0,$$

we have $(T\varphi, \psi) = (\varphi, T^*\psi)$. So, in this example, the operator T is closed, but not self-adjoint. Its deficiency subspaces are one-dimensional. Indeed, all solutions of $(T^* \mp i)\eta_\pm(x) = 0$ coincide with the functions $\eta_\pm(x) = \exp(\pm x)$ to within constants.

Let us consider an arbitrary self-adjoint extension S of the operator T. Let $\varphi \in \operatorname{Dom} S \setminus \operatorname{Dom} T$. Since $\operatorname{Dom} S \subset \operatorname{Dom} T^*$, the vector $\varphi \in AC[0,1]$. Now, the equality $(S\varphi, \varphi) = (\varphi, S\varphi)$ is equivalent to $|\varphi(1)|^2 - |\varphi(0)|^2 = 0$. This equality imposes a nontrivial construct, since in general, $\varphi(1) \neq 0$ and $\varphi(0) \neq 0$, and $\varphi \notin \operatorname{Dom} T$. The condition $|\varphi(1)|^2 - |\varphi(0)|^2 = 0$, is obviously satisfied, if $\varphi(1) - \alpha\varphi(0) = 0$ with some fixed α such that $|\alpha| = 1$. We need to fix a constant α due to the linearity of the set $\operatorname{Dom} S$. Denote $S = T_\alpha$, where $T_\alpha = i\frac{d}{dx}$ with the domain

$$\operatorname{Dom} T_\alpha = \{\varphi \in AC[0,1] \mid \varphi(1) = \alpha\varphi(0),\ |\alpha| = 1\}.$$

Then S is a self-adjoint extension of the original operator T. Thus, the set of all self-adjoint extensions of the symmetric operator T is parameterized by the complex number α with $|\alpha| = 1$:

$$\{T \subset S = S^* \subset T^*\} = \{S = T_\alpha,\ |\alpha| = 1\}.$$

In other words, the set $\{T_\alpha\}_{|\alpha|=1}$ is parameterized by unit circle.

Example 2.3.11. Let us construct the operator T_α from the previous example, using the von Neumann theory and formulas (2.6).

Let us describe the deficiency subspaces $\mathcal{K}_\pm = \mathcal{N}_{\pm i}$ of the symmetric operator T. It is clear that the function $\psi \in \mathcal{K}_+$ is a solution of the equation

$$T^*\psi = i\psi' = i\psi,$$

only if it is absolutely continuous. Hence, ψ belongs to the subspace $\mathcal{K}_+$, since ψ is absolutely and arbitrary times differentiable due to $\psi' = \psi$. Now, it is easy to understand that the deficiency subspaces of the operator T are one-dimensional and have the form

$$\mathcal{K}_+ = \{ce^x | c \in \mathbb{C}\}, \quad \mathcal{K}_- = \{ce^{-x} | c \in \mathbb{C}\}.$$

The functions $\psi_+ = c_+e^x$ and $\psi_- = c_-e^{-x}$ with $c_+ = \sqrt{\frac{2}{e^2-1}}$ and $c_- = e\sqrt{\frac{2}{e^2-1}}$, respectively, are normalized to one. It is obvious that all partially isometric mappings between $\mathcal{K}_+$ and $\mathcal{K}_+$ are parameterized by functions γ satisfying $|\gamma| = 1$:

$$U_\gamma : \mathcal{K}_+ \longrightarrow \mathcal{K}_-, \quad U_\gamma\psi_+ = \gamma\psi_-.$$

According to Theorem 2.3.3, each self-adjoint extension A_γ of the operator T is specified by a number γ, $|\gamma| = 1$, and is given by the restriction of the adjoint operator $T^* = i\frac{d}{dx}$ to the domain

$$\mathrm{Dom}A_\gamma = \{\varphi + ce^x + \gamma ce^{-x} \mid \varphi \in \mathrm{Dom}\, T,\ c \in \mathbb{C}\}.$$

It is easily verified that $A_\gamma = T_\alpha$ with $\gamma = (\alpha - e)/(1 - \alpha e)$.

The following example shows how much can be different the spectra of symmetric operators depending on the boundary conditions.

Example 2.3.12. Consider the operators T_1 and T_2, which act in $L_2(0, 1)$ as $i\frac{d}{dx}$ with domains:

$$\mathrm{Dom}\, T_1 = \{\varphi \in AC[0, 1]\}, \quad \mathrm{Dom}\, T_2 = \{\varphi \in AC[0, 1] \mid \varphi(0) = 0\}$$

(recall that $AC[0, 1]$ denotes the set of absolutely continuous functions on $[0, 1]$ belonging to L_2 together with their derivatives).

It is easy to see that both operators T_1 and T_2 are densely defined and closed. Moreover T_2 is a restriction of T_1: $T_2 \subset T_1$. We can show that the spectrum of the operator T_2 is purely point, consists of eigenvalues, and fills the whole complex plane:

$$\sigma(T_1) = \sigma_{\mathrm{pp}}(T_1) = \mathbb{C},$$

and $\rho(T_1) = \varnothing$. Indeed, $\eta_\lambda(x) = e^{-i\lambda x}$ is an eigenfunction for each $\lambda \in \mathbb{C}$, namely $T_1\eta_\lambda(x) = \lambda\eta_\lambda(x)$. The condition $\varphi(0) = 0$, $\varphi \in \mathrm{Dom}\, T_2$, does not hold for all $\eta_\lambda(x)$. Then the restriction of T_1 to T_2 leads to loss of all eigenvectors. More precisely, the resolvent set occupies the whole space $\rho(T_2) = \mathbb{C}$, and the spectrum of T_2 is empty. Let us note that the corresponding resolvent $R_\lambda(T_2) = (T_2 - \lambda)^{-1}$ acts as (for details, see [169])

$$R_\lambda(T_2)g(x) = i \int_0^1 e^{-i\lambda(x-y)} g(y)dy.$$

2.4 Closable quadratic forms in a Hilbert space

Let Φ be an abstract linear space over the field of complex numbers $\mathbb{C}$. The complex-valued function $\gamma(\varphi, \psi)$, $\varphi, \psi \in \Phi$, defined on $\Phi \times \Phi$ is called a *sesquilinear form* if it is linear in the first variable φ and antilinear in the second one ψ, in particular $\gamma(\varphi, \lambda\psi) = \bar{\lambda}\gamma(\varphi, \psi)$, $\lambda \in \mathbb{C}$. The mapping $\Phi \ni \varphi \mapsto \gamma[\varphi] \equiv \gamma(\varphi, \varphi)$ is called a *quadratic form.*

By the polarization identity a sesquilinear form is restored from the quadratic form:

$$\gamma(\varphi, \psi) = \frac{1}{4}(\gamma[\varphi + \psi] - \gamma[\varphi - \psi] + i\gamma[\varphi + \imath\psi] - i\gamma[\varphi - \imath\psi]).$$

For this reason, we use the term "the quadratic form" for $\gamma[\varphi]$, and also for $\gamma(\varphi, \psi)$.

Usually quadratic forms are considered in complex Hilbert spaces and have additional properties. In particular, each quadratic form $\gamma[\varphi]$, regarded as a mapping from $\mathcal{H}$ into $(-\infty, \infty]$, satisfies

$$\begin{aligned} \gamma[\varphi + \psi] + \gamma[\varphi - \psi] &= 2\gamma[\varphi] + 2\gamma[\psi], \\ \gamma[\lambda\varphi] &= |\lambda|^2\gamma[\varphi], \quad \lambda \in \mathbb{C}. \end{aligned} \tag{2.15}$$

The *domain* of the quadratic form denoted $\operatorname{Dom}\gamma \equiv Q(\gamma) =: \Phi$ is the set of all $\varphi \in \mathcal{H}$ such that $\gamma[\varphi] < \infty$. Φ is a linear space. If Φ is dense in $\mathcal{H}$, then the form γ is said to be *densely defined.*

If $\gamma(\varphi, \psi) = \overline{\gamma(\psi, \varphi)}$ for all $\psi, \varphi \in \Phi$, then the form γ is called *Hermitian*; we write $\gamma = \gamma^*$.

Let $\mathcal{H}$ be a Hilbert space with the norm $\|\varphi\|$. If $\gamma = \gamma^*$ and

$$m\|\varphi\|^2 \leq \gamma[\varphi]$$

for all $\varphi \in Q(\gamma)$, where the number $m > -\infty$ independent of φ, then γ is said to be *bounded from below* and we write $\gamma \geq m$. The largest number among all m's is called the *lower bound* of γ, and it is denoted by m_γ. If $\gamma \geq 0$, then the form is called *positive.*

It is known that each linear operators A, on a Hilbert space $\mathcal{H}$ generates a quadratic form by

$$\gamma_A(\varphi, \psi) = (A\varphi, \psi), \quad \varphi, \psi \in \operatorname{Dom} A = Q(\gamma). \tag{2.16}$$

If the operator A is bounded and defined on the whole space, $\operatorname{Dom} A = \mathcal{H}$, then the formula (2.16) establishes bijective correspondence between bounded operators and quadratic forms γ with $Q(\gamma) = \mathcal{H}$ (see Theorem 1.3.11).

A densely defined and bounded from below quadratic form γ on $\mathcal{H}$ is called *closed*, if the following implication holds true:

$$\begin{aligned} &\varphi_n \in Q(\gamma), \quad \varphi_n \longrightarrow \varphi \in \mathcal{H}, \quad \gamma[\varphi_n - \varphi_m] \longrightarrow 0 \\ &\qquad \Longrightarrow \varphi \in Q(\gamma) \ \text{ and } \ \gamma[\varphi_n - \varphi] \longrightarrow 0, \quad n, m \longrightarrow \infty. \end{aligned} \tag{2.17}$$

In this a case we write $\gamma = \gamma^{\mathrm{cl}}$ (cl stands for a closure). If the form γ has a closed extension, $\gamma \subseteq \gamma^{\mathrm{cl}}$, then it is called *closable.*

The minimal closed extension γ^{cl} of a closable form γ is called its *closure.*

Let $\gamma \geq 0$. We denote by $\mathcal{H}_{\gamma+\chi}$ the Hilbert space obtained by completing $Q(\gamma)$ with respect to the norm

$$\|\varphi\|_{\gamma+\chi} = (\gamma[\varphi] + \chi[\varphi])^{1/2}, \quad \chi[\varphi] := \|\varphi\|^2. \tag{2.18}$$

In the sequel, the space $\mathcal{H}_{\gamma+\chi}$ well be often denoted by $\mathcal{H}_\tau, \tau = \chi + \gamma$, or $\mathcal{H}_+$.

There is a well-known criterion for a quadratic form to be closed (see [107], Theorem VI.1.11).

A densely defined positive quadratic form γ on a Hilbert space $\mathcal{H}$ is *closed* if and only if its domain $Q(\gamma)$ coincides with the Hilbert space $\mathcal{H}_\tau$, $\tau = \gamma + \chi$ with respect to the norm (2.18).

It is easy to extend this criterion to bounded from below quadratic forms.

From the definition of the form (2.18) it follows that $\mathcal{H} \supset \mathcal{H}_{\gamma+\chi}$ and that this mapping is continuous. Hence, the closability criterion for the form $\gamma = \gamma^* \geq 0$ can be written as follows:

$$\gamma \subseteq \gamma^{\mathrm{cl}} \Longleftrightarrow \mathcal{H} \sqsupset \mathcal{H}_{\gamma+\chi}, \tag{2.19}$$

where $\sqsupset$ stands for dense and continuous embedding.

Due to linearity, from (2.19) it follows that a positive densely defined quadratic form γ on the Hilbert space $\mathcal{H}$ is closable if and only if each fundamental sequence φ_n in $\mathcal{H}_{\gamma+\chi}$ converging to zero in $\mathcal{H}$, converges to zero with respect to the form γ, i.e., $\gamma[\varphi_n] \to 0$.

2.4.1 Operator representation of closed quadratic form

There is a close connection between quadratic forms and linear operators on Hilbert space. This connection is bijective under certain conditions.

Let γ be a densely defined ($Q(\gamma)^{\mathrm{cl}} = \mathcal{H}$), symmetric quadratic form. According to the definition, the operator A_γ^{as} *associated* with the quadratic form γ, is constructed in the following way. A vector ψ from $Q(\gamma)$ belong to $\operatorname{Dom} A_\gamma^{\mathrm{as}}$, if the linear functional $l_\psi(\varphi) := \gamma(\varphi, \psi)$ is continuous in φ on $\mathcal{H}$. Then by the Riesz theorem $l_\psi(\varphi) := (\varphi, \psi^*)$ for some $\psi^* \in \mathcal{H}$. Let us put $A_\gamma^{\rm as}\psi = \psi^*$. Hence,

$$\gamma(\varphi, \psi) = (\varphi, A_\gamma^{\rm as}\psi), \quad \psi \in \operatorname{Dom} A_\gamma^{\rm as} \subset Q(\gamma).$$

Since the form γ is symmetric, the operator $A_\gamma^{\rm as}$ is obviously, Hermitian, but in general, may not be densely defined. For example, the operator $A_{\gamma_\delta}^{\rm as}$ on L_2 associated with the quadratic form γ_δ, generated by the delta-function, is defined only on the zero vector, $\operatorname{Dom} A_{\gamma_\delta}^{\rm as} = \{0\}$.

But, if a symmetric form γ is closed and bounded from below, then the associated operator is self-adjoint.

The following result is known as the first representation theorem [107, 170].

Theorem 2.4.1. *Each symmetric, closed, bounded from below quadratic form $\gamma = \gamma^{\rm cl} = \gamma^* \geq m_\gamma > -\infty$ on the Hilbert space $\mathcal{H}$ has an operator representation:*

$$\gamma(\varphi, \psi) = (A\varphi, \psi), \quad \varphi, \psi \in \operatorname{Dom} A \subseteq Q(\gamma), \tag{2.20}$$

where $A = A^* \geq m_A = m_\gamma$ is a self-adjoint operator on $\mathcal{H}$ associated with the form γ, i.e., $A = A_\gamma^{\rm as}$. Moreover, the correspondence (2.20) *between forms and operators is one-to-one:*

$$\gamma = \gamma^{\rm cl} = \gamma^* \geq m_\gamma > -\infty \Longleftrightarrow A \equiv A_\gamma^{\rm as} \geq m_A, \ m_A = m_\gamma. \tag{2.21}$$

Proof. Suppose that the lower bound of the quadratic form is

$$m_\gamma = \inf_{\|\varphi\|=1} \gamma[\varphi] = 1, \quad \varphi \in \operatorname{Dom} \gamma,$$

i.e., $1 \leq \gamma$. Then it is easy to show that $\operatorname{Dom}\gamma$ is a Hilbert space with an inner product $(\varphi, \psi)_1 := \gamma(\varphi, \psi)$. We denote this space by $\mathcal{H}_1$. Clearly $\mathcal{H}_1$ is densely and continuously embedded into $\mathcal{H}$; we write $\mathcal{H} \sqsupset \mathcal{H}_1$. We introduce the *rigged space*

$$\mathcal{H}_{-1} \sqsupset \mathcal{H} \sqsupset \mathcal{H}_1,$$

where $\mathcal{H}_{-1}$ is the space adjoint to $\mathcal{H}_1$ with respect to $\mathcal{H}$ (rigged spaces are discussed in more detail in Chapter 4). By the properties of the *Berezansky canonical isomorphism*,

$$D_{-1,1} : \mathcal{H}_1 \ni \psi \longmapsto \psi^* \in \mathcal{H}_{-1},$$

(the corresponding definition is given in Chapter 4), the linear functional

$$l_\psi(\varphi) := \langle \varphi, \psi^* \rangle_{-1,1} = (\varphi, \psi)_1, \quad \varphi \in \mathcal{H}_1, \tag{2.22}$$

is continuous on $\mathcal{H}_1$. Here $\langle \cdot, \cdot \rangle_{-1,1}$ denotes the duality pairing for $\mathcal{H}_{-1}$ and $\mathcal{H}_1$.

The operator A on $\mathcal{H}$ is defined as follows:

$$A = D_{-1,1} \restriction \operatorname{Dom} A, \quad \operatorname{Dom} A := \{\psi \in \mathcal{H}_1 \mid D_{-1,1}\psi = \psi^* \in \mathcal{H}\}.$$

It is easy to see that $\operatorname{Ran} A = \mathcal{H}$. Since the mappings $\mathcal{H}_{-1} \sqsupset \mathcal{H}$ and $D_{-1,1}$ are unitary, we conclude that the domain $\operatorname{Dom} A$ is dense in $\mathcal{H}_1$. Hence, the operator A is densely defined on $\mathcal{H}$. For all $\psi \in \operatorname{Dom} A$, the functional $l_\psi(\varphi)$, defined in (2.22) is, obviously, continuous on $\mathcal{H}$. Consequently $A = A_\gamma^{\mathrm{as}}$.

The following relations show that the operator A is symmetric:

$$(A\varphi, \psi) = \langle D_{-1,1}\varphi, \psi\rangle_{-1,1} = (\varphi, \psi)_1 = \gamma(\varphi, \psi)$$
$$= \overline{\gamma(\psi, \varphi)} = \overline{(A\psi, \varphi)} = (\varphi, A\psi), \quad \varphi, \psi \in \operatorname{Dom} A.$$

Since $\operatorname{Ran} A = \mathcal{H}$, the operator A is self-adjoint and $A = A^* \geq 1$ by the Hellinger–Toeplitz theorem.

In the case $m_\gamma < 1$ we consider the form $\gamma' = \gamma + (1 + m_\gamma)\chi$ and construct a corresponding self-adjoint operator $A' \geq 1$. Then it is obvious that

$$\gamma(\phi, \psi) = (A\varphi, \psi), \quad A = A' - (m_\gamma + 1)\mathbf{1}.$$

We leave to the reader the task of verifying that for each operator $A = A^* \geq m_A > -\infty$, the quadratic form

$$\gamma_A(\varphi, \psi) := (A\varphi, \psi), \quad \varphi, \psi \in \operatorname{Dom} A,$$

is symmetric, closed, and bounded from below. In addition, the operator, associated with its closure coincides with A and also with the operator, constructed in accordance with the procedure described above. □

Example 2.4.2 (Laplace operators and Dirichlet forms). Let

$$\mathcal{H} = L_2(\mathbb{R}^d, dx), \quad A = -\Delta, \quad \Delta = \sum_{i=1}^{d} \frac{\partial^2}{\partial x_i^2}, \quad d \geq 1,$$

where $-\Delta$ denotes the Laplace operator. The function $f \in L_2$ belongs to $\operatorname{Dom} A$ if $\Delta f \in L_2$. It is well known that the Laplace operator is essentially self-adjoint on $C_0^\infty(\mathbb{R}^d)$. It is also known that the operator$-\Delta$ is associated with the *Dirichlet form* :

$$\gamma_{-\Delta}(f, g) = \int_{\mathbb{R}^d} \left(\sum_{i=1}^{d} \frac{\partial f(x)}{\partial x_i} \frac{\partial \overline{g(x)}}{\partial x_i} \right) dx.$$

The Dirichlet form with the domain

$$Q(\gamma_{-\Delta}) = \{f \in L_2 \mid \int_{\mathbb{R}^d} |k|^2 |(\mathcal{F}f)(k)|^2 dk < \infty, \ |k|^2 = \sum_{i=1}^{d} |k_i|^2\},$$

is strictly positive and closed. Here $\mathcal{F}$ denotes the Fourier transform. Note that the domain of the operator $-\Delta$ can be described in terms of the Fourier transform:

$$\operatorname{Dom}(-\Delta) = \{f \in L_2 \mid |k|^2 |(\mathcal{F}f)(k)|^2 \in L_2\}.$$

Example 2.4.3 (Schrödinger operators). Consider $\mathcal{H} = L_2(\mathbb{R}^d, dx)$, $d \geq 1$. The expression $A = -\Delta + V(x)$, is called the Schrödinger operator, and V is called the potential. Usually, V is a locally integrable nonnegative function. The operator A has the domain

$$\operatorname{Dom} A = \{f \in L_2 \mid -\Delta f + Vf \in L_2\}.$$

In the literature (see, e.g., [107, 171, 172]), one can find numerous results, which establish conditions on the potential V (not necessarily nonnegative) ensuring the essential self-adjointness of the Schrödinger operator on $C_0^\infty(\mathbb{R}^d)$. In addition, the construction of the Schrödinger operator often begins with the study of the quadratic forms

$$\gamma_{-\Delta+V}(f,g) = \gamma_{-\Delta}(f,g) + \int_{\mathbb{R}^d} V(x) f(x) \bar{g}(x) dx,$$

checking its boundedness from below and its closability.

For positive quadratic forms there is a known, so-called second operator representation theorem.

Theorem 2.4.4. *Let $\gamma \geq 0$ be a closed form on $\mathcal{H}$ and let A be the associated self-adjoint operator: $\gamma = \gamma_A$. Then the domain $Q(\gamma_A) = \operatorname{Dom} A^{1/2}$ and*

$$\gamma_A(\varphi, \psi) = (A^{1/2}\varphi, A^{1/2}\psi), \quad \varphi, \psi \in \operatorname{Dom} A^{1/2}.$$

2.5 The spectral theorem

The proofs of many theorems in analysis are greatly simplified by using the spectral theorem.

Let us consider an operator-valued function $\mathfrak{B} \ni \triangle \mapsto \mathbb{E}(\triangle) \in \mathcal{B}(\mathcal{H})$, where $\mathfrak{B}$ is the Borel σ-algebra of subsets on $\mathbb{R}^1$, and $\mathcal{B}(\mathcal{H})$ is the set of bounded operators defined on the whole space $\mathcal{H}$. The function $\mathbb{E}(\cdot)$ is called an *operator-valued spectral measure*, if the following conditions are satisfied:

(i) σ-additivity in the strong sense:

$$\lim_{N \to \infty} \sum_{i=1}^{N} \mathbb{E}(\triangle_i) = \mathbb{E}\left(\bigcup_{i=1}^{\infty} \triangle_i\right);$$

(ii) self-adjointness and positivity:

$$\mathbb{E}(\triangle) = \mathbb{E}^*(\triangle) \geq 0, \quad \triangle \in \mathfrak{B};$$

(iii) orthogonality and normalization

$$\mathbb{E}(\triangle_1)\mathbb{E}(\triangle_2) = \mathbb{E}(\triangle_1 \cap \triangle_2), \quad \triangle_1, \triangle_2 \in \mathfrak{B},$$
$$\mathbb{E}(\mathbb{R}^1) = \mathbf{1},$$

where $\mathbf{1}$ denotes the identity operator on the space $\mathcal{H}$.

In accordance with [32, 42, 56], to the self-adjoint operator $A = A^*$ on the separable Hilbert space $\mathcal{H}$ there corresponds a projection-valued spectral measure $\mathbb{E}(\triangle)$ associated with the resolution of the identity $\{\mathbb{E}_\lambda\}_{\lambda\in\mathbb{R}^1}$, which forms a one-parameter family of orthogonal projections on $\mathcal{H}$ [32]: $\mathbb{E}_\lambda = \mathbb{E}((-\infty, \lambda))$. The spectral theorem for self-adjoint operators can be stated as follows.

Theorem 2.5.1 ([48, 168, 170]). *The selfadjoint operator $A = A^*$ on $\mathcal{H}$ corresponds bijectively to the operator-valued spectral measure $\mathbb{E}(\triangle)$:*

$$A = \int_{\mathbb{R}} \lambda d\mathbb{E}_\lambda,$$

with

$$\operatorname{Dom} A = \left\{ \psi \in \mathcal{H} \mid \int_{\mathbb{R}} \lambda^2 d(\mathbb{E}_\lambda \psi, \psi) < \infty \right\}.$$

Using the spectral theorem, we can calculate functions of the operator A for a wide class of scalar-valued functions $F(\lambda)$ on $\mathbb{R}^1$:

$$F(A) = \int_{\mathbb{R}} F(\lambda) d\mathbb{E}_\lambda,$$

$$\operatorname{Dom} F(A) = \left\{ \psi \in \mathcal{H} \mid \int_{\mathbb{R}} |F(\lambda)|^2 d(\mathbb{E}_\lambda \psi, \psi) < \infty \right\}.$$

The measure $\mu_\psi(\triangle) = \int_\triangle d(\mathbb{E}_\lambda \psi, \psi)$, $\triangle \in \mathfrak{B}$ is called the *spectral measure* of the operator A associated with the vector $\psi \in \mathcal{H}$. The vector ψ is called *cyclic* for the operator A, if $\psi \in \operatorname{Dom} A^n$, $n \in \mathbb{N}_0$ and $\operatorname{span}\{A^n \psi\}_{n=0}^{\infty}$ is dense in $\mathcal{H}$. The existence of a cyclic vector for A means that the spectrum of the operator is simple. Only in this case the operator A is unitarily equivalent to the operator of multiplication by the independent variable on the space $L_2(\mathbb{R}, d\mu_\psi)$. Then there exists a unitary operator $U\colon \mathcal{H} \to L_2(\mathbb{R}, d\mu_\psi)$ such that

$$(UAU^{-1}f)(\lambda) = \lambda f(\lambda), \quad f \in \operatorname{Dom}(UA).$$

Of course, the spectral theorem is a generalization of the well-known decomposition of the finite-dimensional or compact self-adjoint operators into a sum of operators, each of which is a scalar multiple of the projection P_i onto an eigen-subspace:

$$A = \sum_i \lambda_i P_i, \quad A\psi_i = \lambda_i \psi_i, \quad \psi_i \in P_i \mathcal{H}.$$

It is worth mentioning that, in the general case of an arbitrary self-adjoint operator A in $\mathcal{H}$, there is the well-known developed theory due to Berezansky, Gelfand, and Kostyuchenko [42, 159]. This theory provides the procedure of eigenfunctions expansion, namely, the expansion into generalized eigenvectors of an operator A. According to this theory, there is a scalar-valued spectral measure $\rho(\triangle)$

such that the resolution of the identity $\mathbb{E}(\triangle)$ of the operator A can be differentiated with respect to this measure. So, one has a representation

$$\mathbb{E}(\triangle) = \int_{\triangle} P(\lambda) d\rho(\lambda),$$

where the operator-valued function $P(\lambda)$ is defined almost everywhere with respect to ρ and acts from $\mathcal{H}_+$ into $\mathcal{H}_-$ in the rigged Hilbert space $\mathcal{H}_- \supset \mathcal{H} \supset \mathcal{H}_+$. In addition, the embedding $\mathcal{H}_+$ into $\mathcal{H}$ should be quasinuclear and under some additional conditions on the construction of the rigged space, the range $\operatorname{Ran} P(\lambda)$ consists of generalized eigenvectors of the operator A. The last statement means (see details in [42]) that for all $\varphi \in \operatorname{Ran} P(\lambda)$, the equality

$$\langle \varphi, (A - \lambda)u \rangle_{-,+} = 0, \quad u \in \operatorname{Dom} A \cap \mathcal{H}_+, \quad Au \in \mathcal{H}_+.,$$

holds true.

Chapter 3

Self-adjoint Extensions of Symmetric Operators

There are numerous works devoted to the theory of extension of symmetric operators, were properties of extended operators are described. Here we refer only to some sources that have influenced our research in this area: [28, 32, 34, 35, 39, 45, 54, 55, 64, 65, 68, 70, 71, 73, 75, 82, 91, 92, 110, 147, 148, 152, 167, 179].

In this chapter we present needed further facts from the theory of self-adjoint extensions of symmetric densely defined operators.

The construction of self-adjoint extensions of symmetric operators is an important and active field of research due mainly to its physical applications. It is known that the energy operator (Hamiltonian) in physical theories is one of the main objects of study. In particular, this operator is the generator of the time evolution for the corresponding dynamical system. According to axioms of quantum mechanics, the state space of every physical system is described by the appropriate Hamiltonian on a certain Hilbert space.

It is a highly nontrivial problem to construct the Hamiltonian as a self-adjoint operator starting from physical considerations. Often, as a variant of the energy operator, one can write a formal mathematical expression (usually in the form of a differential operator), for which we need to construct a meaningful self-adjoint operator. However, a formal mathematical expression has only the meaning of a densely defined symmetric operator or a Hermitian quadratic form. Since the time evolution of a physical system should be given by a self-adjoint operator, a standard problem arises: to construct a self-adjoint extension of a formally given symmetric operator.

This problem leads to a more complex task: to describe the whole family self-adjoint extensions and to study their spectral properties.

For physical reasons, usually only positive or bounded from below symmetric operators and their extensions are considered. Here we present one of the productive approaches, in which the parameter describing the extensions is an auxiliary self-adjoint operator acting on deficiency subspace of the considered symmetric operator.

Without loss of generality, we assume that the initial symmetric operator is strictly positive, and its lower bound is equal 1. We can reduce other cases to such operator by shifting, if necessary, the initial operator by the identity operator multiplied on a constant.

3.1 The operator parametrization

In this section we establish the operator parametrization of the whole family of bounded from below self-adjoint extensions $\tilde{A}$ of a symmetric operator $\mathbf{A}$, in terms of positive operators B acting in the deficiency subspace.

Let $\mathbf{A}$ be a bounded from below symmetric operator on the Hilbert space $\mathcal{H}$. The number

$$-\infty < m_{\mathbf{A}} = \inf_{\|\varphi\|=1} (\mathbf{A}\varphi, \varphi),$$

is called the *lower bound* of the operator $\mathbf{A}$. Since $\mathbf{A} \subseteq \mathbf{A}^*$, the operator $\mathbf{A}$ is closable, since the adjoint operator $\mathbf{A}^*$ is closed by the construction. So we can assume that $\mathbf{A}$ is already closed, $\mathbf{A} = \mathbf{A}^{\mathrm{cl}}$, although in specific examples, the initial symmetric operator is not closed on its natural domain.

In the sequel, we always assume that $\mathbf{A}$ is unbounded:

$$\sup_{\|\varphi\|=1} (\mathbf{A}\varphi, \varphi) = \infty, \quad \varphi \in \operatorname{Dom} \mathbf{A}$$

and *simple*. The last means that there are no closed proper subspaces in $\mathcal{H}$ invariant under $\mathbf{A}$.

Without loss of generality, we can assume that $\mathbf{A}$ is a positive operator. Indeed, if the lower bound $m_{\mathbf{A}}$ of the original operator is negative, then instead of $\mathbf{A}$ we take $\mathbf{A}' = \mathbf{A} + a\mathbf{1}$ ($\mathbf{1}$ denotes the identity operator), where a is chosen, so that $m_{\mathbf{A}'} \geq 0$. Of course, it is necessary to make the inverse shift by $-a\mathbf{1}$ after constructing and describing the self-adjoint extensions of the operator $\mathbf{A}'$. Moreover, it is convenient to assume that the lower bound $m_{\mathbf{A}} = 1$, namely $\mathbf{A} \geq 1$. Then 0 is a point of regular type for $\mathbf{A}$ and therefore there exists the bounded operator $\mathbf{A}^{-1}$, defined on the subspace:

$$\mathcal{M}_0 := \operatorname{Dom} \mathbf{A}^{-1} = \operatorname{Ran} \mathbf{A}, \quad \mathcal{M}_0 \subset \mathcal{H}.$$

Of course, if $\mathbf{A}$ would be essentially self-adjoint, then as a consequence of a basic criterion of a self-adjointness, we would have the obvious equality $\operatorname{Ran} \mathbf{A} = \operatorname{Dom} \mathbf{A}^{-1} = \mathcal{H}$. But we consider there a symmetric operator $\mathbf{A}$ which is not essentially self-adjoint, $\mathbf{A}^{\mathrm{cl}} \neq \mathbf{A}^*$. Therefore, the space $\mathcal{H}$ is decomposed into a nontrivial orthogonal sum

$$\mathcal{H} = \mathcal{M}_0 \oplus \mathcal{N}_0, \quad \mathcal{N}_0 = \mathcal{M}_0^{\perp}, \quad 0 < \dim \mathcal{N}_0 = n \leq \infty.$$

Let us recall that the deficiency subspace $\mathcal{N}_0$ can be defined by means of the adjoint operator:

$$\mathcal{N}_0 = \{h \in \mathcal{H} \mid \mathbf{A}^* h = 0\} = \operatorname{Ker} \mathbf{A}^*.$$

Here we consider the following problem. We want to describe the set of all bounded from below self-adjoint extensions of the operator $\mathbf{A}$. Let us denote this set by $\mathcal{A}(\mathbf{A})$. However, we analyze mostly bounded from below extensions $\tilde{A} \in \mathcal{A}(\mathbf{A})$ which have inverse operators $\tilde{A}^{-1}$ defined and bounded on the whole space $\mathcal{H}$. This set of extensions we denote by $\mathcal{A}_{\neq 0}(\mathbf{A})$. Namely,

$$\mathcal{A}_{\neq 0}(\mathbf{A}) := \{\tilde{A} \geq m > -\infty \mid \mathbf{A} \subset \tilde{A} = \tilde{A}^* \subset \mathbf{A}^*,\ \operatorname{Dom} \tilde{A}^{-1} = \mathcal{H}\}.$$

The subset of $\mathcal{A}_{\neq 0}(\mathbf{A})$ consisting of positive extensions we denote by $\mathcal{A}_+(\mathbf{A})$.

One of the creators of the theory of operators and quantum mechanics, J. von Neumann, claimed that there exists at least one self-adjoint extension $\tilde{A}$ such that for each bounded from below symmetric operator $\mathbf{A}$ we have $m_{\tilde{A}} = m_{\mathbf{A}}$ [161]. Later, Stone [178] and Friedrichs [86] proved this fact, and Kreĭn [148] described the set of all positive self-adjoint extensions of a positive symmetric operator.

If the operator $\mathbf{A} \geq 1$, then the method of rigged spaces gives an appropriated way for the construction of its self-adjoint extension preserving the lower bound. To this end, we consider the quadratic form

$$\gamma_{\mathbf{A}}[\varphi] = (\mathbf{A}\varphi, \varphi), \quad \varphi \in \operatorname{Dom} \mathbf{A}.$$

We note that the quadratic form $\gamma_{\mathbf{A}}$ on $Q(\gamma_{\mathbf{A}}) = \operatorname{Dom} \mathbf{A}$ is not closable, even if $\mathbf{A}$ is a closed operator. The reason is that the operator $\mathbf{A}$ is unbounded. We denote the closure of the form $\gamma_{\mathbf{A}}$ by γ_∞: $\gamma_\infty = \gamma_{\mathbf{A}}^{\mathrm{cl}}$. It is clear that $\gamma_\infty \geq 1$. We define $A_\infty \geq 1$ to be the operator associated with the quadratic form γ_∞ (see Theorem 2.4.1). The construction of A_∞ by the method of rigged Hilbert space is carried out as follows.

By using the quadratic form γ_∞ we introduce the Hilbert space $\mathcal{H}_1$ with the norm $\|\varphi\|_1 = (\gamma_\infty[\varphi])^{1/2}$. This space coincides with the domain of the form γ_∞, since it is closed:

$$\mathcal{H}_1 = \operatorname{Dom} \gamma_\infty \equiv Q(\gamma_\infty).$$

Further, since $\gamma_\infty \geq 1$, the Hilbert space $\mathcal{H}_1$ is positive with respect to $\mathcal{H}$ in the sense of Berezansky (see [42, 44]), i.e., $\|\cdot\| \leq \|\cdot\|_1$. Then the couple $\mathcal{H} \sqsupset \mathcal{H}_1$ forms a pre-rigged space (see Chapter 4). It has a unique extension to a rigged space:

$$\mathcal{H}_{-1} \sqsupset \mathcal{H} \sqsupset \mathcal{H}_1. \tag{3.1}$$

Finally, A_∞ is constructed as a restriction of the Berezansky canonical isomorphism $D_{-1,1} : \mathcal{H}_1 \to \mathcal{H}_{-1}$. That is,

$$A_\infty = D_{-1,1} \upharpoonright \operatorname{Dom} A_\infty,\ \operatorname{Dom} A_\infty = \{\varphi \in \mathcal{H}_1 \mid D_{-1,1}\varphi \in \mathcal{H}\}. \tag{3.2}$$

The operator A_∞ is called the *Friedrichs extension* of the operator $\mathbf{A}$. In the terminology of M. Kreĭn, it is the hard extension. It is clear that each bounded from below symmetric operator has the Friedrichs extension.

In the fundamental work of M. Kreĭn, [148] a number of important results concerning positive self-adjoint extensions of a symmetric positive operator were established (see also [28, 54, 82, 179]. In the next theorem we formulate some of these results that will be used for the descriptions of the set $\tilde{A} \in \mathcal{A}_+(\mathbf{A})$.

Theorem 3.1.1 (Kreĭn). *Let $\mathbf{A} \subset \mathbf{A}^*$, $\mathbf{A} \geq 1$ be an unbounded closed symmetric operator on $\mathcal{H}$. Then in the set $\mathcal{A}_+(\mathbf{A})$ of all positive self-adjoint extension of the operator $\mathbf{A}$, there exists two distinguished ones, A_0 and A_∞. The operator A_0 is the minimal operator and A_∞ is the maximal one, and coincides with the Friedrichs extension. All operators from $\mathcal{A}_+(\mathbf{A})$ satisfy the Kreĭn inequality*

$$A_0 \leq \tilde{A} \leq A_\infty, \quad \tilde{A} \in \mathcal{A}_+(\mathbf{A}). \tag{3.3}$$

In addition, the operator A_∞ is unique in the set $\mathcal{A}_+(\mathbf{A})$ and has the property

$$\mathrm{Dom} A_\infty \subseteq \mathrm{Dom} \gamma_\infty. \tag{3.4}$$

The proof of this theorem can be found in [28, 32, 54, 148]. Let us note that as a corollary of Theorem 3.1.1,

$$\mathrm{Dom}\, A_\infty^{1/2} = \mathrm{Dom}\, \gamma_\infty \tag{3.5}$$

Thus, according to Kreĭn's theory, one can introduce a partial order in the set $\mathcal{A}_+(\mathbf{A})$,. Let us recall that for a couple of operators $A_1, A_2 \in \mathcal{A}_+(\mathbf{A})$ one writes $A_1 \leq A_2$, if for some (and also for all) common regular point a from $\mathbb{R}$,

$$((A_1 - a)^{-1}h, h) \geq ((A_2 - a)^{-1}h, h), \quad \forall h \in \mathcal{H}. \tag{3.6}$$

In particular, if $\mathbf{A} \geq 0$, then similarly to Theorem 3.1.1, in the set of its positive self-adjoint extensions, there exist two of extreme operators: A_0 is the minimal one and A_∞ is the maximal one in the sense of the ordering introduced above. We should point out that sometime for $\mathbf{A} \geq 0$ it may happen that $A_0 = A_\infty$. This means that there is one only non-negative self-adjoint extension. Of course, in the case $\mathbf{A} \geq 1$, there is a continuum of positive extensions. In the literature (see, e.g., [28]), the operator A_0 is called the Kreĭn extension. This operator is fixed by the equality

$$A_0 \mathcal{N}_0 = 0.$$

The second extreme operator A_∞ is usually called the Friedrichs extension [86], although it was independently introduced by many authors (see [107]). According to the construction, A_∞ is the largest operator in the set $\mathcal{A}_+(\mathbf{A})$, i.e., $\tilde{A} \leq A_\infty$ for all $\tilde{A} \in \mathcal{A}_+(\mathbf{A})$. It is easy to see that A_∞ has the same lower bound as the operator $\mathbf{A}$.

In the sequel we continue to investigate the self-adjoint extensions of a simple symmetric operator $\mathbf{A} \subset \mathbf{A}^*$, $\mathbf{A} \geq 1$ on $\mathcal{H}$. Let us recall (see Chapter 2) that for every real number $a < 1$, the subspace

$$\mathcal{N}_a = \mathcal{M}_a^{\perp}, \quad \mathcal{M}_a := \operatorname{Ran}(\mathbf{A} - a)$$

consists of eigenvectors of the adjoint operator $\mathbf{A}^*$

$$\mathbf{A}^* \eta_a = a\eta_a, \quad \forall \eta_a \in \mathcal{N}_a. \tag{3.7}$$

Proposition 3.1.2. *Let the operator $\mathbf{A} \subset \mathbf{A}^*$, $\mathbf{A} \geq 1$ be given. Then for every real $a < 1$, there is a self-adjoint extension $\tilde{A} = A_a \in \mathcal{A}_+(\mathbf{A})$ with the lower bound $m_{\tilde{A}} = a$. Moreover,*

$$A_a \eta_a = a\eta_a, \quad \eta_a \in \mathcal{N}_a. \tag{3.8}$$

Proof. Let us put

$$A_a \varphi = \mathbf{A}\varphi, \quad \varphi \in \operatorname{Dom} \mathbf{A}, \quad A_a \eta_a = a\eta_a, \ \eta_a \in \mathcal{N}_a, \tag{3.9}$$

and define the operator $\tilde{A} = A_a$ on

$$\operatorname{Dom} A_a = \operatorname{span}\{\operatorname{Dom} \mathbf{A}, \mathcal{N}_a\},$$

by linearity. According to this construction, $A_a \subset \mathbf{A}^*$ (see (3.7)). It is easy to verify that A_a is a self-adjoint operator, $A_a = A_a^*$. Indeed, from (3.9) it follows that A_a has an orthogonal decomposition into two self-adjoint components: the operator $A_a P_{\mathcal{M}_a}$ acting on the space $\mathcal{M}_a$ according to the rule

$$A_a P_{\mathcal{M}_a} \varphi = \mathbf{A}\varphi - a P_{\mathcal{N}_a} \varphi, \quad \varphi \in \operatorname{Dom}(\mathbf{A}),$$

and the multiplication operator on $\mathcal{N}_a$. The first component is self-adjoint on $\mathcal{M}_a$ because $\operatorname{Ran}(\mathbf{A} - a) = \mathcal{M}_a$. It is easy to show that the operator A_a is the smallest self-adjoint extension such that $\tilde{A} \geq a$. □

Thus, the symmetric operator $\mathbf{A} \geq 1$, and hence each bounded from below symmetric operator $\mathbf{A}$, has a continuum of different self-adjoint extensions A_a, $a < m_{\mathbf{A}}$. On the other hand, it is clear that the set $\{\mathbf{A} \subset A_a = A_a^* \subset \mathbf{A}^* \mid a < m_{\mathbf{A}}\}$ does not contain of all self-adjoint extensions of the operator $\mathbf{A}$. In particular, this set does not contain even the Friedrichs extension A_∞.

Let us describe this operator in the case of an arbitrary lower bound $m_{\mathbf{A}}$. A characteristic property of the operator A_∞ is that its lower bound is the same as for the operator $\mathbf{A}$:

$$m_{A_\infty} = m_{\mathbf{A}}. \tag{3.10}$$

In general, the operator $\mathbf{A}$ can have other self-adjoint extensions that preserve the lower bound. Kreĭn [148] (see also [32]) found a criterion for (3.10) to hold true only for A_∞.

In the case where $m_{\mathbf{A}} < 1$, for the construction A_∞, we introduce a positive form on $\operatorname{Dom} A$ by

$$\chi_1(\varphi, \psi) = \gamma_{\mathbf{A}}(\varphi, \psi) + (1 - m_{\mathbf{A}})\chi(\varphi, \psi), \quad \left(\chi[\cdot] = \|\cdot\|^2_{\mathcal{H}}\right).$$

The closability of the form χ_1 is proved by checking the corresponding criterion (see Theorem VII.17 in [107]). Let $\varphi_n \in \operatorname{Dom} \mathbf{A}$, $\varphi_n \to 0$ on $\mathcal{H}$, and $\chi_1[\varphi_n - \varphi_m] \to 0$. Then, by using the operator representation $\gamma_{\mathbf{A}}(\cdot, \cdot) = (\mathbf{A}\cdot, \cdot)$, we conclude that $\chi_1[\varphi_n] \to 0$. Now according to Theorem 2.2.1, the self-adjoint operator $A' \geq 0$ is associated with the form χ^{cl}. The operator A_∞ is defined as $A' - (1 - m_{\mathbf{A}})\mathbf{1}$. By this construction, the lower bound m_{A_∞} is the same as the lower bound of the operator $\mathbf{A}$. Thus, $\gamma_\infty[\cdot] := (\gamma_{\mathbf{A}}[\cdot])^{\mathrm{cl}}$. Finally, if we assume that there exists another self-adjoint extension $\tilde{A}$, with $\operatorname{Dom} \tilde{A} \subseteq \operatorname{Dom} \gamma_\infty$, then it is easy to see that $\tilde{A} \subseteq A_\infty$. Due to the self-adjointness, this means that $\tilde{A} = A_\infty$.

Now, the task of describing the operators $\tilde{A} \in \mathcal{A}_+(\mathbf{A})$ (that is, their parametrization) is reduced to establishing of the connection between operators $\tilde{A}$ and A_∞ based on the partial ordering (3.3). Further, for a simplification, we reduce this problem and consider not all operators from the set $\mathcal{A}_+(\mathbf{A})$, but only the ones that are mutually simple to A_∞ with respect to $\mathbf{A}$, meaning that the symmetric operator $\mathbf{A}$ is a maximal common part for each couple $\tilde{A}$ and A_∞. Namely, the set

$$\mathfrak{D} = \{\varphi \in \operatorname{Dom} \tilde{A} \cap \operatorname{Dom} A_\infty \mid \tilde{A}\varphi = A_\infty\varphi\}$$

coincides with the domain $\operatorname{Dom} \mathbf{A}$:

$$\tilde{A} \restriction \mathfrak{D} = A_\infty \restriction \mathfrak{D} = \mathbf{A}. \tag{3.11}$$

This reduction of the problem practically does not diminish the generality, because if it happens that the domain $\operatorname{Dom} \mathbf{A}$ is a proper subspace in $\mathfrak{D}$, then instead of $\mathbf{A}$, one should consider the symmetric operator $\mathbf{A}' = \tilde{A} \restriction \mathfrak{D} = A_\infty \restriction \mathfrak{D}$, so that $\tilde{A}$ and A_∞ will be mutually simple operators.

Some variants of the next theorem were published in [116, 120, 121, 126, 135]. It is one of the main result which will be used in the subsequent considerations.

Theorem 3.1.3. *Let $\mathbf{A} \geq \mathbf{1}$ be an unbounded, closed symmetric operator on $\mathcal{H}$, and A_∞ be its Friedrichs extension. Suppose that $\tilde{A} \in \mathcal{A}_+(\mathbf{A})$ and satisfies the condition* (3.11). *Then one has the representation*

$$\tilde{A}^{-1} = A_\infty^{-1} + B^{-1}P_{\mathcal{N}_0}, \tag{3.12}$$

where $P_{\mathcal{N}_0}$ is the orthogonal projection on $\mathcal{N}_0 = \operatorname{Ker} \mathbf{A}^*$, and $B = B^*$ is a bounded self-adjoint operator on $\mathcal{N}_0$ such that B^{-1} exists and is defined on the whole subspace $\mathcal{N}_0$. Moreover, the formula (3.12) *establishes the bijection between the sets containing $\tilde{A}$ and B, correspondingly, with the above-described properties.*

Proof. Let us consider a mutually simple, with respect to A_∞, operator $\tilde{A} \in \mathcal{A}_+(\mathbf{A})$. According to the definition of the class $\mathcal{A}_+(\mathbf{A})$, there exists the bounded inverse operator $\tilde{A}^{-1}$ with the domain $\operatorname{Dom}\tilde{A}^{-1} = \mathcal{H}$. Since $A_\infty \geq 1$, the domain $\operatorname{Dom} A_\infty^{-1} = \mathcal{H}$ too. According to the Kreĭn theorem, $\tilde{A} \leq A_\infty$. Therefore, $A_\infty^{-1} \geq \tilde{A}^{-1}$ and the operator

$$\tilde{B} := \tilde{A}^{-1} - A_\infty^{-1} \tag{3.13}$$

is bounded, self-adjoint and positive, $\tilde{B} = \tilde{B}^* \geq 0$. It is clear that $\tilde{B} = 0$ on the subspace $\mathcal{M}_0 = \operatorname{Ran} A$. This follows from the fact that $\tilde{A}^{-1}$ coincides with A_∞^{-1} and with $\mathbf{A}^{-1}$ on this subspace. Since the operators $\tilde{A}$ and A_∞ are mutually simple, the subspace $\mathcal{M}_0$ is maximal, where $\tilde{A}^{-1}$ coincides with A_∞^{-1}. It means that $\operatorname{Ker}\tilde{B} = \mathcal{M}_0$. We can define the operator $B = (\tilde{B} \upharpoonright \mathcal{N}_0)^{-1}$. It is obvious that $B = B^* > 0$, since $\tilde{B}$ is positive. Hence $\tilde{B} = B^{-1} P_{\mathcal{N}_0}$. Let us remark that the boundedness of B^{-1} follows from the boundedness of $\tilde{A}^{-1}$.

Conversely, let $B > 0$ be a positive self-adjoint operator on $\mathcal{N}_0$. We assume B^{-1} exists and is bounded. Let us define the operator $\tilde{B} = B^{-1} P_{\mathcal{N}_0}$. It is obvious that $\tilde{B}^* = \tilde{B} \geq 0$ is a bounded operator on $\mathcal{H}$. Hence, $\tilde{A}^{-1} := A_\infty^{-1} + \tilde{B}$ is also a bounded self-adjoint operator on $\mathcal{H}$ such that $\operatorname{Ker}\tilde{A}^{-1} = \{0\}$. So, there is $\tilde{A} = (\tilde{A}^{-1})^{-1}$. Moreover, $\tilde{A}$ is not only strictly positive, $\tilde{A} > 0$, but also positively defined, since $\operatorname{Dom}\tilde{A}^{-1} = \mathcal{H}$. It is clear that this operator is self-adjoint and coincides with $\mathbf{A}$ on $\operatorname{Dom}\mathbf{A}$, since on $\operatorname{Ran}\mathbf{A} = \mathcal{M}_0 = \operatorname{Ker}\tilde{B}$, the operator A_∞^{-1} acts as $\mathbf{A}^{-1}$. Thus, the operator $\tilde{A}$ belongs to the set $\mathcal{A}_+(\mathbf{A})$. The operators $\tilde{A}$ and A_∞ are mutually simple, since $\operatorname{Ker} B^{-1} = \{0\}$.

Finally, we note that, according to the formula (3.12), the correspondence between the operators $B > 0$ for which $\operatorname{Dom} B^{-1} = \mathcal{N}_0$ and $\tilde{A}$ from the set $\mathcal{A}_+(\mathbf{A})$ is bijective. □

Sometimes the operators $\tilde{A} \in \mathcal{A}_+(\mathbf{A})$ which have the representation (3.12) will denoted by A_B.

Let us note that the boundedness of the operator B^{-1} on $\mathcal{N}_0$, which is equivalent to the positive definitness of B, is a necessary and sufficient condition for the positive definitness of a self-adjoint extensions $\tilde{A} = A_B$. If we now take the strictly positive operator $B = B^* > 0$ on the space $\mathcal{N}_0$, as an extension parameter, then the formula (3.12) is also true, although B^{-1} and $\tilde{A}^{-1}$ are, in general, unbounded positive operators. However, A_B will be not positive definite, but only strictly positive, $A_B > 0$.

Let us denote by $\mathcal{A}_{>0}(\mathbf{A})$ the set of all strictly positive self-adjoint extensions of the operator $\mathbf{A}$, and by $\mathcal{B}_{>0}(\mathcal{N}_0)$ the set of operators $B = B^* > 0$ that are self-adjoint and strictly positive on the subspace $\mathcal{N}_0 = \operatorname{Ker}\mathbf{A}^*$:

$$\mathcal{B}_{>0}(\mathcal{N}_0) = \{B = B^* > 0 \mid \operatorname{Dom} B \subseteq \mathcal{N}_0\}.$$

Theorem 3.1.4. *Between the subset of operators $\tilde{A} \in \mathcal{A}_{>0}(\mathbf{A})$, $\tilde{A} > 0$ which are mutually simple to A_∞ with respect to $\mathbf{A}$, and the set of self-adjoint on $\mathcal{N}_0$ operators*

$B \in \mathcal{B}_{>0}(\mathcal{N}_0)$, *there exists a bijective correspondence given by the formula*

$$\tilde{A}^{-1} = A_B^{-1} = A_\infty^{-1} + B^{-1}P_{\mathcal{N}_0}. \tag{3.14}$$

Proof. Let $B = B^* > 0$ on $\mathcal{N}_0$. It is possible that the lower bound $m_B = 0$. Let us consider the operator

$$A_B^{-1} := A_\infty^{-1} + \tilde{B}, \quad \tilde{B} = B^{-1}P_{\mathcal{N}_0},$$

where, recall that A_∞^{-1} is bounded and strictly positive. Clearly A_B^{-1} is self-adjoint and strictly positive, $A_B^{-1} > 0$, although it can be unbounded if the lower bound of the operator B is zero. Hence, there exists the inverse positive self-adjoint operator, $A_B = (\tilde{A}^{-1})^{-1} > 0$. It is clear that $A_B \equiv \tilde{A} \in \mathcal{A}_{>0}(\mathbf{A})$, since

$$A_B^{-1} \restriction \mathcal{M}_0 = A_\infty^{-1} \restriction \mathcal{M}_0 = \mathbf{A}^{-1}.$$

In particular, by Kreĭn's theorem, $A_B \leq A_\infty$.

Conversely, let the operator $\tilde{A} \in \mathcal{A}_{>0}(\mathbf{A})$, $\tilde{A} > 0$ (possibly, with the lower bound $m_{\tilde{A}} = 0$) which satisfies the condition (3.11) be given. Then, there exist the operator $\tilde{A}^{-1} > 0$, which in general is an unbounded and satisfies the inequality $\tilde{A}^{-1} \geq A_\infty^{-1}$ in the sense of quadratic forms:

$$(\tilde{A}^{-1}h, h) \geq (A_\infty^{-1}h, h), \quad h \in \operatorname{Ran} \tilde{A}.$$

This inequality in turn follows from the inequality $\tilde{A} \leq A_\infty$, which holds for each positive self-adjoint extension thanks Kreĭn's theorem.

Let us consider the quadratic form

$$\gamma_{\tilde{B}}[h] = (\tilde{A}^{-1}h, h) - (A_\infty^{-1}h, h). \tag{3.15}$$

It is densely defined and nonnegative. It is easy to verify that it is also closable. Hence, its closure is associated with a nonnegative self-adjoint operator $\tilde{B}$. According to the construction, $\operatorname{Ker} \tilde{B} = \mathcal{M}_0$, since the operator $\tilde{A} \in \mathcal{A}_{>0}(\mathbf{A})$ is mutually simple with A_∞. Then there exists an operator $B^{-1} := \tilde{B} \restriction \mathcal{N}_0$, which is strictly positive, $B^{-1} > 0$ and self-adjoint on $\mathcal{N}_0$. So, we have the equality

$$(\tilde{A}^{-1}h, h) = (A_\infty^{-1}h, h) + (B^{-1}P_{\mathcal{N}_0}h, h), \quad h \in \operatorname{Ran} \tilde{A}.$$

It is easy to show that it this has a meaning in terms of operators:

$$\tilde{A}^{-1} = A_\infty^{-1} + B^{-1}P_{\mathcal{N}_0}.$$

Further, it is clear that for different operators $\tilde{A}_1, \tilde{A}_2 \in \mathcal{A}_{>0}(\mathbf{A})$ the corresponding operators B_1, B_2 are also different. This means that each operator $\tilde{A} \in \mathcal{A}_{>0}(\mathbf{A})$ has a unique representation (3.14). On the other hand, we have shown that each operator $B = B^* > 0$ on $\mathcal{N}_0$ defines some $A_B \in \mathcal{A}_{>0}(\mathbf{A}), A_B > 0$ by the same formula (3.14). Hence, the correspondence between $\tilde{A} = A_B \in \mathcal{A}_{>0}(\mathbf{A})$, $\tilde{A} > 0$ and $B \in \mathcal{B}_{>0}(\mathcal{N}_0)$, $B > 0$, is both injective and surjective, i.e., it is bijective. The theorem is completely proved. □

To extend this theorem to all operators $\tilde{A} \geq 0$, we have to use the whole set of positive operators $B \in \mathcal{B}_+(\mathcal{N}_0)$, which parameterize them.

Let us consider a mutually simple with A_∞ operator $\tilde{A} \geq 0$, which is not strictly positive. Moreover, we assume $\operatorname{Ker} \tilde{A} =: \mathcal{N}_{00} \neq \{0\}$. Then $\mathcal{N}_0 = \operatorname{Ker} \mathbf{A}^*$ contains the subspace $\mathcal{N}_{00}$. Since $\tilde{A}$ is a self-adjoint, it has the orthogonal sum decomposition $\tilde{A} = \tilde{A} \restriction (\mathcal{H} \ominus \mathcal{N}_{00}) \oplus 0_{\mathcal{N}_{00}}$, where $0_{\mathcal{N}_{00}}$ denotes the zero operator on $\mathcal{N}_{00}$. The operator $\tilde{A} \restriction (\mathcal{H} \ominus \mathcal{N}_{00})$ is strictly positive, and, according to Theorem 3.1.4, its inverse admits the representation

$$(\tilde{A} \restriction (\mathcal{H} \ominus \mathcal{N}_{00})^{-1} = A_\infty^{-1} \restriction (\mathcal{H} \ominus \mathcal{N}_{00}) + B^{-1} P_{\mathcal{N}_0 \ominus \mathcal{N}_{00}}, \tag{3.16}$$

where B is the self-adjoint nonnegative operator on $\mathcal{N}_0$ with $\operatorname{Ker} B = \mathcal{N}_{00}$.

Conversely, if we start with the self-adjoint on $\mathcal{N}_0$ operator $B = B^* \geq 0$ for which $\mathcal{N}_{00} := \operatorname{Ker} B \neq \{0\}$, then denoting $\mathcal{N}_0' = \mathcal{N}_0 \ominus \mathcal{N}_{00}$, $B' = B \restriction \mathcal{N}_0' > 0$ and introducing the subspace $\mathcal{H}' = \mathcal{M}_0 \oplus \mathcal{N}_0'$, one can define the operator

$$\mathbf{A}' : P'\varphi \longmapsto \mathbf{A}\varphi \in \mathcal{M}_0, \quad \varphi \in \operatorname{Dom} \mathbf{A},$$

where P' denotes the orthogonal projection in $\mathcal{H}$ onto $\mathcal{H}'$. It is obvious that $\mathbf{A}'$ is a closed symmetric operator on $\mathcal{H}'$, $\mathbf{A}' \subset (\mathbf{A}')^*$, $\mathbf{A}' \geq 1$. Its Friedrichs extension is denoted by A_∞'. According to Theorem 3.1.4, the equality

$$(\tilde{A}')^{-1} = (A_\infty')^{-1} + (B')^{-1} P_{\mathcal{N}_0'} \tag{3.17}$$

defines an operator $\tilde{A}' \in \mathcal{A}_{>0}(\mathbf{A}')$, $\tilde{A}' > 0$. We denote the extension of the operator $\tilde{A}'$ by zero on $\mathcal{N}_{00}$ by A_B. One verifies directly that $A_B \geq 0$. Given $\tilde{A} \geq 0$ we can recover the operator B by the method already described above.

Formally, we can set B^{-1} equal to ∞ on $\mathcal{N}_{00}$. Then $\tilde{A}^{-1} = \infty$ on $\mathcal{N}_{00}$, which is equivalent to $\operatorname{Ker} \tilde{A} = \mathcal{N}_{00}$. In particular, if $\operatorname{Ker} B = \mathcal{N}_0$, i.e., $B = 0$, then, formally, $B^{-1} = \infty$ on whole $\mathcal{N}_0$ and also $\tilde{A}^{-1} = \infty$ on whole $\mathcal{N}_0$. This means that $\operatorname{Ker} \tilde{A} = \mathcal{N}_0$. In such a case $\tilde{A} = A_0$, where A_0 is the Kreĭn extension of the operator $\mathbf{A}$.

Now we will establish a bijective correspondence between positive not necessarily mutually simple to A_∞ self-adjoint extensions $\tilde{A}$ of the operator $\mathbf{A}$ and positive operators B that are not necessary densely defined on $\mathcal{N}_0$. We denote by $\mathcal{B}_+(\mathcal{N} \subseteq \mathcal{N}_0)$ the set of operators of the last kind.

Let us consider the situation, where $\tilde{A} \in \mathcal{A}_+(\mathbf{A})$ is not mutually simple to A_∞, but $\operatorname{Ker} \tilde{A} = \{0\}$. It means that the subspace

$$\mathcal{N}_{0\infty} := \{\eta \in \mathcal{N}_0 \mid \tilde{A}^{-1}\eta = A_\infty^{-1}\eta\} \neq \{0\}.$$

In such a case, we introduce the space $\mathcal{N}_0' = \mathcal{N}_0 \ominus \mathcal{N}_{0\infty}$, $\mathcal{H}' = \mathcal{H} \ominus \mathcal{N}_{0\infty}$ and $\mathcal{M}_0' = \mathcal{M}_0 \oplus \mathcal{N}'_0$. Instead of $\tilde{A}$, $\mathbf{A}$, and A_∞ on $\mathcal{H}'$, one can consider the self-adjoint operator $\tilde{A}'$ and symmetric operators $\mathbf{A}'$ and A_∞', defined through the restriction of corresponding inverse operators on $\mathcal{H}'$. Now $\tilde{A}'$ is mutually simple to A_∞' with respect to $\mathbf{A}'$, since on $\mathcal{N}_0'$ the operators $(\tilde{A}')^{-1}$ and $(A_\infty')^{-1}$ are nowhere equal. According to Theorem 3.1.4 we construct the operator $B' = B'^* > 0$ on $\mathcal{N}_0'$. To B' there corresponds on $\mathcal{N}_0$ a non-densely defined operator B. Namely, $B = B^* > 0$ acts on the subspace $(\mathrm{Dom}\, B')^{\mathrm{cl}} = \mathcal{N}_B \equiv \mathcal{N}_0'$. Then, formally $B^{-1} = 0$ on $\mathcal{N}_{0\infty}$. This is equivalent to the fact that $\tilde{A}^{-1}$ equals A_∞^{-1} on $\mathcal{N}_{0\infty}$.

Let the operator $\tilde{A} \geq 0$ be constructed by means of a not densely defined operator B on $\mathcal{N}_0$. It is self-adjoint and strictly positive on the subspace $\mathcal{N}_B := (\mathrm{Dom} B)^{\mathrm{cl}} \equiv \mathcal{N}_0'$, and it is not mutually simple to A_∞.

Finally, if $\tilde{A} \geq 0$, $\mathrm{Ker}\, \tilde{A} \neq \{0\}$, and $\tilde{A}$ are not mutually simple to A_∞, then by means of the construction described above and Theorem 3.1.4, we establish a bijective correspondence between $\tilde{A}$ and $B = B^* \geq 0$, which is not densely defined on $\mathcal{N}_0$, although it is self-adjoint on the subspace $\mathcal{N}_B \subseteq \mathcal{N}_0$.

These arguments prove the following result.

Theorem 3.1.5. *Between the set of all self-adjoint positive extensions of the symmetric operator* $\mathbf{A}$ *and the set of all operators of the form* $B \in \mathcal{B}_+(\mathcal{N} \subseteq \mathcal{N}_0)$, *there exists a bijective correspondence which is defined on the subspace* $\tilde{\mathcal{H}} = \mathcal{H} \ominus \mathrm{Ker}\, B$ *by the operator formula*

$$A_B^{-1} = (A_\infty^{-1} \restriction \tilde{\mathcal{H}}) + B^{-1} P_{\mathcal{N}}, \tag{3.18}$$

where $P_{\mathcal{N}}$ *denotes the orthogonal projection onto the subspace* $\mathcal{N} = \mathcal{N}_B \ominus \mathcal{N}_{00} \subseteq \mathcal{N}_0$, *where* $\mathcal{N}_{00} := \mathrm{Ker}\, B$. *Moreover,* $\tilde{A} = (A_B^{-1})^{-1} \oplus 0_{\mathcal{N}_{00}}$, *where* $0_{\mathcal{N}_{00}}$ *denotes the zero operator on the subspace* $\mathcal{N}_{00}$.

3.2 Description of extensions in terms of quadratic forms

Here we will reformulate the bijective correspondence established in Theorem 3.1.5 between the set of all positive self-adjoint extensions $\tilde{A}$ of a symmetric operator $\mathbf{A} \geq 1$ and the set of positive (in general not densely defined) operators $B \geq 0$ on $\mathcal{N}_0$ in terms of quadratic forms (for more details, see [28, 82]).

To formulate the results precisely, we need some preparatory material, which is also of an independent interest. To simplify the formulation we take as a basis Theorem 3.1.4. So, we are dealing with operators $\tilde{A} \in \mathcal{A}_{>0}(\mathbf{A})$ and $B \in \mathcal{B}_+(\mathcal{N}_0)$, $B > 0$, where $\mathcal{N}_0 = \mathrm{Ker}\, \mathbf{A}^*$.

Let us recall that the Friedrichs extension A_∞ of a symmetric operator $\mathbf{A}$ is associated with the closure of the quadratic form $\gamma_{\mathbf{A}}[\cdot] = (\mathbf{A}\cdot, \cdot)$. We denote this form by γ_∞. Thus,

$$(A_\infty \varphi, \psi) = \gamma_\infty(\varphi, \psi), \quad \varphi, \psi \in \mathrm{Dom}\, A_\infty \subset Q(\gamma_\infty).$$

We also know that the domain of γ_∞ is with a Hilbert space with respect to the norm $\|\cdot\|_1^2 = \gamma_\infty[\cdot]$:

$$Q(\gamma_\infty) = \mathrm{Dom}\, A_\infty^{1/2} =: \mathcal{H}_1.$$

Proposition 3.2.1. *Let* $\mathbf{A} \geq 1$ *and* $\gamma_\infty = (\gamma_{\mathbf{A}})^{\mathrm{cl}}$. *Then the deficiency subspace* $\mathcal{N}_0 = \operatorname{Ker} \mathbf{A}^*$ *of the operator* $\mathbf{A}$ *has null intersection with* $Q(\gamma_\infty) = \mathcal{H}_1$*:*

$$Q(\gamma_\infty) \cap \mathcal{N}_0 = \{0\}. \tag{3.19}$$

In particular

$$\operatorname{Dom} A_\infty \cap \mathcal{N}_0 = \{0\}. \tag{3.20}$$

Moreover, for each $a < 1$

$$Q(\gamma_\infty) \cap \mathcal{N}_a = \{0\}, \quad \mathcal{N}_a = \operatorname{Ker}(\mathbf{A}^* - a); \tag{3.21}$$

in particular,

$$\operatorname{Dom} A_\infty \cap \mathcal{N}_a = \{0\}. \tag{3.22}$$

Proof. Let $\varphi \in Q(\gamma_\infty) \cap \mathcal{N}_a$. Consider a sequence $\varphi_n \in \operatorname{Dom} \mathbf{A}$ such that $\varphi_n \to \varphi$ in $\mathcal{H}$ and $\gamma_\infty[\varphi_n - \varphi] \to 0, n \to \infty$. Then

$$\gamma_\infty[\varphi] = \lim_{n\to\infty} (\mathbf{A}\varphi_n, \varphi) = \lim_{n\to\infty} (\varphi_n, \mathbf{A}^*\varphi) = a \lim_{n\to\infty} (\varphi_n, \varphi) = a\|\varphi\|^2,$$

where we used that $\varphi \in \mathcal{N}_a = \operatorname{Ker}(\mathbf{A}^* - a)$. But $\gamma_\infty[\varphi] \geq \|\varphi\|^2$. Since $a < 1$, we conclude $\varphi = 0$. □

Proposition 3.2.2. *Let* $\mathbf{A} \geq 1$. *Then the domain of the adjoint operator* $\mathbf{A}^*$ *has a representation as the direct sum of* $\operatorname{Dom} A_\infty$ *and the deficiency subspace* $\mathcal{N}_a = \operatorname{Ker}(\mathbf{A}^* - a)$ *with any* $a < 1$*:*

$$\operatorname{Dom}\mathbf{A}^* = \operatorname{Dom} A_\infty \dot{+} \mathcal{N}_a = \operatorname{Dom} A_\infty \dot{+} \mathcal{N}_0. \tag{3.23}$$

Proof. Let $\psi \in \operatorname{Dom} \mathbf{A}^*$. Then the vector

$$\varphi = (A_\infty - a)^{-1}(\mathbf{A}^* - a)\psi \in \operatorname{Dom} A_\infty.$$

Here we used the fact that $(A_\infty - a)^{-1}$ is defined on the whole $\mathcal{H}$. We claim that the vector $\eta_a = \psi - \varphi \in \mathcal{N}_a$. Indeed,

$$(\mathbf{A}^* - a)\eta_a = (\mathbf{A}^* - a)\psi - (\mathbf{A}^* - a)(A_\infty - a)^{-1}(\mathbf{A} - a)\psi = 0,$$

since $\operatorname{Dom} A_\infty \subset \operatorname{Dom} \mathbf{A}^*$. Hence $\psi = \varphi + \eta_a$, $\varphi \in \operatorname{Dom} A_\infty$, $\eta_a \in \mathcal{N}_a$. By Proposition 3.2.1, this is a direct sum. □

Theorem 3.2.3. *The domain of an operator* $\tilde{A} \in \mathcal{A}_{>0}(\mathbf{A})$ *has the following description:*

$$\operatorname{Dom} \tilde{A} = \{g \in \mathcal{H} \mid g = f + B^{-1}P_{\mathcal{N}_0}A_\infty f,\ f \in \operatorname{Dom} A_\infty\}, \tag{3.24}$$

where $B = B^* > 0$ *is the operator on the subspace* $\mathcal{N}_0 = \operatorname{Ker} \mathbf{A}^*$ *which bijectively corresponds to* $\tilde{A} \equiv A_B$. *Moreover,*

$$\tilde{A}g = A_\infty f. \tag{3.25}$$

Proof. The operators $\tilde{A}$ and A_∞ are strictly positive and

$$\operatorname{Ran}\tilde{A} \subseteq \operatorname{Ran} A_\infty = \mathcal{H}.$$

Then, for all $h \in \operatorname{Ran}\tilde{A}$ one has that

$$h = \tilde{A}g = A_\infty f, \quad g \in \operatorname{Dom}\tilde{A}, \ f \in \operatorname{Dom} A_\infty.$$

Now, from (3.14) it follows that

$$g = \tilde{A}^{-1}h = (A_\infty^{-1} + B^{-1}P_{\mathcal{N}_0})h = f + B^{-1}P_{\mathcal{N}_0}A_\infty f, \ f \in \operatorname{Dom} A_\infty,$$

where we recall that $P_{\mathcal{N}_0}$ is the orthogonal projection of $\mathcal{N}_0$ onto $\mathcal{H}$. □

We are ready to obtain the direct sum representation for the quadratic form $\tilde{\gamma} = (\gamma_{\tilde{A}})^{\mathrm{cl}}$ (cf. with [28]).

Theorem 3.2.4. *The quadratic form $\tilde{\gamma}[\cdot] = (\tilde{A}\cdot,\cdot)^{\mathrm{cl}}$ associated with $\tilde{A} \in \mathcal{A}_{>0}(\mathbf{A})$, has the direct sum representation:*

$$\tilde{\gamma} = \gamma_\infty \dotplus \gamma_B, \quad Q(\tilde{\gamma}) = Q(\gamma_\infty) \dotplus Q(\gamma_B), \tag{3.26}$$

where γ_B denotes the closure of the form $(B\eta,\eta)$, $\eta \in \operatorname{Dom} B \subseteq \mathcal{N}_0$ (the operator B corresponds bijectively to $\tilde{A}$, according to Theorem 3.1.4). The direct sum $\dotplus$ means that

$$Q(\tilde{\gamma}) = \{\psi \in \mathcal{H} \mid \psi = \varphi + \eta, \varphi \in Q(\gamma_\infty), \ \eta \in Q(\gamma_B) \subseteq \mathcal{N}_0\} \tag{3.27}$$

and

$$\tilde{\gamma}[\psi] = \gamma_\infty[\varphi] + \gamma_B[\eta]. \tag{3.28}$$

Proof. Let $\tilde{A} \in \mathcal{A}_{>0}(\mathbf{A})$. We claim that $\tilde{\gamma} \subseteq \gamma_\infty \dotplus \gamma_B$. Indeed, since any $g \in \operatorname{Dom}\tilde{A}$ admits the representation (see (3.24)

$$g = f + \eta, \quad f \in \operatorname{Dom} A_\infty, \quad \eta = B^{-1}P_{\mathcal{N}_0}A_\infty f, \quad \eta \in \operatorname{Dom} B,$$

then (3.24) yields the equality

$$(\tilde{A}g, g) = (A_\infty f, f) + (B\eta,\eta), \tag{3.29}$$

because

$$\begin{aligned}(\tilde{A}g, g) &= (A_\infty f, g) + (A_\infty f, f + \eta)\\ &= (A_\infty f, f) + (P_{\mathcal{N}_0}A_\infty f, B^{-1}P_{\mathcal{N}_0}A_\infty f)\\ &= (A_\infty f, f) + (B\eta,\eta).\end{aligned}$$

Let us verify that the equality (3.29) remains valid under closure. Let the sequence $g_n \in \operatorname{Dom}\tilde{A}$ be such that $g_n \to 0$ in $\mathcal{H}$ and $\tilde{\gamma}[g_n - g_m] \to 0$, $n, m \to$

∞. Then, thanks to the representation $g_n = f_n + \eta_n$, $f_n \in \operatorname{Dom} A_\infty$, $\eta_n = B^{-1}P_{\mathcal{N}_0}A_\infty f_n$, (3.29) yields

$$\tilde{\gamma}[g_n] = \gamma_\infty[f_n] + \gamma_B[\eta_n] \geq \gamma_\infty[f_n] \geq \|f_n\|^2.$$

From the closability of the form $(\tilde{A}\cdot,\cdot)$ it follows that $\tilde{\gamma}[g_n] \to 0$. Hence $f_n \to 0$ as well. But then $\eta_n \to 0$. Now it is clear that for the last sequence $\gamma_B[\eta_n - \eta_m] \to 0$, $\gamma_\infty[f_n] \to 0$. This means that

$$\tilde{\gamma}[\psi] = \gamma_\infty[\varphi] + \gamma_B[\eta], \; \psi = \varphi + \eta \in Q(\tilde{\gamma}), \; \varphi \in Q(\gamma_\infty), \; \eta \in Q(\gamma_B).$$

Hence, $\tilde{\gamma} \subseteq \gamma_\infty \dot{+} \gamma_B$, as claimed.

On the other hand, it is easy to prove (see, e.g., [28]) that the quadratic form $\tilde{\gamma}' = \gamma_\infty \dot{+} \gamma_B$ is positive and closed. Let $\tilde{A}' > 0$ be the self-adjoint operator associated with $\tilde{\gamma}'$. According to the construction, $(\tilde{A}'f, f) = (\mathbf{A}f, f)$, $f \in \operatorname{Dom} \mathbf{A}$. Therefore, $\tilde{A}'$ is an extension of $\mathbf{A}$. Since $\tilde{\gamma} \subseteq \tilde{\gamma}'$, then $\tilde{A} = \tilde{A}'$, and so $\tilde{\gamma} = \tilde{\gamma}'$. □

From Theorem 3.2.4, we can get the following important corollary which we formulate as an independent result.

Theorem 3.2.5. *Formula* (3.26) *establishes a bijective correspondence between the quadratic form $\tilde{\gamma}$ generated by the operator $\tilde{A} \in \mathcal{A}_{>0}(\mathbf{A})$, and the quadratic form γ_B generated by the operator $B = B^* > 0$ acting on the subspace $\mathcal{N}_0 = \operatorname{Ker} \mathbf{A}^*$.*

This correspondence can be continued to all positive extensions $\tilde{A}$ of the operator $\mathbf{A}$, if one uses the forms γ_B generated by operators $B = B^* \geq 0$ acting on subspaces $\mathcal{N}_B \subseteq \mathcal{N}_0$, where $\mathcal{N}_B = (\operatorname{Dom} B)^{\mathrm{cl}}$. It is convenient to write $\gamma_B[h] = \infty$, if $h \notin Q(\gamma_B)$. Then each operator B on $\mathcal{N}_B \subseteq \mathcal{N}_0$ generates the form γ_B with the domain $Q(\gamma_B)$ which, in general, is not dense in $\mathcal{N}_0$. Now, the equality (3.26) establishes the bijective correspondence between $\tilde{\gamma} = \gamma_{\tilde{A}}$, $\tilde{A} \geq 0$ and γ_B, $B \in \mathcal{B}_+(\mathcal{N}_0)$, $B = B^* \geq 0$ on $\mathcal{N}_B \subseteq \mathcal{N}_0$.

Let us note that one cannot establish a bijective correspondence between all self-adjoint extensions $\tilde{A} \geq 0$ and all positive forms on the subspace $\mathcal{N}_0$ (see Theorem 2.9 in [28]), because in the case where $\dim \mathcal{N}_0 = \infty$ there exist positive singular forms which cannot be associated with any self-adjoint operator $B = B^* \geq 0$ on $\mathcal{N}_B \subset \mathcal{N}_0$.

3.3 On operators $\tilde{A} \in \mathcal{A}_+^1(\mathrm{A})$

In this section we study in more details self-adjoint extensions of symmetric operators with deficiency indices $(1,1)$.

Let us consider a couple of Hilbert spaces $\mathcal{H}$ and $\mathcal{H}_+$, which form a so-called pre-rigged space [42]:

$$\mathcal{H} \sqsupset \mathcal{H}_+.$$

By this we mean that $\mathcal{H}_+$ is densely and continuously embedded into $\mathcal{H}$. Moreover, the norms of these spaces satisfy the inequality

$$\|\cdot\| \leq \|\cdot\|_+. \tag{3.30}$$

Let us denote the inner products in $\mathcal{H}$ and $\mathcal{H}_+$ by the quadratic forms χ and χ_+, correspondingly

$$\chi(\cdot,\cdot) = (\cdot,\cdot) \equiv (\cdot,\cdot)_{\mathcal{H}},\ \chi_+(\cdot,\cdot) = (\cdot,\cdot)_+ \equiv (\cdot,\cdot)_{\mathcal{H}_+}.$$

Then (3.30) can be recast as

$$\chi_+[\varphi] \geq \chi[\varphi], \quad \varphi \in \mathcal{H}_+. \tag{3.31}$$

It is obvious that the quadratic form χ_+ is densely defined in $\mathcal{H}$, positive, and closed. It also admits the second representation [107]:

$$\chi_+(\varphi,\psi) = (A\varphi, A\psi), \quad \varphi,\psi \in \operatorname{Dom} A, \tag{3.32}$$

where $A = A^*$ is a positive self-adjoint operator on $\mathcal{H}$ with, $\operatorname{Ker} A = 0$. Moreover, it is easy to see that $A \geq 1$. Indeed, because of the inequality (3.30),

$$(A\varphi, A\varphi) = \chi_+[\varphi] = \|\varphi\|_+^2 \geq \|\varphi\|^2.$$

Therefore $A^2 \geq 1$, and $A \geq 1$ also. Obviously,

$$\operatorname{Dom} A = \mathcal{H}_+, \quad \operatorname{Ran} A = \mathcal{H},$$

since from $\{\varphi_n\} \in \Sigma_+$ it follows that $\{A\varphi_n\} \in \Sigma$ (membership of a sequence either to Σ_+ or to Σ means it is fundamental either in $\mathcal{H}_+$ or $\mathcal{H}$). Thus, for any vector $\varphi = \mathcal{H}_+$ there exists a sequence $\lim_{n\to\infty} \varphi_n \in \operatorname{Dom} A$ such that

$$A\varphi = \mathcal{H}\text{-}\lim_{n\to\infty} A\varphi_n \in \mathcal{H}.$$

Moreover, the mappings

$$A : \mathcal{H}_+ \longrightarrow \mathcal{H}, \quad A^{-1} : \mathcal{H} \longrightarrow \mathcal{H}_+$$

are isometric, since $A^{-1} \leq 1$ and

$$\|\varphi\|_+ = \|A\varphi\|, \quad \|A^{-1}h\|_+ = \|h\|, \quad \varphi \in \mathcal{H}_+,\ h \in \mathcal{H}.$$

Let us fix a vector $\eta \in \mathcal{H}\backslash\mathcal{H}_+$. Since $\mathcal{H} = \operatorname{Ran} A$, $\eta = A\eta_+$ for some $\eta_+ \in \mathcal{H}_+$. Let us denote

$$\mathcal{N}_+ = \{c\eta_+\}_{c\in\mathbb{C}}, \quad \mathcal{M}_+ = \mathcal{N}_+^{\perp}.$$

Then

$$\mathcal{H}_+ = \mathcal{M}_+ \oplus \mathcal{N}_+.$$

Lemma 3.3.1. *The subspace $\mathcal{M}_+$ is dense in $\mathcal{H}$ if and only if the vector η does not belong to the domain of the operator A:*

$$\mathcal{H} \sqsupset \mathcal{M}_+ \Longleftrightarrow \eta \in \mathcal{H} \setminus \operatorname{Dom} A. \tag{3.33}$$

Proof. Sufficiency. Let $\eta \notin \mathcal{H}_+ = \operatorname{Dom} A$. Suppose the contrary namely that $\mathcal{M}_+$ is not dense in $\mathcal{H}$. Then there exist a vector $0 \neq h \in \mathcal{H}$, orthogonal to $\mathcal{M}_+$: $(h, \varphi) = 0$ for all $\varphi \in \mathcal{M}_+$. Then

$$0 = (h, \varphi) = (A^{-1}h, A^{-1}\varphi)_+ = (A^{-2}h, \varphi)_+, \quad \varphi \in \mathcal{M}_+.$$

This means that $A^{-2}h = c\eta_+$, with some $c \in \mathbb{C}$. Since in fact $\eta_+ = A^{-1}\eta$, we conclude that $A^{-1}h = c\eta$. Hence, $\eta \in \operatorname{Dom} A = \mathcal{H}_+$, which contradicts the initial assumption.

Necessity. Let $\mathcal{M}_+$ be dense in $\mathcal{H}$. Then, there exists a sequence $\varphi_n \in \mathcal{M}_+$ that converges to η on $\mathcal{H}$. In particular, for the inner product $(\varphi_n, \eta) \to \|\eta\|^2 > 0$. It follows that $\eta = A\eta_+ \notin \operatorname{Dom} A$. Indeed, if we suppose that $\eta \in \operatorname{Dom} A$, then

$$(\varphi_n, A\eta) = (A\varphi_n, \eta) = 0,$$

since $A\varphi_n \perp \eta$. On the other hand, since $A \geq 1$, it should hold that, $(\varphi_n, A\eta) \to (\eta, A\eta) \geq \|\eta\|^2 > 0$. Hence, $\eta \notin \operatorname{Dom} A$. □

The proved result can be formulated in a slightly different form.

Let $\eta_+ \in \operatorname{Dom} A$. Recall that $\operatorname{Dom} A = \mathcal{H}_+$. Denote

$$\mathcal{M}_+ = \{\varphi \in \operatorname{Dom} A \mid (A\varphi, A\eta_+) = 0\}.$$

We claim that $\mathcal{M}_+$ being dense in $\mathcal{H}$ is equivalent to

$$\eta = A\eta_+ \notin \operatorname{Dom} A.$$

Let an unbounded simple (without invariant subspaces) self-adjoint operator $A = A^* \geq 1$ be given on $\mathcal{H}$.

According to our constructions above, the domain $\operatorname{Dom} A$ of this operator is a (complete) Hilbert space $\mathcal{H}_+$ with respect to the inner product

$$(\varphi, \psi)_+ = (A\varphi, A\psi), \quad \varphi, \psi \in \operatorname{Dom} A.$$

Obviously, the couple $\mathcal{H} \sqsupset \mathcal{H}_+$ forms a pre-rigged space in the sense that $\mathcal{H}_+$ is a dense subset in $\mathcal{H}$ and, in addition, the corresponding norms satisfy the inequality

$$\|\varphi\| \leq \|\varphi\|_+, \quad \varphi \in \mathcal{H}_+. \tag{3.34}$$

It is important to remark that the operator A can be recovered from the couple $\mathcal{H} \sqsupset \mathcal{H}_+$ with the condition (3.34). Indeed, the quadratic form

$$\chi_+(\varphi, \psi) := (\varphi, \psi)_+, \quad \varphi, \psi \in \mathcal{H}_+$$

is densely defined, positive, and closed in $\mathcal{H}$. Therefore, in accordance with the second representation theorem, there exists a positive operator A in $\mathcal{H}$ such that

$$\chi_+(\varphi,\psi) = (A\varphi, A\psi), \quad \varphi,\psi \in \operatorname{Dom} A = \mathcal{H}_+.$$

Because of (3.34), the operator A satisfies the inequality $A \geq 1$. Hence $A^{-1} \leq 1$, $\operatorname{Dom} A^{-1} = \mathcal{H}$, and the operators A and A^{-1} regarded as mappings

$$A : \mathcal{H}_+ \longrightarrow \mathcal{H}, \quad A^{-1} : \mathcal{H} \longrightarrow \mathcal{H}_+, \tag{3.35}$$

are isometric. We emphasize that by our assumption the operator A is unbounded, since otherwise, $\operatorname{Dom} A = \mathcal{H}$, and the norms and $\|\cdot\|$ and $\|\cdot\|_+$ are equivalent. In such a case $\mathcal{H}_+$ is not a proper subspace in $\mathcal{H}$.

So, we have the following result.

Theorem 3.3.2. *Each couple of subspaces $\mathcal{H}$ and $\mathcal{H}_+$ which forms a pre-rigged space*

$$\mathcal{H} \sqsupset \mathcal{H}_+, \quad \|\cdot\| \leq \|\cdot\|_+, \tag{3.36}$$

is uniquely associated with a self-adjoint operator $A = A^* \geq 1$ on $\mathcal{H}$ such that $\operatorname{Dom} A = \mathcal{H}_+$, $\operatorname{Ran} A = \mathcal{H}$, and the mapping $A : \mathcal{H}_+ \to \mathcal{H}$ is isometric. Moreover, the set of all operators $A = A^* \geq 1$ on $\mathcal{H}$ and the set of couples of spaces that form pre-rigged spaces (3.36) *are in a bijective correspondence. This correspondence is established by the equality*

$$\|A\varphi\| = \|\varphi\|_+, \quad \varphi \in \operatorname{Dom} A = \mathcal{H}_+. \tag{3.37}$$

Let us fix a pre-rigged space (3.36) and the associated operator A. We will carry out the following construction. Taking a vector $\eta \in \mathcal{H}$, $\|\eta\| = 1$, we introduce a decomposition of $\mathcal{H}_+$ onto the orthogonal sum of subspaces:

$$\mathcal{H}_+ = \mathcal{M}_+ \oplus \mathcal{N}_+, \quad \mathcal{N}_+ = \{c\eta_+\}_{c\in\mathbb{C}}, \quad \mathcal{M}_+ = \mathcal{N}_+^\perp,$$

where $\eta_+ = A^{-1}\eta$. By Lemma 3.3.1, the subspace $\mathcal{M}_+$ is dense in $\mathcal{H}$, if and only if $\eta \notin \mathcal{H}_+$.

Proceeding in this way, we take the vector $\eta \in \mathcal{H} \setminus \mathcal{H}_+$, so that $\eta = A\eta_+$, where $\eta_+ \in \mathcal{H}_+ = \operatorname{Dom} A$. Now we define $\mathbf{A} := A \restriction \mathcal{M}_+$. Then, the operator $\mathbf{A}$ is densely defined, closed, and symmetric. In addition,

$$\operatorname{Ran} \mathbf{A} = \mathcal{M}_0 = \mathcal{N}_0^\perp, \quad \mathcal{N}_0 = \{c\eta\}_{c\in\mathbb{C}}, \quad \dim \mathcal{N}_0 = 1,$$

and the deficiency indices of the operator $\mathbf{A}$ are equal to $(1,1)$.

We start the description of operators $\tilde{A} \in \mathcal{A}^1_+(\mathbf{A})$, $\tilde{A} \geq 1$, with the construction of the maximal self-adjoint extension $A_\infty \in \mathcal{A}^1_+(\mathbf{A})$, i.e., the Friedrichs extension (see [82, 107, 148]). This operator can be obtained as follows. Let us denote by γ_∞ the closure of the positive on $\mathcal{H}$ quadratic form generated by the symmetric operator $\mathbf{A}$. Let $\mathcal{H}_{+,\infty}$ denote the completion of $\mathcal{M}_+$ with respect to the inner product $(\cdot,\cdot)_{+,\infty} := \gamma_\infty(\cdot,\cdot)$. Obviously, $\mathcal{H} \sqsupset \mathcal{H}_{+,\infty}$, $\|\cdot\| \leq \|\cdot\|_{+,\infty}$, since $\mathbf{A} \geq 1$. The operator $A_\infty^{1/2}$ is defined as the operator associated with this pre-rigged couple. Therefore,

$$\gamma_\infty(\cdot,\cdot) = (A_\infty^{1/2}\cdot, A_\infty^{1/2}\cdot) = (A_\infty\cdot,\cdot)^{\mathrm{cl}}, \quad \operatorname{Dom} A_\infty^{1/2} = \mathcal{H}_{+,\infty}.$$

The operator A_∞ can be also constructed in a slightly different way. We extend the given couple $\mathcal{H} \sqsupset \mathcal{H}_{+,\infty}$ to the rigged space $\mathcal{H}_{-,\infty} \sqsupset \mathcal{H} \sqsupset \mathcal{H}_{+,\infty}$ and then introduce the Berezansky canonical isomorphism $D_{-+} : \mathcal{H}_{+,\infty} \mapsto \mathcal{H}_{+,\infty}$. Then

$$A_\infty := D_{-+} \upharpoonright \{\varphi \in \mathcal{H}_{+,\infty} \mid D_{-+}\varphi \in \mathcal{H}\}.$$

Let us address the following question. Under what conditions $A_\infty = A$? Since both $A_\infty, A \in \mathcal{A}^1_+(\mathbf{A})$, the equality $A_\infty = A$ holds true if and only if $\operatorname{Dom} A_\infty = \operatorname{Dom} A = \mathcal{H}_+$. This is obviously equivalent to the equality $\mathcal{H}_{+,\infty} = \mathcal{H}_1$, where $\mathcal{H}_1$ is the completion of $\operatorname{Dom} A$ with respect to the inner product $(\cdot,\cdot)_1 = (A\cdot,\cdot) \equiv \gamma_A(\cdot,\cdot)$. Further, $\mathcal{H}_{+,\infty} = \mathcal{H}_1$ if and only if $\mathcal{M}_+$ is dense in $\mathcal{H}_1$, because the form γ_∞ coincides with the form γ_A on $\mathcal{M}_+$. Since $\mathcal{H} \sqsupset \mathcal{H}_1 \sqsupset \mathcal{H}_+$, Lemma 3.3.1 shows that the subspace $\mathcal{M}_+$ is dense in $\mathcal{H}_1$ if and only if the vector

$$\eta_1 := A^{-1/2}\eta = A^{1/2}\eta_+ \notin \mathcal{H}_+. \tag{3.38}$$

Thus, the validity of the next assertion is established.

Proposition 3.3.3. *The Friedrichs extension of the symmetric operator*

$$\mathbf{A} := A \upharpoonright \mathcal{M}_+,\ \mathcal{M}_+ = \mathcal{N}_+^\perp,\ \mathcal{N}_+ = \{c\eta_+ | c \in \mathbb{C}\}, \eta_+ = A^{-1}\eta$$

coincides with the original operator $A_\infty = A$ if and only if the condition (3.38) *is fulfilled.*

Remark 3.3.4. The condition (3.38) is equivalent to the fact that $\eta \notin \mathcal{H}_1$, since $A^{1/2} : \mathcal{H}_1 \to \mathcal{H}$ is an isometric operator. Hence,

$$\eta_1 \in \mathcal{H}_1 \setminus \mathcal{H}_+ \Longleftrightarrow A^{1/2}\eta_1 = \eta \in \mathcal{H} \setminus \mathcal{H}_1. \tag{3.39}$$

Remark 3.3.5. If a vector η from $\mathcal{H}$ belongs to $\mathcal{H}_+$, i.e., $\eta \in \operatorname{Dom} A$, then it does not make sense to ask about the existence of a densely defined symmetric operator $\mathbf{A} = A \upharpoonright \mathcal{M}_+$, since in such a case the subspace $\mathcal{M}_+ = \mathcal{N}_+^\perp$, ($\mathcal{N}_+ = \{cA^{-1}\eta\}_{c\in\mathbb{C}}$) is not dense in $\mathcal{H}$. Therefore, further we assume that $\eta \notin \mathcal{H}_+$ and, moreover, $\eta \notin \operatorname{Dom} A^{1/2}$. Then $A_\infty = A$ and this equality is basic for the description of all operators $\tilde{A} \in \mathcal{A}^1_+(\mathbf{A})$.

Let us consider one more question. Do there exist self-adjoint extensions $\tilde{A} \geq 1$ of the operator $\mathbf{A} \geq 1$, different from A_∞, but with the same lower bound. In other words, do there exist operators $\tilde{A} \neq A_\infty, \tilde{A} \geq 1$? If yes, then how to describe them?

Let us recall that the Friedrichs extension A_∞ is a maximal positive extension of the operator $\mathbf{A}$. Therefore, for any strictly positive extension $\tilde{A}$ it holds that $0 < \tilde{A} \le A_\infty$. Now, we want to describe all $\tilde{A}$, for which the inequality

$$1 \le \tilde{A} \le A_\infty = A \tag{3.40}$$

holds. The condition (3.40), is obviously equivalent to

$$0 < A^{-1} \le \tilde{A}^{-1} \le 1. \tag{3.41}$$

We introduce the bounded self-adjoint operator $\tilde{B} := \tilde{A}^{-1} - A^{-1}$. It is clear that

$$\operatorname{Ker} \tilde{B} = \mathcal{M}_0 = \mathbf{A}\mathcal{M}_+,$$

since the operators A, $\tilde{A}$, and $\mathbf{A}$ all coincide on $\mathcal{M}_+$. Since the subspace $\mathcal{N}_0 = \mathcal{M}_0^\perp$ is one-dimensional and $\tilde{B}$ is self-adjoint, it has the form

$$\tilde{B} = b^{-1} P_{\mathcal{N}_0}, \quad b > 0,$$

where $P_{\mathcal{N}_0}$ is the orthogonal projection onto $\mathcal{N}_0$. Since $\|\eta\| = 1$,

$$\tilde{B}h = b^{-1}(h, \eta)\eta, \quad h \in \mathcal{H}.$$

From $\tilde{A}^{-1} = A^{-1} + \tilde{B}$, $A^{-1} \le 1$, and (3.41) it follows that $\tilde{B} \le 1 - A^{-1} \le 1$. But $\operatorname{Ker} \tilde{B} = \mathcal{M}_0$. Consequently,

$$(\tilde{A}^{-1}\eta, \eta) = (A^{-1}\eta, \eta) + b^{-1}(\eta, \eta) \le 1. \tag{3.42}$$

Theorem 3.3.6. *For an operator $\tilde{A} \in \mathcal{A}_+^1(\mathbf{A})$, $\tilde{A} \ge 1$ if and only if*

$$b \ge \frac{1}{1 - \|\eta\|_{-1}^2}, \quad \|\eta\|_{-1}^2 = (A^{-1}\eta, \eta). \tag{3.43}$$

Proof. It follows from (3.42). We only need to note that the negative norm $\|\eta\|_{-1}^2 < 1$. Indeed, $(A^{-1}\eta, \eta) \le 1$, since $A^{-1} \le 1$ and $\|\eta\| = 1$, but the equality to one, due to the self-adjointness, means that η is an eigen-vector of the operator A. This contradicts the condition $\eta \notin \operatorname{Dom} A$. □

So, in the set of all positive self-adjoint extensions of the operator $\mathbf{A}$ there exist a continuum of different self-adjoint operators $\tilde{A}$ that satisfy the inequality $\tilde{A} \ge 1$. These operators have a parametrization by numbers $\beta = b^{-1}$, $0 < \beta \le 1 - (A^{-1}\eta, \eta)$.

Let us note that, in general case, to ensure the property $\tilde{A} \ge 1$, the conditions

$$0 \le \tilde{B} \le 1 - A^{-1}$$

should be fulfilled. Its extension on the subspace $\mathcal{N}_0$ has a form

$$0 \le (B^{-1}\eta, \eta) \le \|\eta\|^2 - \|\eta\|_{-1}^2, \quad \eta \in \mathcal{N}_0.$$

We remain the proof of this theorem as an exercise for our reader.

Theorem 3.3.7. *Let* $\mathcal{N}_0 = \mathrm{Ker}\mathbf{A}^*$, $\dim\mathcal{N}_0 = 1$, $\eta \in \mathcal{N}_0$, $\|\eta\|_0 = 1$. *Then the set of operators* $\tilde{A} \in \mathcal{A}^1(\mathbf{A})$ *has the following parametrization:*

$$\mathcal{A}^1(\mathbf{A}) \ni \tilde{A} = A_b \longleftrightarrow b \in \mathbb{R}^1 \cup \{\infty\},$$

where $\tilde{A} = A_\infty$, $b = \infty$, $A_b^{-1} = A_\infty^{-1} + b^{-1}P_{\mathcal{N}_0}$, $b \neq 0$, $\tilde{A} = A_0$, $A_0\eta = 0$, $b = 0$.

3.4 Examples

Here are several examples concerning the previous statements.

Example 3.4.1. Let the operator $\mathbf{A} = -\frac{d^2}{dx^2}$ be defined on the space $L^2(0,\infty)$ in the domain $\mathrm{Dom}\,\mathbf{A} = C_0^\infty(0,\infty)$. It is obvious that the quadratic form

$$\gamma_{\mathbf{A}}[\varphi] = \int\limits_0^\infty |\varphi'(x)|^2 dx \geq 0,$$

i.e., is positive, and hence $\mathbf{A}$ is a positive symmetric operator. The adjoint operator $\mathbf{A}^*$ is defined by the same differential expression $-\frac{d^2}{dx^2}$, but with the domain $\mathrm{Dom}\,\mathbf{A}^* = \{\varphi \in AC^2(0,\infty)\}$, the set of functions that belong to $AC(0,\infty)$ together with their first derivatives (see [170]). It is easy to verify that

$$\mathrm{Ker}\,(A^* - z) = \{c\exp(-(-z)^{1/2}x) \mid c \in \mathbb{C}\}.$$

In particular, the solutions of the equations

$$-\varphi''(x) = \pm i\varphi(x),$$

have, up to constants, the form

$$\begin{aligned} &\exp\{(-1+i)x/\sqrt{2}\}, \qquad &&\exp\{(1-i)x/\sqrt{2}\}, \\ &\exp\{(1+i)x/\sqrt{2}\}, &&\exp\{(-1-i)x/\sqrt{2}\}. \end{aligned}$$

Only the first and the last of these functions (we denote them by $\eta_+(x)$ and $\eta_-(x)$, respectively) belong to $L^2(0,\infty)$. Therefore, the deficiency subspaces $\mathcal{K}_+$ and $\mathcal{K}_-$ are one-dimensional and have the form

$$\mathcal{K}_+ = \{c\eta_+ \mid c \in \mathbb{C}\}, \quad \mathcal{K}_- = \{c\eta_- \mid c \in \mathbb{C}\}.$$

So, the deficiency indices of the operator $\mathbf{A}$ are equal to $(1,1)$. Let us remark that $\eta_\pm(0) = 1$, but $\eta'_\pm(0) = -\exp\{\pm\frac{\pi}{4}i\}$. All self-adjoint extensions $\tilde{A} = A_\alpha$ of the operator $\mathbf{A}$ are restrictions of the operator $\mathbf{A}^* = -\frac{d^2}{dx^2}$ and are parameterized by the number $a \in \mathbb{R}^1 \cup \infty$:

$$\begin{aligned} \mathrm{Dom}\,(A_a) &= \{\psi \in AC^2[0,\infty) \mid \psi'(0) + a\psi(0) = 0,\ a \in \mathbb{R}^1\}, \\ \mathrm{Dom}\,A_\infty &= \{\psi \in AC^2[0,\infty) \mid \psi(0) = 0\}. \end{aligned} \tag{3.44}$$

To verify that (3.44) is valid, we consider the boundary form $B(g,h)$ of the operator $\mathbf{A}$. Let us fix the vector $g \in \mathcal{K}_+ \oplus \mathcal{K}_-$ as a sum $g = \eta_+ + \exp(i\theta)\eta_-$, $0 \le \theta < 2\pi$, where $\exp(i\theta)$ corresponds to an arbitrary isometric transformation from $\mathcal{K}_+$ into $\mathcal{K}_-$. Then the boundary condition $B(g,h) = 0$ is equivalent to the equality

$$\cos\frac{\theta}{2}\psi'(0) + \cos\left(\frac{\theta}{2} - \frac{\pi}{4}\right)\psi(0) = 0,$$

where we used the fact that

$$B(\eta_\pm, \psi) = \psi'(0) + \exp\left(\pm i\frac{\pi}{4}\right)\psi(0).$$

Thus, in the most general form, the boundary condition has the form

$$\alpha\psi'(0) + \beta\psi(0) = 0,$$

where real numbers α, and β are not equal to zero simultaneously. Now, each self-adjoint extension of the operator $\mathbf{A}$ is determined by the value of one parameter $a = \alpha/\beta$. We obtain (3.44).

Example 3.4.2. Let the operator $\mathbf{A} = -\frac{d^2}{dx^2}$ be defined on the domain C_0^∞ in $L^2(0,\pi)$. It is obvious $\mathbf{A}$ is a positive symmetric operator. The adjoint operator $\mathbf{A}^* = -\frac{d^2}{dx^2}$ that is naturally defined for all $\varphi \in AC^2[0,\pi]$. The Friedrichs extension $\mathbf{A}_\infty$ is defined by the Dirichlet boundary conditions

$$\varphi(0) = \varphi(1) = 0.$$

The operator $\mathbf{A}$ has continuum self-adjoint extensions, which have the same lower bound. One of these extensions is defined by the condition

$$\varphi(0) = -\varphi(\pi), \quad \varphi'(0) = -\varphi'(\pi).$$

Example 3.4.3. Let the operator $\mathbf{A} = -\frac{d^2}{dx^2} + 1$ be defined on the domain $C_0^\infty(0,\infty)$ in $L^2(0,\infty)$. It is obvious that $\mathbf{A} \ge 1$, and that the point $0 \in \rho(\mathbf{A})$. It is easy to verify that the deficiency subspace

$$\mathcal{N}_0 = \operatorname{Ker} \mathbf{A}^* = \{ce^{-x} \mid c \in \mathbb{C}\}.$$

The Friedrichs extension $A_\infty = -\frac{d^2}{dx^2} + 1$ is determined by the domain

$$\operatorname{Dom} A_\infty = \{\psi, \psi' \in AC[0,\infty) \mid \psi(0) = 0\},$$

The domain of the quadratic form $\gamma_\infty = \gamma_{\mathbf{A}}^{\mathrm{cl}}$,

$$\gamma_{\mathbf{A}}[\varphi] = \int_0^\infty (|\varphi'(x)|^2 + |\varphi(x)|^2)dx, \ \varphi \in C_0^\infty,$$

coincides with the Sobolev space $\overset{\circ}{W}{}_1^2(0,\infty)$. By Proposition 3.2.2,

$$\operatorname{Dom}\mathbf{A}^* = \operatorname{Dom}A_\infty \dot{+} \mathcal{N}_0 = \operatorname{Dom}A_\infty \dot{+} \mathcal{N}_a, \quad a<1.$$

It is clear that for functions that belong to the domains of self-adjoint extensions distinct from A_∞, instead of the condition $\varphi(0)=0$ will satisfy another boundary condition. In particular, for each $a<1$, the operator $A_a = -\frac{d^2}{dx^2}+1$, with the domain

$$\operatorname{Dom}A_a = \operatorname{Dom}\mathbf{A} \dot{+} \{ce^{-(\sqrt{a-1})x} \mid c\in\mathbb{C}\},$$

is the self-adjoint extension with the lover bound $m_{A_a}=a$, which is an eigenvalue of the operator A_a, since $\mathcal{N}_a = \{ce^{-(\sqrt{a-1})x}\}$.

On the other hand, it is easy to show that the subset of all positive self-adjoint extensions of the operator $\mathbf{A}$ is described by the boundary condition

$$\psi'(0) = (b-1)\psi(0), \quad b>0.$$

Indeed, let $\tilde{A} = A_B \in \mathcal{A}_+^1(\mathbf{A})$, where the operator B acts on $\mathcal{N}_0$ as multiplication by the number $b>0$. Then the quadratic form of the operator $\tilde{A}$, which, in accordance with Theorem 3.2.4, is written as a direct sum of forms, $\gamma_{\tilde{A}} = \gamma_\infty \dot{+} \gamma_B$, can be expressed as

$$\gamma_{\tilde{A}}[\psi] = \int_0^\infty (|\psi'(x)|^2 + |\psi(x)|^2)dx + (b-1)|\psi(0)|^2,$$

where we used the fact that each function $\psi\in\operatorname{Dom}\tilde{A}$ has a unique representation (see Proposition 3.2.1) as a sum $\psi = \varphi+\eta$ with $\varphi\in\operatorname{Dom}A_\infty$ and $\eta\in\mathcal{N}_0$. Since the subspace $\mathcal{N}_0$ is one-dimensional, the representation has an explicit form:

$$\psi(x) = [\psi(x) - \psi(0)e^{-x}] + \psi(0)e^{-x}.$$

Taking into account now that $\psi\in\operatorname{Dom}\tilde{A}$ if and only if the functional $l_g(h) = (\tilde{A}\psi, h)$ is continuous with respect h, and that

$$(\tilde{A}\psi,h) = -(\psi'',h) + (\psi,h) + \bar{h}(0)((b-1)\psi(0) - \psi'(0)),$$

we conclude that $\psi'(0) = (b-1)\psi(0)$.

Let us note that the boundary conditions of this type establish a one-to-one correspondence between positive self-adjoint extensions and positive operators B in the deficiency subspace $\mathcal{N}_0$. This correspondence can be extended as to establish a connection between operators B and singular quadratic forms that act in rigged spaces (see Chapter 7).

3.5 Abstract boundary conditions. The Weyl function

The method of abstract boundary conditions (see [91, 92] and references therein) is extremely fruitful from the standpoint of the analysis of spectral properties of singularly perturbed operators. This method is based on the concept of boundary triplets, which was developed by many authors (see, e.g., [34, 36, 37, 63, 70–74, 156] and is closely associated with the use of an abstract analog of the Weyl function. We briefly sketch this method.

In what follows A denotes a closed simple symmetric operator on $\mathcal{H}$ with deficiency indices $\mathbf{n}^- = \mathbf{n}^+ \neq 0$.

A *boundary triplet*, for the adjoint operator A^*, consists of a Hilbert space $\mathfrak{H}$ and a couple of linear mappings:

$$\Gamma_i : \operatorname{Dom} A^* \longmapsto \mathfrak{H}, \quad i = 0, 1,$$

such that the mapping

$$\operatorname{Dom} A^* \ni \varphi \longmapsto \Gamma_0\varphi \oplus \Gamma_1\varphi \in \mathfrak{H} \oplus \mathfrak{H}$$

is surjective and the Green formula holds:

$$(A^*\varphi, \psi) - (\varphi, A^*\psi) = (\Gamma_1\varphi, \Gamma_0\psi) - (\Gamma_0\varphi, \Gamma_1\psi), \quad \varphi, \psi \in \operatorname{Dom} A^*,$$

where the inner product in $\mathfrak{H}$ is denoted as in the space $\mathcal{H}$: $(\cdot,\cdot)_{\mathfrak{H}} = (\cdot,\cdot)_{\mathcal{H}}$.

It is clear that for a fixed operator A^* there are many boundary triplets $\Pi = \{\mathcal{H}, \Gamma_0, \Gamma_1\}$, But this is a problem to write some of the triplet in a certain given case.

The extension A' of the operator A is said to be *proper*, (and we write $A' \in \operatorname{Ext} A$), if $A \subset A' \subset A^*$. A couple of proper extensions A', A'' is called *disjunctive*, if $\operatorname{Dom} A' \cap \operatorname{Dom} A'' = \operatorname{Dom} A$ and *transversal*, if additionally $\operatorname{Dom} A' + \operatorname{Dom} A'' = \operatorname{Dom} A^*$. In [74] it is shown that for each couple of transversal extensions A_i, $i = 0, 1$ there exists a boundary triplet $\Pi = \{\mathfrak{H}, \Gamma_0, \Gamma_1\}$ such that

$$\operatorname{Dom} A_i = \operatorname{Ker} \Gamma_i, \quad i = 0, 1,$$

and, in addition,

$$\dim \mathfrak{H} = \mathbf{n}^- = \mathbf{n}^+.$$

Conversely (see [156]), each boundary triplet Π is associated with a couple of extensions $A_i := A^* \restriction \operatorname{Ker}\Gamma_i$, $i = 0, 1$, which are self-adjoint.

It is easy to check that for a boundary triplet $\Pi = \{\mathfrak{H}, \Gamma_0, \Gamma_1\}$, each bounded self-adjoint operator B on $\mathfrak{H}$ defines a new boundary triplet $\Pi_B = \{\mathfrak{H}, \Gamma_0^B, \Gamma_1^B\}$, where $\Gamma_0^B := B\Gamma_0 - \Gamma_1$, $\Gamma_1^B = \Gamma_0$.

In accordance with [174], a closed subspace of $\mathfrak{H} \oplus \mathfrak{H}$ is called a closed *relation*, and is denoted by θ. The relation is called *symmetric* if

$$(\psi_1, \varphi_2) - (\varphi_1, \psi_2) = 0, \quad \{\varphi_1, \psi_1\}, \{\varphi_2, \psi_2\} \in \theta,$$

and *self-adjoint* if θ has no closed symmetric extensions.

Let us remark that the graph of any self-adjoint operator B on $\mathfrak{H}$ is a self-adjoint relation and conversely, each self-adjoint relation is a graph of some self-adjoint operator on $\mathfrak{H}$.

If for A^* the boundary triplet $\Pi = \{\mathfrak{H}, \Gamma_0, \Gamma_1\}$ is fixed, then we can associate each self-adjoint relation θ with an operator

$$A^\theta := A^* \upharpoonright \mathfrak{D}^\theta,$$
$$\operatorname{Dom} A^\theta = \mathfrak{D}^\theta := \{\varphi \in \operatorname{Dom} A^* \mid \{\Gamma_0\varphi, \Gamma_1\varphi\} \in \theta\}.$$

Theorem 3.5.1 ([73, 92]). *The correspondence $\theta \leftrightarrow A^\theta$ between self-adjoint relations θ on $\mathfrak{H}$ and self-adjoint extensions of a symmetric operator A is bijective.*

Fix a boundary triplet $\Pi = \{\mathfrak{H}, \Gamma_0, \Gamma_1\}$ for A^*. Denote A_0 an extension of A such that

$$\operatorname{Dom} A_0 = \operatorname{Ker} \Gamma_0.$$

In accordance with [73], the *Weyl function* of the operator A is the bounded operator-valued function $M(z)$, $z \in \rho(A_0)$, defined by the equality

$$M(z)\Gamma_0 f_z = \Gamma_1 f_z, \quad f_z \in \mathcal{N}_z, \ z \in \rho(A_0),$$

where $\mathcal{N}_z = \operatorname{Ker}(A^* - zI)$.

It turns out that the Weyl function has excellent analytical properties, namely, it is always a Herglotz (or Nevanlinna) function, meaning that it is holomorphic on $\mathbb{C}_+$ and dissipative:

$$\operatorname{Im}(M(z)) = \frac{M(z) - \bar{M}(z)}{2i} \geq 0.$$

Therefore, the Weyl functions provide a powerful tool for establishing various, in particular spectral, properties of symmetric and self-adjoint extensions (see, e.g., [36, 37, 63, 73, 74]). A typical result in this direction reads:

Theorem 3.5.2 ([73]). *Let A^θ be a self-adjoint extension of a simple symmetric operator A, where θ denotes a self-adjoint relation on $\mathfrak{H}$. The point $\lambda \in \rho(A_0)$ belongs to $\rho(A^\theta)$ (to the point or continuous spectrum of the operator A^θ) if and only if 0 belongs to $\rho(\theta - M(\lambda))$, (resp., to the point or continuous spectrum of the operator $\theta - M(\lambda)$).*

Chapter 4

Rigged Hilbert Spaces

A considerable part of functional analysis, including the theory of linear operators, particularly the spectral theory, cannot be presented successively without the notion of a rigged Hilbert space. The use of the method of rigged spaces allows one to beyond the setting with two starting objects, a single Hilbert space $\mathcal{H}$ and an operator A on $\mathcal{H}$. To provide a complete picture of the spectral properties for A in the general case, it is necessary to equip the space $\mathcal{H}$ with an additional couple of Hilbert (or topological) spaces in a specific way. So, in a natural way there appear the triples of embedding spaces of a view $\mathcal{H}_- \supset \mathcal{H} \supset \mathcal{H}_+$, (or $\Phi^* \supset \mathcal{H} \supset \Phi$). In this book we will restrict our attention to the Hilbert equipping which form the so-called rigged Hilbert spaces.

In the most advanced approach (see [42] and references therein) it is assumed that $\operatorname{Dom} A$ is a part of some positive Hilbert space $\mathcal{H}_+$, which is continuously embedded into $\mathcal{H}$. Then the generalized eigen-functions of the operator A belong to the negative space $\mathcal{H}_-$ which extends $\mathcal{H}$. This approach proved its success in a wide class of various problems of mathematical physics, the theory of differential operators, and, especially, in the infinite-dimensional analysis (see, e.g., [45]).

In this chapter we describe the standard construction of the rigged Hilbert space following to the detail presentation in Berezansky's books [42, 44] (see also [48]). In addition, we will briefly analyze the properties of the A-scale of Hilbert spaces which will be used in that follows.

4.1 Construction of a rigged Hilbert space

By definition (see [42, 44]), a triple of Hilbert spaces

$$\mathcal{H}_- \supset \mathcal{H}_0 \supset \mathcal{H}_+$$

forms a *rigged Hilbert space*, if $\mathcal{H}_+$ is a proper subset of $\mathcal{H}_0$, and, in turn, $\mathcal{H}_0$ is a proper subset of $\mathcal{H}_-$, and the following three conditions are fulfilled (these conditions are not independent):

(a) both embedding $\mathcal{H}_0 \supset \mathcal{H}_+$ and $\mathcal{H}_- \supset \mathcal{H}_0$ are continuous and dense, we denote,

$$\mathcal{H}_- \sqsupset \mathcal{H}_0 \sqsupset \mathcal{H}_+; \tag{4.1}$$

(b) the norms in $\mathcal{H}_-$, $\mathcal{H}_0$, and $\mathcal{H}_+$ satisfy the inequalities

$$\|\cdot\|_- \leq \|\cdot\|_0 \leq \|\cdot\|_+; \tag{4.2}$$

(c) the spaces $\mathcal{H}_-$ and $\mathcal{H}_+$ are dual to one another with respect to $\mathcal{H}_0$.

By conditions (a) and (b), the identity mappings

$$\mathcal{H}_+ \ni \varphi \longmapsto \varphi \in \mathcal{H}_0, \quad \mathcal{H}_0 \ni f \longmapsto f \in \mathcal{H}_-$$

are continuous. Condition (c) means that for each vector $\varphi \in \mathcal{H}_+$, the linear functional

$$l_\varphi(f) := (f, \varphi)_0, \quad f \in \mathcal{H}_0$$

has an extension by continuity on the whole space $\mathcal{H}_-$. This extension defines the duality (inner) pairing between $\mathcal{H}_-$ and $\mathcal{H}_+$:

$$\langle \omega, \varphi \rangle_{-,+}, \quad \omega \in \mathcal{H}_-, \ \varphi \in \mathcal{H}_+.$$

It is clear that this duality pairing extends the usual inner product in $\mathcal{H}_0$, i.e., the following equality holds true:

$$\langle \omega, \varphi \rangle_{-,+} = (f, \varphi)_0, \quad \text{if } \omega = f \in \mathcal{H}_0.$$

In addition, the duality pairing between $\mathcal{H}_+$ and $\mathcal{H}_-$ in the rigged Hilbert space (4.1), as an extension of the inner product $(\cdot, \cdot)_0$ in $\mathcal{H}_0$, satisfies the symmetry condition

$$\langle \omega, \varphi \rangle_{-,+} = \overline{\langle \varphi, \omega \rangle_{+,-}}, \quad \omega \in \mathcal{H}_-, \ \varphi \in \mathcal{H}_+.$$

Of course, the latter product also is continuous in $\omega \in \mathcal{H}_-$ and $\varphi \in \mathcal{H}_+$. In particular, if we fix $\omega \in \mathcal{H}_-$, then $\langle \omega, \varphi \rangle_{-,+}$ defines an conjugate-linear and continuous functional on $\mathcal{H}_+$. Therefore, the spaces $\mathcal{H}_-$, $\mathcal{H}_+$ are dual to one another with respect to $\mathcal{H}_0$. In accordance with [42], we name these spaces *negative* and *positive*, respectively.

We note that the rigged space (4.1) can be constructed starting with a couple of embedded Hilbert spaces $\mathcal{H}$ and $\mathcal{K}$ if they constitute a *pre-rigged space*, i.e., if one of them is a dense subset of the other, for example $\mathcal{H} \supset \mathcal{K}$. In addition, the inequality

$$\|\varphi\|_\mathcal{H} \leq \|\varphi\|_\mathcal{K}, \quad \forall \varphi \in \mathcal{K}$$

holds true. Hence, we can write $\mathcal{H} \sqsupset \mathcal{K}$. We denote $\mathcal{H} = \mathcal{H}_0$, and $\mathcal{K} = \mathcal{H}_+$. Then

$$\|\varphi\|_0 \leq \|\varphi\|_+, \quad \forall \varphi \in \mathcal{H}_+.$$

Now we can introduce the negative norm

$$\|f\|_- := \sup_{\|\varphi\|_+=1} |(\varphi, f)_0|, \quad \varphi \in \mathcal{H}_+$$

for vectors $f \in \mathcal{H}_0$. It is clear that this norm satisfies the inequality

$$\|f\|_- \leq \|f\|_0, \quad \forall f \in \mathcal{H}_0.$$

The completion of $\mathcal{H}_0$ with respect to the negative norm gives the space $\mathcal{H}_-$, which contains $\mathcal{H}_0$ and is dual to $\mathcal{H}_+$. Thus, beginning with a pre-rigging $\mathcal{H}_0 \sqsupset \mathcal{H}_+$, we obtain a rigged Hilbert space of the form (4.1).

It should be noted that starting with a pair $\mathcal{H} \sqsupset \mathcal{K}$, we can put $\mathcal{H} = \mathcal{H}_-$, $\mathcal{K} = \mathcal{H}_0$, and construct a rigged space in a slightly different way. Specifically, the pre-rigging couple $\mathcal{H}_- \sqsupset \mathcal{H}_0$ can be extended to the right side. Let us describe the corresponding procedure in more detail.

We will construct the positive space $\mathcal{H}_+$ by means of the linear functionals

$$l_\varphi(f) := (f, \varphi)_0, \quad f \in \mathcal{H}_0,$$

defined for every fixed $\varphi \in \mathcal{H}_0$. It is clear that $l_\varphi(f)$ is continuous on $\mathcal{H}_0$. But, in general, it is not continuous on $\mathcal{H}_-$. We form the positive space $\mathcal{H}_+$ by taking only those $\varphi \in \mathcal{H}_0$ for which the functional $l_\varphi(f)$ has a continuous extension on the whole space $\mathcal{H}_-$.

It is easy to show (see [42, 44]) that above φ's form a linear, dense in $\mathcal{H}_0$ set. The positive norm $\|\varphi\|_+$ is defined by the obvious formula

$$\|\varphi\|_+ = \sup_{\|f\|_-=1} |(f, \varphi)_0|, \quad f \in \mathcal{H}_0.$$

The space $\mathcal{H}_+$ consisting of the vectors described above is complete with respect to this norm. The inner product in $\mathcal{H}_+$ is determined by the polarization identity:

$$(\varphi, \psi)_+ = \frac{1}{4}(\|\varphi + \psi\|_+^2 - \|\varphi - \psi\|_+^2 + i\|\varphi + i\psi\|_+^2 - i\|\varphi - i\psi\|_+^2).$$

Clearly, $\|\varphi\|_0 \leq \|\varphi\|_+$. Thus, starting with $\mathcal{H}_- \sqsupset \mathcal{H}_0$, we construct the rigged Hilbert space

$$\mathcal{H}_- \sqsupset \mathcal{H}_0 \sqsupset \mathcal{H}_+.$$

The important role in the theory of rigged spaces is played by the canonical identification operators, which we call the Berezansky canonical isomorphisms. They arise as follows.

Let us consider for a fixed $\varphi \in \mathcal{H}_+$ the functional

$$l_\varphi(f) = (f, \varphi)_0, \quad f \in \mathcal{H}_0.$$

It has a continuous extension to $\mathcal{H}_-$ which can be written via the duality pairing: $l_\varphi(\omega) = \langle \omega, \varphi \rangle_{-,+}$. Now, according to the Riesz theorem, l_φ, as a functional on

$\mathcal{H}_-$, has the representation $l_\varphi(\omega) = \langle \omega, \varphi^* \rangle_-$ with some $\varphi^* \in \mathcal{H}_-$. It is understood that $\|\varphi\|_+ = \|\varphi^*\|_-$. Therefore, the mapping

$$D_{-,+} : \mathcal{H}_+ \ni \varphi \longmapsto \varphi^* \in \mathcal{H}_- \tag{4.3}$$

is isometric. The operators

$$D_{-,+} : \mathcal{H}_+ \longrightarrow \mathcal{H}_-, \quad I_{+,-} = D_{-,+}^{-1} : \mathcal{H}_- \longrightarrow \mathcal{H}_+, \tag{4.4}$$

are called the *Berezansky canonical isomorphisms*. It is a simple exercise to prove the validity of the following relations:

$$\begin{aligned}
\langle \omega, \varphi \rangle_{-,+} &= (\omega, \varphi^*)_- = (\omega, D_{-,+}\varphi)_- = (I_{+,-}\omega, \varphi)_+, \\
\langle \varphi, D_{-,+}\varphi \rangle_{+,-} &= \|\varphi\|_+^2 \geq 0, \\
\langle I_{+,-}\omega, \omega \rangle_{+,-} &= \|\omega\|_-^2 \geq 0, \\
\|D_{-,+}\varphi\|_- &= \|\varphi^*\|_- = \|\varphi\|_+, \quad \omega \in \mathcal{H}_-, \ \varphi \in \mathcal{H}_+.
\end{aligned}$$

For further considerations it is important that each rigged space of the form (4.1) can be continued any number times in both directions, left and right. In particular, by the above-described procedure we can extend the rigged space (4.1) to a chain containing five spaces,

$$\mathcal{H}_{--} \sqsupset \mathcal{H}_- \sqsupset \mathcal{H}_0 \sqsupset \mathcal{H}_+ \sqsupset \mathcal{H}_{++}. \tag{4.5}$$

For example, to obtain $\mathcal{H}_{--}$, we need to continue the pre-rigging couple $\mathcal{H}_- \sqsupset \mathcal{H}_0$ to the rigged space

$$\mathcal{H}_{--} \sqsupset \mathcal{H}_- \sqsupset \mathcal{H}_0,$$

where $\mathcal{H}_{--}$ is the dual to the space $\mathcal{H}_0$ space with respect to $\mathcal{H}_-$. In turn, the space $\mathcal{H}_{++}$ is dual to $\mathcal{H}_0$ with respect to $\mathcal{H}_+$. At the same time, $\mathcal{H}_{++}$ is dual to $\mathcal{H}_{--}$ with respect to $\mathcal{H}_0$. The indicated procedure can be iterated countably many times, yielding the discrete scale of Hilbert spaces

$$\cdots \sqsupset \mathcal{H}_{-k} \sqsupset \cdots \sqsupset \mathcal{H}_- \sqsupset \mathcal{H}_0 \sqsupset \mathcal{H}_+ \sqsupset \cdots \sqsupset \mathcal{H}_k \sqsupset \cdots, \tag{4.6}$$

where $k \in \mathbb{N}$, $\mathcal{H}_+ := \mathcal{H}_1$, and $\mathcal{H}_{++} := \mathcal{H}_2$.

Exercise 4.1.1. Show that a pair of Hilbert spaces $\mathcal{H}$ and $\mathcal{K}$ whose norms satisfy the inequality

$$\|\varphi\|_{\mathcal{H}} \leq \|\varphi\|_{\mathcal{K}}, \quad \forall \varphi \in \mathcal{K},$$

form a pre-rigging couple $\mathcal{H} \sqsupset \mathcal{K}$ only if the kernel of the identity mapping $\mathcal{K} \ni \varphi \mapsto \varphi \in \mathcal{H}$ is zero.

Example 4.1.2. Let $\gamma \geq 0$ be a positive, densely defined, and closed quadratic form on $\mathcal{H}$. It is well known (see Chapter 2) that the domain $Q(\gamma)$ of such a form is a complete Hilbert space with respect to the inner product

$$(\varphi, \psi)_+ = (\varphi, \psi) + \gamma(\varphi, \psi), \quad \varphi, \psi \in Q(\gamma).$$

We denote this space by $\mathcal{H}_+$. It is obvious that $\|\varphi\|_+ \geq \|\varphi\|$ for all $\varphi \in Q(\gamma)$. It follows that $\mathcal{H}_0 \sqsupset \mathcal{H}_+$, where $\mathcal{H}_0 \equiv \mathcal{H}$. Thus, the pre-rigging couple $\mathcal{H}_0 \sqsupset \mathcal{H}_+$ can be continued by means of the standard procedure described above to the rigged space

$$\mathcal{H}_- \sqsupset \mathcal{H}_0 \sqsupset \mathcal{H}_+.$$

Example 4.1.3. Let T be a closed linear operator on $\mathcal{H}$ with dense domain $\mathfrak{D}(T)$. Then $\mathcal{H}_+ = \mathfrak{D}(T)$ is a complete Hilbert space with respect to the inner product

$$(\varphi, \psi)_+ = (\varphi, \psi) + (T\varphi, T\psi), \quad \varphi, \psi \in \mathfrak{D}(T).$$

It is clear that $\mathcal{H} \sqsupset \mathcal{H}_+$ and these spaces form a pro-rigging couple which has an extension to the rigged space $\mathcal{H}_- \sqsupset \mathcal{H} \sqsupset \mathcal{H}_+$. We can introduce the positive norm on $\mathfrak{D}(T)$ in a slightly different way. Namely, let us define an equivalent norm by $\|\varphi\|_+ = \|(1+T)\varphi\|$. Then $\mathcal{H}_-$ arises as the completion of $\mathcal{H}$ with respect to the negative norm $\|f\|_- = \|(1+T)^{-1}f\|$, $f \in \mathcal{H}$.

4.2 Connections with self-adjoint operators

There is a well-known connection between triplets of the kind (4.1) and positive self-adjoint operators in $\mathcal{H}_0$. The following theorem describes this connection in a general setting.

Theorem 4.2.1. *Each unbounded self-adjoint operator $A = A^* \geq 1$ on $\mathcal{H}_0$ is uniquely associated with the rigged Hilbert space*

$$\mathcal{H}_-(A) \sqsupset \mathcal{H}_0 \sqsupset \mathcal{H}_+(A) \tag{4.7}$$

where the space $\mathcal{H}_+ = \mathcal{H}_+(A)$ coincides with the domain $\mathfrak{D}(A)$, equipped with the norm $\|\varphi\|_+ = \|A\varphi\|_0$, $\varphi \in \mathfrak{D}(A)$.

Conversely, for each rigged Hilbert space of the form (4.1), *there exists the self-adjoint operator $A = A^* \geq 1$ on $\mathcal{H}_0$ such that the triple* (4.7) *coincides with* (4.1).

Proof. Let us start with an abstract rigged Hilbert space

$$\mathcal{H}_- \sqsupset \mathcal{H}_0 \sqsupset \mathcal{H}_+.$$

Consider the Berezansky canonical isomorphism

$$D_{-,+} : \mathcal{H}_+ \ni \varphi \longmapsto \varphi^* \in \mathcal{H}_-.$$

Its restriction to $\mathcal{H}_0$ we denote by L_A, i.e.,

$$L_A := D_{-,+} \restriction \mathfrak{D}(L_A) = \{\varphi \in \mathcal{H}_+ \mid D_{-,+}\varphi \in \mathcal{H}_0\}.$$

We claim that L_A is a symmetric operator on $\mathcal{H}_0$. Indeed, for all vectors $\varphi, \psi \in \mathfrak{D}(L_A) \subset \mathcal{H}_+$ we have

$$\begin{aligned}(L_A\varphi, \psi)_0 &= (D_{-,+}\varphi, \psi)_0 = \langle \varphi^*, \psi\rangle_{-,+} = (\varphi, \psi)_+ \\ &= \langle \varphi, \psi^*\rangle_{+,-} = (\varphi, D_{-,+}\psi)_0 = (\varphi, L_A\psi)_0,\end{aligned} \tag{4.8}$$

where $\varphi^* = D_{-,+}\varphi$ and $\psi^* = D_{-,+}\psi$. In fact, L_A is self-adjoint on $\mathcal{H}_0$ since by construction, its range coincides with the whole space $\mathcal{H}_0$, i.e., $\mathrm{Ran}A = \mathcal{H}_0$. It is obvious, L_A is unbounded and positive. So, we can define the operator $A := L_A^{1/2}$. From (4.8) it follows that

$$(L_A^{1/2}\varphi, L_A^{1/2}\psi)_0 = (A\varphi, A\psi)_0 = (\varphi, \psi)_+.$$

Therefore, $\mathfrak{D}(A) = \mathcal{H}_+$ and $A \geq 1$ since $\|\cdot\|_+ \geq \|\cdot\|_0$.

Conversely, let $A = A^* \geq 1$ be an unbounded self-adjoint operator with the domain $\mathfrak{D}(A)$ on $\mathcal{H}_0$. Starting with A we can to construct the rigged Hilbert space (4.7). To this end we notice that the domain $\mathfrak{D}(A)$ forms a complete Hilbert space with respect to the inner product

$$(\varphi, \psi)_+ := (A\varphi, A\psi)_0, \quad \varphi, \psi \in \mathfrak{D}(A).$$

We denote this space by $\mathcal{H}_+(A)$. Further, since $\mathcal{H}_0 \sqsupset \mathcal{H}_+(A)$, we can continue this chain to the rigged space (4.7). The corresponding procedure was described above. Now we can consider the Berezansky canonical isomorphism $D_{-,+}(A)$ in (4.7). It is clear that the operator L_A related to $D_{-,+}(A)$ coincides with A. Thus, we proved that there exists a bijective connection between rigged Hilbert spaces (4.7) and self-adjoint unbounded operators $A \geq 1$ on $\mathcal{H}_0$. □

The next two propositions describe the above connection more precisely.

Proposition 4.2.2. *Given A as in Theorem 4.2.1, we consider the pre-rigging couple $\mathcal{H}_- \sqsupset \mathcal{H}_0$, where $\mathcal{H}_-$ is the completion of $\mathcal{H}_0$ with respect to the norm*

$$\|h\|_- = \|A^{-1}h\|_0, \quad h \in \mathcal{H}_0. \tag{4.9}$$

Then the positive space $\mathcal{H}_+$ corresponding to $\mathcal{H}_- \sqsupset \mathcal{H}_0$ coincides with $\mathfrak{D}(A)$ in the norm $\|f\|_+ := \|Af\|$, i.e.,

$$\mathcal{H}_+ = \mathcal{H}_+(A).$$

Proof. According to the definition of the positive space, $\mathcal{H}_+$ consists of those $f \in \mathcal{H}$ for which the linear continuous functional $l_f(h) = (h, f)_0$, $h \in \mathcal{H}_0$ has an extension by continuity on the whole space $\mathcal{H}_-$. The latter requirement is equivalent to the property

$$\|f\|_+ = \sup_{\|h\|_-=1} |l_f(h)| < \infty.$$

In other words, it means that f belongs to $\mathfrak{D}(A)$. Indeed, by the construction of the rigged Hilbert space, we have

$$l_f(h) = (h, f)_0 = \langle h, f\rangle_{-,+} = \langle D_{-,0}\varphi, f\rangle_{-,+} = (A\varphi, f)_0, \quad \varphi \in \mathfrak{D}(A).$$

Since $D_{-,0}$ is isometric, the continuity of the functional $l_f(h)$ in $h \in \mathcal{H}_-$ is equivalent to the continuity of this functional in $\varphi = A^{-1}h \in \mathcal{H}_0$. Therefore, $l_f(h) = (\varphi, f^*)_0$. This means that $f \in \mathfrak{D}(A^*)$ and $f^* = A^*f$. Since $A = A^*$, we get $f \in \mathfrak{D}(A)$. Thus, $\mathcal{H}_+ = \mathcal{H}_+(A)$. □

Proposition 4.2.3. *Given $A = A^* \geq 1$ on $\mathcal{H}$, let $\mathcal{H}_0 \sqsupset \mathcal{H}_+$ be the pre-rigging couple, where $\mathcal{H}_+ = \mathfrak{D}(A)$ with respect to the norm $\|f\|_+ := \|Af\|_0$. Then the negative space $\mathcal{H}_-$ corresponding to $\mathcal{H}_0 \sqsupset \mathcal{H}_+$ coincides with the completion of $\mathcal{H}_0$ with respect to the norm $\|h\|_-$, i.e.,*

$$\mathcal{H}_- = \mathcal{H}_-(A).$$

Proof. According to the construction of the rigged space, the negative space $\mathcal{H}_-$ is defined as the completion of $\mathcal{H}_0$ with respect to the norm

$$\|h\|_- := \sup_{\|f\|_+=1} |(f, h)_0|, \quad h \in \mathcal{H}_0.$$

We have to verify that this norm is equivalent to (4.9). Really, since A is self-adjoint and $\operatorname{Ran} A = \mathcal{H}_0$, we can write for each $h \in \mathcal{H}_0$:

$$(f, h)_0 = (f, A\varphi)_0 = (Af, \varphi)_0 = (g, \varphi)_0$$

with some $\varphi \in \mathcal{H}_+$. Hence,

$$\|h\|_- = \sup_{\|f\|_+=1} |(f, h)_0| = \sup_{\|g\|_0=1} |(g, \varphi)_0| = \|\varphi\|_0 = \|A^{-1}h\|_0,$$

where we used the equality $\|g\|_0 = \|Af\|_0 = \|f\|_+$ with $g = Af$. □

Exercise 4.2.4. Construct $\mathcal{H}_0$ starting with a pre-rigging couple $\mathcal{H}_- \sqsupset \mathcal{H}_+$.

Example 4.2.5. Let $\mathcal{H} = L_2(\mathbb{R}^n, dx)$ and A be the multiplication operator defined by a real-valued function. For example,

$$(Af)(x) = \sqrt{\rho(x)}f(x), \quad \rho(x) = 1 + |x|^2,$$

where $|x|^2 := \sum_{i=1}^{n} |x_i|^2$, $x = (x_1, \ldots, x_n) \in \mathbb{R}^n$. Then the triplet

$$L_2(\mathbb{R}^n, \rho^{-1}dx) \sqsupset L_2(\mathbb{R}^n, dx) \sqsupset L_2(\mathbb{R}^n, \rho dx)$$

forms the rigged Hilbert space associated with the operator A.

Example 4.2.6. Let $A = 1 - \Delta$ on $\mathcal{H} = L_2(\mathbb{R}^3, dx)$, where $-\Delta$ is the Laplace operator. Then $\mathcal{H}_+ = \mathcal{H}_+(A)$ coincides with the Sobolev space $W_2^2 = W_2^2(\mathbb{R}^3)$ (see Subsection 1.2.7) and $\mathcal{H}_- = \mathcal{H}_-(A) = W_2^{-2}(\mathbb{R}^3)$. Thus, one arrives at the well-known rigged space

$$W_2^{-2} \sqsupset L_2 \sqsupset W_2^2.$$

4.3 A-scales of Hilbert spaces

In what follows we will use the doubly-infinite chain of Hilbert spaces associated with an operator A. Such a chain generalizes the rigging (4.6) and is called the *A-scale.* Below we describe its construction.

Let $A = A^* \geq 1$ be a self-adjoint unbounded operator with the domain $\mathfrak{D}(A)$ in the complex separable Hilbert space $\mathcal{H}$ with the inner product $(\cdot, \cdot)$.

For each $k \geq 0$ we define the Hilbert space $\mathcal{H}_k \equiv \mathcal{H}_k(A)$ which coincides as a set with the domain $\mathfrak{D}(A^{k/2})$ equipped with the norm $\|\cdot\|_k$ corresponding to the inner product

$$(\varphi, \psi)_k := (A^{k/2}\varphi, A^{k/2}\psi), \quad \varphi, \psi \in \mathfrak{D}(A^{k/2}). \tag{4.10}$$

We put $\mathcal{H}_0 \equiv \mathcal{H}$ and define $\mathcal{H}_{-k}$ as the completion of $\mathcal{H}_0$ with respect to the negative norm $\|\cdot\|_{-k}$ generated by the inner product

$$(f, g)_{-k} := (A^{-k/2}f, A^{-k/2}g), \quad f, g \in \mathcal{H}. \tag{4.11}$$

It is easy to see that for each fixed $k > 0$ the triplet

$$\mathcal{H}_{-k} \sqsupset \mathcal{H}_0 \sqsupset \mathcal{H}_k \tag{4.12}$$

is a rigged Hilbert space in the sense of Berezansky's definition [42, 44].

The chain of Hilbert spaces infinite in both sides

$$\cdots \sqsupset \mathcal{H}_{-k} \sqsupset \cdots \sqsupset \mathcal{H}_0 \sqsupset \cdots \sqsupset \mathcal{H}_k \cdots \tag{4.13}$$

is called the *A-scale.* We will denote it by $\{\mathcal{H}_k\}_{k\in\mathbb{R}}$.

Example 4.3.1. A typical example of a scale of Hilbert spaces is provided by the Sobolev scale

$$W_2^{-k} \sqsupset L_2 \sqsupset W_2^k, \quad k > 0,$$

which can be considered as the A-scale for the operator $A = 1-\Delta$. The spaces $W_2^{\pm k}$ are defined by the powers of the operator $A = 1-\Delta$ or by their Fourier transforms, which are the operators of multiplication by $(1 + |x|^2)^{\pm k}$ (see Subsection 1.2.7).

4.3.1 Properties of the A-scale

The infinite chain of spaces (4.13) has a series of interesting properties. We will briefly discuss some of them, for more details, see [2, 51, 58, 143].

An important property is the invariance of the structure of rigged triplet (4.12) under shifts along the A-scale, i.e., shifts of the index k. We will call this property the *first invariance principle* of the A-scale. Its essence is that for any fixed $k > 0$ and an arbitrary q, the triple of spaces, $\mathcal{H}_{-k+q}, \mathcal{H}_q, \mathcal{H}_{k+q}$ form a new rigged Hilbert space, i.e., we can write

$$\mathcal{H}_{-k+q} \sqsupset \mathcal{H}_q \sqsupset \mathcal{H}_{k+q}, \; q \in \mathbb{R}^1. \tag{4.14}$$

This property follows from the general procedure of construction for the rigged Hilbert space applied to one of the pre-rigging couples, $\mathcal{H}_q \sqsupset \mathcal{H}_{k+q}$ or $\mathcal{H}_{-k+q} \sqsupset \mathcal{H}_q$, taken from the A-scale.

It should be remarked that by using the first invariance principle of the A-scale, we can extend a notion of the Berezansky canonical isomorphism for each ordered pair of spaces $\mathcal{H}_k$ and $\mathcal{H}_l$, $k > l$ and define the operator $D_{l,k} : \mathcal{H}_k \to \mathcal{H}_l$. We denote $I_{k,l} := D_{l,k}^{-1}$. It is clear that these operators are unitary mappings from $\mathcal{H}_k$ to $\mathcal{H}_l$ and from $\mathcal{H}_l$ to $\mathcal{H}_k$, respectively.

The next important property we call the *second invariance principle* of the A-scale. It essentially asserts that A is unitarily equivalent to its image under any shift produced by the operator $D_{l,0}$ or $I_{0,l}$, $l > 2$ along the A-scale, i.e., under the transition to any space in the scale.

Let us explain this principle in more detail. To this end, we fix $k \geq 0$ and define the operator $A_k := D_{k,k+2}$. It is easy to check that A_k coincides with the restriction of A to $\mathcal{H}_k$. So, A_k is self-adjoint in $\mathcal{H}_{k+2}$. In accordance with this definition,

$$A \equiv A_{k=0} = D_{0,2}.$$

In a similar way we define the operator

$$A_{-k} := D_{-k,2-k}, \quad k > 0.$$

It is self-adjoint on $\mathcal{H}_{2-k}$ and coincides with the closure of A in $\mathcal{H}_{-k}$. It is important that all operators $A_{\pm k}$, $k \in \mathbb{R}^1_+$, are unitary images of the original operator A on $\mathcal{H}_0$. In particular,

$$A_k = I_{k-2,0} A D_{2,k}, \quad A_{-k} = D_{-(k+2),0} A I_{2,-k}.$$

At the same time, the operator A is essentially self-adjoint in each space $\mathcal{H}_{-k}$, $k > 0$.

Thus, the second invariance principle of the A-scale can be formulated as follows. For any $\alpha \in \mathbb{R}$ the operator $A_\alpha := D_{\alpha,\alpha+2}$ is self-adjoint in $\mathcal{H}_\alpha$ and is unitarily equivalent to the original operator A. It is obvious that if $\alpha > 0$, then A_α coincides with the restriction of A to $\mathcal{H}_{\alpha+2}$, and if $\alpha < 0$, then A_α is the closure of A in $\mathcal{H}_{\alpha+2}$. Moreover, it is easy to see that A_α is essentially self-adjoint in each space $\mathcal{H}_\beta$, with $\beta < \alpha$.

It is convenient for the sequel to recall once more the construction the A-scale and rewrite its properties in slightly different notations.

Starting with a self-adjoint operator $A = A^* \geq 1$, we define the Hilbert space $\mathcal{H}_\alpha \equiv \mathcal{H}_\alpha(A)$ for each $\alpha > 0$. This space coincides, as a set, with the domain $\mathfrak{D}(A^{\alpha/2})$, and is equipped with the norm $\|\cdot\|_\alpha$ corresponding to the inner product

$$(\varphi, \psi)_\alpha := (A^{\alpha/2}\varphi, A^{\alpha/2}\psi)_0, \quad \varphi, \psi \in \mathfrak{D}(A^{\alpha/2}). \tag{4.15}$$

The space $\mathcal{H}_{-\alpha}$ is obtained as the completion of $\mathcal{H}_0$ with respect to the negative norm $\|\cdot\|_{-\alpha}$ generated by the inner product

$$(f,g)_{-\alpha} := (A^{-\alpha/2}f, A^{-\alpha/2}g)_0, \quad f,g \in \mathcal{H}_0. \tag{4.16}$$

The chain of spaces

$$\cdots \sqsupset \mathcal{H}_{-\alpha} \sqsupset \cdots \sqsupset \mathcal{H}_0 \sqsupset \cdots \sqsupset \mathcal{H}_\alpha \sqsupset \cdots \tag{4.17}$$

with dense embeddings forms the A-scale of Hilbert spaces $\{\mathcal{H}_\alpha\}_{\alpha\in\mathbb{R}}$. It is easy to see that for each fixed $\alpha > 0$ the triplet

$$\mathcal{H}_{-\alpha} \sqsupset \mathcal{H}_0 \sqsupset \mathcal{H}_\alpha \tag{4.18}$$

forms a rigged Hilbert space.

Let

$$D_{-\alpha,\alpha} : \mathcal{H}_\alpha \longrightarrow \mathcal{H}_{-\alpha}, \quad I_{\alpha,-\alpha} := D^{-1}_{-\alpha,\alpha} : \mathcal{H}_{-\alpha} \longrightarrow \mathcal{H}_\alpha$$

denote the Berezansky canonical isomorphisms. In accordance with the previous considerations,

$$D_{-\alpha,\alpha} = (A^{\alpha/2})^{\mathrm{cl}}(A^{\alpha/2}),$$

where cl denotes a closure of the mapping $A^{\alpha/2} : \mathcal{H}_0 \to \mathcal{H}_{-\alpha/2}$. And what is more, we can introduce the Berezansky canonical isomorphisms between arbitrary pair of spaces $\mathcal{H}_\alpha$, $\mathcal{H}_\beta$ from the A-scale:

$$D_{\alpha,\beta} : \mathcal{H}_\beta \longrightarrow \mathcal{H}_\alpha, \quad I_{\beta,\alpha} = D^{-1}_{\alpha,\beta} : \mathcal{H}_\alpha \longrightarrow \mathcal{H}_\beta, \quad \alpha < \beta. \tag{4.19}$$

For the further applications we list the following equalities which hold in the A-scale:

$$\begin{aligned}
D_{0,2} &= A : \mathcal{H}_2 \longrightarrow \mathcal{H}_0,\\
I_{2,0} &= A^{-1} : \mathcal{H}_0 \longrightarrow \mathcal{H}_2;\\
D_{0,\alpha} &= A^{\alpha/2} : \mathcal{H}_\alpha \longrightarrow \mathcal{H}_0,\\
D_{-\alpha,0} &= (A^{\alpha/2})^{cl} : \mathcal{H}_0 \longrightarrow \mathcal{H}_{-\alpha}, \quad \alpha > 0;\\
D_{-\alpha,\beta} &= D_{-\alpha,0}D_{0,\beta} : \mathcal{H}_\beta \longrightarrow \mathcal{H}_{-\alpha}, \quad \alpha,\beta > 0;\\
(f,\varphi)_0 &= \langle f,\varphi\rangle_{-\alpha,\alpha}, \quad f \in \mathcal{H}_0,\ \varphi \in \mathcal{H}_\alpha, \quad \alpha > 0;\\
D_{00} &= I_{00} \equiv \mathbf{1}.
\end{aligned} \tag{4.20}$$

In particular, putting $\mathcal{H}_{-\alpha} = \mathcal{H}_-$ and $\mathcal{H}_\alpha = \mathcal{H}_+$, $\alpha > 0$, we have

$$\langle \omega,\varphi\rangle_{-,+} = (\omega, D_{-,+}\varphi)_- = (I_{+,-}\omega,\varphi)_+, \quad \omega \in \mathcal{H}_-,\ \varphi \in \mathcal{H}_+, \tag{4.21}$$

where

$$D_{-,+} \equiv D_{-\alpha,\alpha} \quad I_{+,-} \equiv I_{\alpha,-\alpha},$$

and

$$\langle\cdot,\cdot\rangle_{-,+} \equiv \langle\cdot,\cdot\rangle_{-\alpha,\alpha}$$

denotes the duality pairing between positive and negative spaces. It should also be noted that

$$\langle \omega, \varphi \rangle_{-\alpha,\alpha} = \langle \omega, \varphi \rangle_{-\beta,\beta}, \quad \omega \in \mathcal{H}_{-\alpha}, \quad \varphi \in \mathcal{H}_{\beta},\ 0 \le \alpha < \beta, \tag{4.22}$$

although in general

$$\langle \omega, \varphi \rangle_{-\alpha,\beta} \neq \langle \omega, \varphi \rangle_{-\beta,\alpha}.$$

Of course, when the value of index α is not important, we denote the triplet (4.18) as (4.1).

The next theorem follows from the previous considerations.

Theorem 4.3.2. *All four mappings listed below coincide with the self-adjoint operators $A^{k/2}$, $k > 0$ on $\mathcal{H}_0$ (or with their inverse):*

(a) $D_{0,k} : \mathcal{H}_k \longrightarrow \mathcal{H}_0$,
(b) $D_{-k/2,k/2} \restriction \{f \in \mathcal{H}_{k/2} \mid D_{-k/2,k/2} f \in \mathcal{H}_0\}$,
(c) $D_{-k,0} \restriction \{f \in \mathcal{H}_0 \mid D_{-k,0} f \in \mathcal{H}_0\}$,
(d) $A^{-k/2} = I_{0,-k} \restriction \{\omega \in \mathcal{H}_{-k} \mid I_{0,-k}\omega \in \mathcal{H}_k\}$.

Proof. (a) is true according to the construction of the A-scale. Indeed, $D_{0,k}$ and $A^{k/2}$ obviously coincide as operators acting from $\mathcal{H}_k$ into $\mathcal{H}_0$. To prove (b) it is sufficient to remark that the operator $D_{0,k}$ can be defined as a restriction of $D_{-k/2,k/2}$. That is,

$$D_{0,k} = D_{-k/2,k/2} \restriction \{f \in \mathcal{H}_{k/2} \mid D_{-k/2,k/2} f \in \mathcal{H}_0\}.$$

In particular,

$$\begin{aligned} D_{0,2} = A &= D_{-1,1} \restriction \{f \in \mathcal{H}_1 \mid D_{-1,1} f \in \mathcal{H}_0\} \\ &= D_{-2,0} \restriction \{f \in \mathcal{H}_0 \mid D_{-2,0} f \in \mathcal{H}_0\}. \end{aligned}$$

The validity of (c) follows from the second invariance principle. Therefore, $D_{0,k}$ coincides with the restriction $D_{-k,0}$ to $\mathcal{H}_k$:

$$D_{-k,0} \restriction \mathcal{H}_k = D_{0,k}.$$

Hence, it holds that

$$A^{k/2} = D_{-k,0} \restriction \mathcal{H}_k.$$

Finally, (d) holds true due to the definition of the mapping $I_{0,-k} = D^{-1}_{-k,0}$. By the second invariance principle, $I_{0,-k} \restriction \mathcal{H}_0 = I_{k,0}$. Now it is obvious that

$$I_{k,0} = D^{-1}_{k,0}, \quad A^{-k/2} = D^{-1}_{0,k}.$$

Hence, $A^{-k/2} = I_{0,-k} \restriction \mathcal{H}_0$. Moreover,

$$A^{-1} = I_{0,-2} \restriction \{\omega \in \mathcal{H}_{-2} \mid I_{0,-2}\omega \in \mathcal{H}_0\}.$$ □

Chapter 5

Singular Quadratic Forms

The notion of singular quadratic form is one of the important objects investigated in this book. In an abstract context the term "singular quadratic form" was perhaps introduced first by B. Simon in [177]. It appears in the canonical decomposition $\gamma = \gamma_{\rm r} + \gamma_{\rm s}$ for a positive quadratic form $\gamma \geq 0$ on a Hilbert space $\mathcal{H}$. Here $\gamma_{\rm r}$ is the largest regular (closable) part of γ. The remainder $\gamma_{\rm s} = \gamma - \gamma_{\rm r}$, is non-closable in $\mathcal{H}$, provided that it is non-trivial. This part of γ is an analog of the singular component $\mu_{\rm sing}$ of a positive measure μ under its Lebesgue decomposition $\mu = \mu_{\rm ac} + \mu_{\rm sing}$. A definition of the form $\gamma_{\rm s}$ independent of the decomposition $\gamma = \gamma_{\rm r} + \gamma_{\rm s}$, was not given in [177]. However, the open question was raised of whether there is a canonical description of $\gamma_{\rm s}$ different than $\gamma_{\rm s} = \gamma - \gamma_{\rm r}$?

At the same time the term "singular quadratic (and bilinear) form" was intensively used by V. Koshmanenko [113]–[142]. These works were devoted to scattering and perturbation problems in quantum field theory. Here, the singularity of a quadratic form γ means that its support is a small set of zero Lebesgue measure, and its null subset $\operatorname{Ker}\gamma$ is dense in the underlying Hilbert state space. The relativistic Wick monomials have exactly these properties. However, it recently became clear that in the abstract setting the density property of $\operatorname{Ker}\gamma$ in the state space is only a typical sufficient condition for singularity of γ. Examples of strongly positive singular quadratic forms such that $\operatorname{Ker}\gamma = \{0\}$ have been exhibited.

In fact, a notion of singular quadratic form was independently introduced in [113], through the precise definition appeared only later (see [105, 106, 121]).

It should be emphasized that the properties of singular quadratic forms are very surprising. They often lead to indeterminacies. Since these forms are non-closable, ambiguous expressions appear in the perturbation theory for linear operators. For this reason, similarly to the theory of tempered distributions, it seems relevant to use the rigged Hilbert spaces for treating singular quadratic forms. Exactly this approach was proposed in [118].

We remark that the decomposition problem for a positive operator T, i.e., the question about the representation of T in the form $T = T_{\rm reg} + T_{\rm sing}$, was investigated in [30, 31] by T. Ando and K. Nishio (see also [78, 95, 96, 165]). Here the singular operator T_{sing} can be associated with a singular quadratic form and its properties can be analyzed in the same way as it was done in [131, 165].

In this chapter we only give a brief exposition of some facts from the theory of singular quadratic forms. For more details see [95, 119, 177]. A comprehensive theory of singular quadratic forms is rather far from completion.

5.1 Quadratic forms

Let us recall the main concepts in the theory of quadratic forms.

A quadratic form γ is usually associated with a complex-valued sesquilinear form $\gamma(\varphi, \psi)$, $\varphi, \psi \in \Phi$, where Φ denotes an abstract linear space over the field $\mathbb{C}$ of complex numbers.

As a rule, quadratic forms are considered on a complex Hilbert space $\mathcal{H}$ as mappings from $\mathcal{H} \times \mathcal{H}$ to $\mathbb{C}$ which obey the conditions (2.15).

The domain $\mathrm{Dom}\,\gamma \equiv Q(\gamma)$ of a form γ can be defined as the set of all vectors $\varphi \in \mathcal{H}$ such that $\gamma[\varphi] < \infty$. Clearly, $\mathrm{Dom}\,\gamma$ is a linear manifold. If it is dense in $\mathcal{H}$, then the form γ is said to be *densely defined.* Each sesquilinear form $\gamma(\varphi, \varphi)$ on a complex Hilbert space can be reconstructed from its (diagonal) quadratic part by the polarization principle (see, e.g., [107]):

$$\gamma(\varphi, \psi) = \frac{1}{4}(\gamma[\varphi + \psi] - \gamma[\varphi - \psi] + \imath\gamma[\varphi + \imath\psi] - \imath\gamma[\varphi - \imath\psi]).$$

The *numerical range* $N(\gamma)$ of a quadratic form γ on a Hilbert space is the set of complex numbers $\gamma[\varphi]$, where φ runs over all unit norm vectors:

$$N(\gamma) = \{\gamma[\varphi] \in \mathbb{C} \mid \varphi \in Q(\gamma),\ \|\varphi\| = 1\}.$$

5.1.1 Symmetric quadratic forms, closability

A quadratic form γ is called Hermitian if $\gamma(\varphi, \psi) = \overline{\gamma(\psi, \varphi)}$ for all $\psi, \varphi \in Q(\gamma)$. In this case we write $\gamma = \gamma^*$.

It is not hard to see that a form γ is Hermitian if and only if $\gamma[\varphi] \in \mathbb{R}$, $\forall \varphi \in Q(\gamma)$, that is if and only if its numerical range consists of the real numbers, $N(\gamma) \subseteq \mathbb{R}$ (see [107]).

If $\gamma[\varphi] \geq 0$ for all $\varphi \in Q(\gamma)$, then γ is said to be *positive* (nonnegative). We say that γ is *strictly positive*, and write $\gamma > 0$, if it is positive and, in addition, from $\gamma[\varphi] = 0$ it follows that $\varphi = 0$.

A form γ on a Hilbert space is said to be *symmetric* if it is Hermitian and densely defined. Of course, every symmetric form is Hermitian. However, not every Hermitian form on a Hilbert space is densely defined and symmetric. So, we distinguish Hermitian and symmetric forms.

Let a form γ be given on a Hilbert space $\mathcal{H}$ with the norm $\|\varphi\|$. If it is symmetric, $\gamma = \gamma^*$, and there is an $m > -\infty$ such that

$$\gamma[\varphi] \geq m\|\varphi\|^2,$$

then γ is said to be *bounded from below*, and we write $\gamma \geq m$. The largest such number m is called the *lower bound* of γ. We denote this bound by m_γ. It can be calculated as

$$m_\gamma := \inf_{\varphi \in Q(\gamma), \|\varphi\|=1} \gamma[\varphi].$$

A positive form γ is said be *positive definite* if its lower bound is strictly larger than zero, $\gamma \geq m_\gamma > 0$. Finally, a symmetric form γ is said to be *bounded* if its numerical range $N(\gamma)$ is a finite interval.

Every quasi-inner product on an abstract linear space Φ is a positive form. And each inner product in a Hilbert space $\mathcal{H}$ is a strongly positive form.

Let A be a linear operator on a Hilbert space $\mathcal{H}$. Then the expression $\gamma_A(\varphi, \psi) = (A\varphi, \psi)$, $\varphi, \psi \in \operatorname{Dom} A = Q(\gamma_A)$ generates a quadratic form on $\mathcal{H}$ (see 2.20). Let A be bounded and defined on the whole space $\mathcal{H}$. Then γ_A is also bounded, its numerical range is a bounded set in $\mathbb{C}$. The relation (2.20) establishes a one-to-one correspondence between bounded operators A and bounded quadratic forms γ_A with $Q(\gamma_A) = \mathcal{H} = \operatorname{Dom} A$.

Let the quadratic form γ on the Hilbert space $\mathcal{H}$ be Hermitian, densely defined, and bounded from below. So, γ is symmetric. In accordance to [107], a sequence $\varphi_n \in Q(\gamma)$ is called *γ-convergent* to $\varphi \in \mathcal{H}$ if

$$\varphi_n \longrightarrow \varphi \text{ and } \gamma[\varphi_n - \varphi_m] \longrightarrow 0, \quad n, m \to \infty. \tag{5.1}$$

In this case we write $\varphi_n \stackrel{\mathcal{H},\gamma}{\longrightarrow} \varphi$.

Definition 5.1.1. A symmetric bounded from below quadratic form γ on $\mathcal{H}$ is said to be closed if

$$\varphi_n \stackrel{\mathcal{H},\gamma}{\longrightarrow} \varphi \Longrightarrow \varphi \in Q(\gamma), \quad \gamma[\varphi_n - \varphi] \longrightarrow 0. \tag{5.2}$$

We write $\gamma = \gamma^{\text{cl}}$ (cl= closure) if γ is a closed form.

A form γ' is called an *extension* of γ, denoted by $\gamma \subset \gamma'$, if $Q(\gamma) \subset Q(\gamma')$ and $\gamma[\varphi] = \gamma'[\varphi]$ for all $\varphi \in Q(\gamma)$. A form γ is said to be *closable* if it has a closed extension, $\gamma \subset \gamma^{\text{cl}}$. The smallest closed extension γ^{cl} of a closable form γ is called its *closure*. To obtain γ^{cl} one has to add to $Q(\gamma)$ all vectors $\varphi \in \mathcal{H}$ which are limits of γ-convergent sequences such that values $c_\varphi = \lim_{n\to\infty} \gamma[\varphi_n]$ are independent on the choice of φ_n. Then we put $\gamma^{\text{cl}}[\varphi] = c_\varphi$.

If $\gamma \geq 0$ is a form on a linear space Φ, then $\mathcal{H}_\gamma$ denotes the Hilbert space constructed by the standard procedure using the (quasi)-inner product $(\varphi, \psi)_\gamma = \gamma(\varphi, \psi)$, $\varphi, \psi \in \Phi$.

Given a positive quadratic form $\gamma \geq 0$ on a Hilbert space $\mathcal{H}$, we denote by $\mathcal{H}_{\gamma+\chi}$ the new Hilbert space which is the completion of $Q(\gamma)$ with respect to the norm

$$\|\varphi\|_{\gamma+\chi} = (\gamma[\varphi] + \chi[\varphi])^{1/2}, \quad \chi[\varphi] := \|\varphi\|^2. \tag{5.3}$$

The following is a fundamental result of the theory (see Theorem VI.1.11 in [107]).

Theorem 5.1.2. *A positive densely defined on $\mathcal{H}$ quadratic form γ is closed if and only if its domain $Q(\gamma)$ is a complete Hilbert space in the norm (5.3), i.e., $\gamma = \gamma^{\mathrm{cl}}$ if and only if $Q(\gamma)$ coincides with $\mathcal{H}_{\gamma+\chi}$.*

Proof. Let us put $\tau = \gamma + \chi$. We will use the following notations. For a sequence $\varphi_n \in \Phi = Q(\gamma)$ we write

$$\varphi_n \in \Sigma_\tau \quad (\varphi_n \in \Theta_\tau)$$

if $\tau[\varphi_n - \varphi_m] \to 0$ (respectively, $\tau[\varphi_n] \to 0$).

Let us assume $\gamma = \gamma^{\mathrm{cl}}$ and take a sequence $\varphi_n \in \Sigma_\tau$. Then $\varphi_n \in \Sigma_\gamma$, and also $\varphi_n \in \Sigma_\chi$ since τ is a sum of positive forms, $\tau = \gamma + \chi$, $\chi \geq 0$, $\gamma \geq 0$. It follows that $\gamma[\varphi - \varphi_n] \to 0$ with $\varphi = \mathcal{H}\text{-}\lim \varphi_n$. Thus, since $\gamma = \gamma^{\mathrm{cl}}$, we conclude that $\varphi \in \Phi \cap \mathcal{H}_\tau$ and therefore $\Phi = \mathcal{H}_\tau$.

Conversely, let us assume $Q(\gamma) = \Phi = \mathcal{H}_\tau$ in the norm (5.3). Let $\varphi_n \in \Phi$ be a γ-convergent sequence (see (5.1)). Then it is easily seen that $\varphi_n \in \Sigma_\tau$. So, φ_n is convergent in $\mathcal{H}_\tau$. Hence, there exists a vector $\varphi' = \mathcal{H}_\tau\text{-}\lim \varphi_n$. Since $\Phi = \mathcal{H}_\tau$, this vector belongs to Φ. Further, since $\tau[\varphi' - \varphi_n] \to 0$ and $\tau = \chi + \gamma$, we conclude that $\gamma[\varphi' - \varphi_n] \to 0$ and $\chi[\varphi' - \varphi_n] \to 0$. Therefore, $\varphi' = \varphi$ and $\gamma[\varphi_n] \to \gamma[\varphi]$. Thus, $\gamma = \gamma^{\mathrm{cl}}$. □

The next construction leads to another fundamental property of closable quadratic forms.

Let $\gamma \geq 0$ be densely defined, $Q(\gamma)^{\mathrm{cl}} = \mathcal{H}$. We consider the mapping

$$O_\gamma : \mathcal{H} \supset Q(\gamma) \ni \varphi \longrightarrow \varphi \in Q(\gamma) \subseteq \mathcal{H}_{\gamma+\chi}.$$

By (5.3), we have

$$\|O_\gamma \varphi\|_{\gamma+\chi} \geq \|\varphi\|.$$

Therefore, the inverse mapping

$$O_\gamma^{-1} : \mathcal{H}_{\gamma+\chi} \ni \varphi \longrightarrow \varphi \in \mathcal{H}, \varphi \in Q(\gamma)$$

is contractive, $\|O_\gamma^{-1}\| \leq 1$.

Let $J : \mathcal{H}_{\gamma+\chi} \to \mathcal{H}$ denotes the continuous extension of O_γ^{-1} to the whole space $\mathcal{H}_{\gamma+\chi}$. We assume that $\operatorname{Ker} J = 0$. Then $\mathcal{H}_{\gamma+\chi}$ is dense in $\mathcal{H}$ as a subset. In this a case it is clear that γ is closable and its closure

$$\gamma^{\mathrm{cl}}[\varphi] := \|\varphi\|^2_{\gamma+\chi} - \|\varphi\|^2, \quad \varphi \in Q(\gamma^{\mathrm{cl}}) = \mathcal{H}_{\gamma+\chi}$$

coincides with the smallest closed extension of γ.

However, if $\operatorname{Ker} J \neq 0$, then it is clear that it is impossible to extend γ to a closed form. Indeed, let $0 \neq \psi \in \operatorname{Ker} J$. Then there exists a sequence $\varphi_n \in Q(\gamma)$ such that $\varphi_n \to \psi$ in $\mathcal{H}_{\gamma+\chi}$. For it $\|\varphi_n\| \to 0$, $\gamma[\varphi_n - \varphi_m] \to 0$, but $\gamma[\varphi_n] \nrightarrow 0$, since $\psi \neq 0$. This means that γ is non-closable if $\operatorname{Ker} J \neq 0$.

We state another fundamental result of the theory (see Theorem VI.1.17 in [107]). Let us recall that Θ denotes the set of all sequences $\varphi_n \in Q(\gamma)$ convergent to zero in $\mathcal{H}$ and $\mathcal{H}_\gamma$ denotes the space constructed from $Q(\gamma)$ by the (quasi)inner product $(\varphi, \psi)_\gamma = \gamma(\varphi, \psi)$.

Theorem 5.1.3. *A positive densely defined in $\mathcal{H}$ quadratic form γ is closable if and only if*

$$\varphi_n \xrightarrow{\mathcal{H},\gamma} 0 \Longrightarrow \gamma[\varphi_n] \longrightarrow 0, \tag{5.4}$$

i.e., $\gamma \subset \gamma^{\mathrm{cl}}$ if and only if each γ-convergent to zero sequence $\varphi_n \in Q(\gamma)$ converges to zero in $\mathcal{H}_\gamma$:

$$\varphi_n \in \Theta \cap \Sigma_\gamma \Longrightarrow \varphi_n \in \Theta_\gamma. \tag{5.5}$$

It is clear that a bounded from below form $\gamma \geq m_\gamma > -\infty$ with $m_\gamma < 0$ is closable in $\mathcal{H}$ if and only if a positive form $\gamma' = \gamma + a\chi$ with $a \geq -m_\gamma$ is closable. Here $\chi[\cdot] = \|\cdot\|^2$. That is why we usually deal with positive forms.

Note that in a general case a sum of two forms is most likely non-closable when at least one of them is non-closable. So, the following question arises naturally: Is it possible to derive from a positive, in general non-closable, form its largest closable component (provided, of course, it exists)? We shall show below that this problem has a constructive solution. But first we shall consider a class of purely non-closable forms with no closable components. These forms are called singular. The precise definition is given below.

5.2 Singular quadratic forms on Hilbert space

The notion of singular quadratic form and its precise definition have been introduced in [113, 117, 121] (see also [105, 106]). These forms were connected with applications in the singular perturbation theory. Indeed, they provide a convenient tool for investigating of singularly perturbed self-adjoint operators.

Let $\gamma \neq 0$ be a Hermitian quadratic form, $\gamma = \gamma^*$, with a dense domain $Q(\gamma) \equiv \Phi$ in a Hilbert space, $\Phi^{\mathrm{cl}} = \mathcal{H}$. Thus, γ is a symmetric form. In what follows we assume that γ is bounded from below, $\gamma \geq m_\gamma > -\infty$, or positive, $\gamma \geq 0$.

Let us recall that a symmetric bounded from below quadratic form γ on $\mathcal{H}$ is said to be *regular* if it is closed (or closable):

$$\gamma \subseteq \gamma^{\mathrm{cl}}.$$

We denote this by $\gamma \| \chi$. This notation was introduced in [99–101]. It symbolizes the tangential property of γ with respect to the form χ which generates the inner product in $\mathcal{H}$.

Obviously, each quadratic form $\gamma_A(\cdot,\cdot) = (A\cdot,\cdot)$ associated with a semi-bounded self-adjoint operator $A = A^* \geq m_A > -\infty$ is regular: $\gamma_A \| \chi$. That is, $\operatorname{Dom} A \subseteq Q(\gamma_A^{\mathrm{cl}})$.

A class of quadratic forms with a property opposite to regularity is that of singular forms.

Let $\gamma \geq 0$. As above, $\mathcal{H}_\gamma$ denotes the Hilbert space which is constructed by the standard procedure of factorization and completion of $\operatorname{Dom}\gamma$ with respect to the inner product $(\varphi,\psi)_\gamma = \gamma(\varphi,\psi)$, $\varphi,\psi \in Q(\gamma)$.

Definition 5.2.1. A positive quadratic form $\gamma \geq 0$ on $\mathcal{H}$ is said to be *purely singular* if for any $\varphi \in Q(\gamma)$ there exists a sequence $\varphi_n \in Q(\gamma)$ convergent to φ in $\mathcal{H}$ and convergent to zero in $\mathcal{H}_\gamma$:

$$\forall \varphi \in Q(\gamma),\ \exists \varphi_n \in Q(\gamma), \quad \varphi_n \xrightarrow{\mathcal{H}} \varphi,\ \gamma[\varphi_n] \longrightarrow 0. \tag{5.6}$$

Putting $\psi_n = \varphi - \varphi_n$, we get an equivalent definition.

Definition 5.2.2. A positive quadratic form $\gamma \geq 0$ on $\mathcal{H}$ is said to be *purely singular* if for any $\varphi \in Q(\gamma)$ there exists a sequence $\psi_n \in Q(\gamma)$ convergent to zero in $\mathcal{H}$ and convergent to φ in $\mathcal{H}_\gamma$:

$$\exists \psi_n \in Q(\gamma), \quad \psi_n \xrightarrow{\mathcal{H}} 0,\ \gamma[\psi_n - \varphi] \longrightarrow 0. \tag{5.7}$$

For a purely singular on $\mathcal{H}$ form γ we write $\gamma \perp \chi$, where $\chi(\cdot,\cdot) = (\cdot,\cdot)$ denotes the inner product in $\mathcal{H}$ restricted onto a dense subset $\Phi = Q(\gamma)$. This notation reflects the property of mutual singularity of forms. We treat this relation as their orthogonality.

In the sequel, we say that a quadratic form γ is singular if it is purely singular, keeping in mind that the class of non-regular forms is wider than the set of purely singular forms. In the next Subsection 5.3 it will be shown that any bounded from below symmetric quadratic form has a unique decomposition into a sum of regular and purely singular components.

A simple sufficient condition for singularity of quadratic forms reads as follows:

Proposition 5.2.3. *Let the quadratic form $\gamma = \gamma^* \geq 0$ be densely defined on $\Phi = Q(\gamma)$ in $\mathcal{H}$. Denote*

$$\Phi_0 := \operatorname{Ker}\gamma = \{\varphi \in \Phi \mid \gamma[\varphi] = 0\}.$$

If Φ_0 is dense in $\mathcal{H}$, then γ is singular in $\mathcal{H}$, i.e.,

$$\Phi_0^{\mathrm{cl}} = \mathcal{H} \Longrightarrow \gamma \perp \chi. \tag{5.8}$$

Proof. (5.8) follows directly from Definition 5.2.1. □

The next original example shows that there exist singular forms γ such that $\operatorname{Ker}\gamma = \{0\}$.

Example 5.2.4 (I. Feshchenko [83]). Let us consider in a Hilbert space $\mathcal{H}$ a couple of linear manifolds Φ_1, Φ_2. We assume the following conditions are fulfilled.

Both Φ_1, Φ_2 are dense in $\mathcal{H}$ and their intersection is zero:

$$\Phi_1^{\mathrm{cl}} = \Phi_2^{\mathrm{cl}} = \mathcal{H}, \quad \Phi_1 \cap \Phi_2 = \{0\}. \tag{5.9}$$

Let $\Phi = \Phi_1 \dot{+} \Phi_2$ denote the direct sum of these manifolds. Surely, Φ is also dense in $\mathcal{H}$. Moreover, each element $\varphi \in \Phi$ has a unique representation as $\varphi = \varphi_1 + \varphi_2$, $\varphi_1 \in \Phi_1$, $\varphi_2 \in \Phi_2$.

Let us define on Φ the quadratic form

$$\gamma[\varphi] := \|\varphi_1 - \varphi_2\|^2, \quad \varphi \in \Phi.$$

Clearly, $\gamma \geq 0$. Moreover, γ is strictly positive and $\operatorname{Ker}\gamma = \{0\}$. Indeed, if $\gamma[\varphi] = 0$, then it follows that $\varphi_1 = \varphi_2$. However, due to the second condition in (5.9) this is possible only for $\varphi_1 = 0 = \varphi_2$.

Let us show that γ is singular in $\mathcal{H}$. Indeed, by the first condition, in (5.9) for each $h \in \mathcal{H}$ there exist sequences $\varphi_{1,n}$ and $\varphi_{2,n}$ belonging to Φ_1 and Φ_2, respectively, which converge to h in $\mathcal{H}$. Therefore, for the sequence $\varphi_n = \frac{1}{2}(\varphi_{1,n} + \varphi_{2,n})$ we have

$$\varphi_n \xrightarrow{\mathcal{H}} h, \quad \gamma[\varphi_n] = \left\| \frac{1}{2}\varphi_{1,n} - \frac{1}{2}\varphi_{2,n} \right\|^2 \longrightarrow 0, \quad n \longrightarrow \infty.$$

Thus, γ is singular in $\mathcal{H}$ (see Definition 5.2.1).

This example shows that there exist mutually singular quadratic forms that are strictly positive:

$$\gamma \perp \chi, \quad \gamma > 0,\ \chi > 0.$$

We recall that χ stands for the inner product in $\mathcal{H}$ restricted on Φ.

Example 5.2.5 (The singular quadratic form γ_δ). Let us consider in $\mathcal{H} = L_2(\mathbb{R}^d)$ the quadratic form generated by the Dirac δ-function:

$$\gamma_\delta(\varphi, \psi) := \varphi(x)\overline{\psi(x)}, \quad \varphi, \psi \in \Phi = C(\mathbb{R}^d),$$

where $x \in \mathbb{R}^d$ is a fixed point. Clearly, γ_δ is positive and singular since the set

$$\operatorname{Ker}\gamma_\delta = \Phi_0 = \{\varphi \in C(\mathbb{R}^d) | \varphi(x) = 0\}$$

is dense in $L_2(\mathbb{R}^d)$ (see Proposition 5.2.3).

Example 5.2.6 (Rank-one singular forms). Consider in an abstract rigged Hilbert space

$$\mathcal{H}_- \sqsupset \mathcal{H}_0 \sqsupset \mathcal{H}_+ \tag{5.10}$$

and a fixed vector $\omega \in \mathcal{H}_- \setminus \mathcal{H}_0$. Define the quadratic form γ_ω on $\mathcal{H}_0$ as follows:

$$\gamma_\omega(\varphi, \psi) = \langle \varphi, \omega \rangle_{+-} \langle \omega, \psi \rangle_{-+}, \quad \varphi, \psi \in \mathcal{H}_+ = Q(\gamma_\omega) \equiv \Phi.$$

Obviously, γ_ω is positive, since $\gamma_\omega[\varphi] = |\langle \varphi, \omega \rangle_{+-}|^2 \geq 0$ for all $\varphi \in \Phi$. The standard relations

$$\langle \varphi, \omega \rangle_{+-} = (\varphi, I_{+-}\omega)_+ = (\varphi, \eta_+)_+, \quad \eta_+ = I_{+-}\omega,$$

where $I_{+-} : \mathcal{H}_- \to \mathcal{H}_+$ is the Berezansky canonical isomorphism, show that

$$\Phi_0 = \operatorname{Ker} \gamma_\omega = \mathcal{M}_+, \quad \mathcal{M}_+ = \mathcal{N}_+^\perp, \quad \mathcal{N}_+ = \{c\eta_+ \mid c \in \mathbb{C}\}.$$

By Theorem 6.1.1 from Chapter 6, the subspace $\mathcal{M}_+$ is dense in $\mathcal{H}_0$ if and only if the subspace $\mathcal{N}_- = D_{-+}\mathcal{N}_+ = \{c\omega \mid c \in \mathbb{C}\}$ has a null intersection with $\mathcal{H}_0$. It is obvious that this condition fulfilled since $\omega \notin \mathcal{H}_0$. Thus, the form γ_ω is singular in $\mathcal{H}$ due to Proposition 5.2.3.

Example 5.2.7 (Finite-rank singular forms). Let us consider a finite set of vectors

$$\Omega = \{\omega_i \in \mathcal{H}_- \setminus \mathcal{H}_0\}_{i=1}^n, \quad n < \infty,$$

in the rigged space (5.10) We assume that Ω is linearly independent with respect to $\mathcal{H}_0$, i.e., the subspace $\mathcal{N}_- = \operatorname{span}(\Omega)$ has a null intersection with $\mathcal{H}_0$:

$$\operatorname{span}(\Omega) \cap \mathcal{H} = \{0\}.$$

Then the quadratic form

$$\gamma_\Omega(\varphi, \psi) = \sum_{i=1}^n \lambda_i \gamma_{\omega_i}(\varphi, \psi), \quad \varphi, \psi \in \mathcal{H}_+, \ \lambda_i \geq 0,$$

is singular in $\mathcal{H}_0$ thanks to the same reason as in the previous example.

Example 5.2.8 (Infinite-rank singular forms). Consider now a countable set of vectors

$$\Omega = \{\omega_i \in \mathcal{H}_- \setminus \mathcal{H}_0\}_{i=1}^\infty$$

in the rigged space (5.10). Assume the subspace $\mathcal{N}_- = (\operatorname{span}\Omega)^{\mathrm{cl},-}$ has a null intersection with $\mathcal{H}_0$, where cl, $-$ denotes closure in $\mathcal{H}_-$. Then the quadratic form

$$\gamma_\Omega(\varphi, \psi) = \sum_{i=1}^\infty \lambda_i \gamma_{\omega_i}(\varphi, \psi), \quad \lambda_i \geq 0,$$

with

$$Q(\gamma_\Omega) = \left\{ \varphi \in \mathcal{H}_+ \mid \sum_{i=1}^\infty \lambda_i \gamma_{\omega_i}[\varphi] < \infty \right\}$$

is singular in $\mathcal{H}$ by Proposition 5.2.3 and Theorem 6.1.4 from Chapter 6, since $\mathcal{N}_- \cap \mathcal{H}_0 = \{0\}$.

More information on singular quadratic forms can be found in [135].

5.3 A canonical decomposition for quadratic forms

One of the main tools of the method of rigged spaces in the theory of singular perturbations of self-adjoint operators is the canonical decomposition for any positive quadratic form $\gamma = \gamma_{\rm r} + \gamma_{\rm s}$, where $\gamma_{\rm r}$ is the largest regular part of γ and $\gamma_{\rm s}$ is a singular form in the sense of Definition 5.2.1. The next theorem states the existence of the canonical decomposition.

Theorem 5.3.1. *Each positive densely defined on a Hilbert space $\mathcal{H}$ quadratic form γ admits a unique decomposition into a sum of two mutually singular components,*

$$\gamma = \gamma_{\rm r} + \gamma_{\rm s}, \quad \gamma_{\rm r} \perp \gamma_{\rm s}, \tag{5.11}$$

where $\gamma_{\rm r}$ is positive and regular in $\mathcal{H}$,

$$\gamma_{\rm r} \| \chi, \quad \chi[\cdot] = \| \cdot \|,$$

and $\gamma_{\rm s}$ is positive and singular in $\mathcal{H}$,

$$\gamma_{\rm s} \perp \chi.$$

That is, $\gamma_{\rm r}$ is the largest closable component of γ, $\gamma_{\rm r} \subseteq \gamma_{\rm r}^{\rm cl} \leq \gamma$.

Proof. Let us denote $\Phi = Q(\gamma)$. Let $\mathcal{H}_+ = \mathcal{H}_{\chi+\gamma}$ be the Hilbert space constructed by the standard procedure as the completion of Φ with respect to the inner product

$$(\varphi, \psi)_+ \equiv (\varphi, \psi)_{\gamma+\chi} = \gamma(\varphi, \psi) + \chi(\varphi, \psi), \tag{5.12}$$

where $\chi[\varphi] = \|\varphi\|^2, \varphi \in \Phi$. Note that (5.12) implies the inequality

$$\| \cdot \| \leq \| \cdot \|_+. \tag{5.13}$$

However, in the general case this inequality does not ensure that the embedding of $\mathcal{H}_+$ into $\mathcal{H}$ is injective, in spite of the fact that the mapping

$$j : \mathcal{H}_+ \ni \varphi \longmapsto \varphi \in \mathcal{H}, \quad \varphi \in Q(\gamma)$$

is well defined and bounded, $\|j\| \leq 1$. Moreover, the domain of this mapping is dense in $\mathcal{H}_+$ and its range is dense in $\mathcal{H}$.

Let $J : \mathcal{H}_+ \to \mathcal{H}$ denote the contraction operator which is the extension by continuity of the mapping j.

If $\operatorname{Ker} J = \{0\}$, then by Theorem 5.1.3 γ is closable in $\mathcal{H}$. In this case $\gamma_{\rm r} = \gamma$ and $\gamma_{\rm s} = 0$.

Otherwise $\operatorname{Ker} J$ is a closed subspace in $\mathcal{H}_+$. Denote $\mathcal{H}_{\rm s} := \operatorname{Ker} J$. Then

$$\mathcal{H}_+ = \mathcal{H}_{\rm r} \oplus \mathcal{H}_{\rm s}, \quad \text{with} \;\; \mathcal{H}_{\rm r} := \mathcal{H}_{\rm s}^{\perp}.$$

Now we can define the required forms by

$$\gamma_{\mathrm{r}}(\varphi,\psi) := (P_{\mathrm{r}}\varphi,\chi)_+ - (\varphi,\psi), \quad \varphi,\psi \in \Phi, \tag{5.14}$$

and

$$\gamma_{\mathrm{s}}(\varphi,\psi) := (P_{\mathrm{s}}\varphi,\psi)_+, \quad \varphi,\psi \in \Phi, \tag{5.15}$$

where P_{r} and P_{s} denote the orthogonal projections in $\mathcal{H}_+$ onto the subspaces $\mathcal{H}_{\mathrm{r}}$ and $\mathcal{H}_{\mathrm{s}}$, respectively. For the form γ_{r} defined in (5.14) one can write

$$\gamma_{\mathrm{r}} = \tilde{\gamma}_{\mathrm{r}} - \chi, \quad \tilde{\gamma}_{\mathrm{r}}[\,\cdot\,] := \|P_{\mathrm{r}}\cdot\|_+, \quad Q(\tilde{\gamma}_{\mathrm{r}}) = \Phi.$$

In general it may happen that $\gamma_{\mathrm{r}} = 0$.

Let us show that the form γ_{s} defined in (5.14) is singular in $\mathcal{H}$, i.e., $\gamma_{\mathrm{s}} \perp \chi$. Let $P_{\mathrm{s}}\varphi \neq 0$ and $\gamma_{\mathrm{s}}[\varphi] \neq 0$ for some $\varphi \in \Phi$. Since $\mathcal{H}_+$ coincides with the completion of Φ with respect to the norm $\|\cdot\|_+ = (\chi[\,\cdot\,]+\gamma[\,\cdot\,])^{1/2}$, there exists a sequence $\varphi_n \in \Phi$ convergent to $P_{\mathrm{s}}\varphi$ in the norm of $\mathcal{H}_+$. Clearly, $\{\varphi_n\}$ it is a Cauchy sequence in $\mathcal{H}$. Thus, since χ is positive, we have $\{\varphi_n\} \in \Sigma_\gamma$, $\gamma_{\mathrm{r}}[\varphi_n] \to 0$, and $\gamma_{\mathrm{s}}[\varphi_n - \varphi] \to 0$. Now, from the equality $\operatorname{Ker} J = \mathcal{H}_{\mathrm{s}} = P_{\mathrm{s}}\mathcal{H}_+$ and due to (5.7), it follows that $\varphi_n \to 0$ in $\mathcal{H}$. This means that $\gamma_{\mathrm{s}} \perp \chi$.

Let us show that $\gamma_{\mathrm{r}} \geq 0$. The equality $\mathcal{H}_{\mathrm{s}} = \operatorname{Ker} J$ implies that $JP_{\mathrm{s}}\varphi = 0$ for all $\varphi \in \Phi$. Thus,

$$\|\varphi\|^2 = \|J\varphi\|^2 = \|JP_{\mathrm{r}}\varphi\|^2 \leq \|P_{\mathrm{r}}\varphi\|_+^2 = \gamma_{\mathrm{r}}[\varphi] + \|\varphi\|^2,$$

and therefore $\gamma_{\mathrm{r}}[\varphi] \geq 0$, $\varphi \in \Phi$.

Next, let us verify that γ_{r} is closable in $\mathcal{H}$. According to our construction, the subspace $\mathcal{H}_{\mathrm{r}}$ in $\mathcal{H}_+$ can be identified with $\mathcal{H}_{\gamma_{\mathrm{r}}+\chi}$ since $\gamma_{\mathrm{r}} + \chi = \tilde{\gamma}_{\mathrm{r}}$, where recall that $\tilde{\gamma}_{\mathrm{r}}[\varphi] = (P_{\mathrm{r}}\varphi,\varphi)_+$. Therefore, one can identify the contraction operators $J_{\gamma_{\mathrm{r}}} : \mathcal{H}_{\gamma_{\mathrm{r}}+\chi} \to \mathcal{H}$ and $J \restriction \mathcal{H}_{\mathrm{r}}$, since $JP_{\mathrm{s}}\varphi = 0$, $\varphi \in \Phi$. This proves that operator $J_{\gamma_{\mathrm{r}}}$ is injective. Therefore, $\operatorname{Ker} J_{\gamma_{\mathrm{r}}} = \{0\}$. By Theorem 5.1.3, it follows that γ_{r} is closable.

The uniqueness of the canonical decomposition for γ into regular and singular components follows from the uniqueness of decomposition for the positive space:

$$\mathcal{H}_+ = \mathcal{H}_{\mathrm{r}} \oplus \mathcal{H}_{\mathrm{s}}, \quad \mathcal{H}_{\mathrm{s}} := \operatorname{Ker} J.$$

The mutual singularity of the forms γ_{r}, γ_{s}, i.e., that

$$\gamma_{\mathrm{r}} \perp \gamma_{\mathrm{s}}, \tag{5.16}$$

follows from the orthogonality of the forms $\tilde{\gamma}_{\mathrm{r}}$ and γ_{s}. In turn, the latter fact follows from the orthogonality of the spaces $\mathcal{H}_{\mathrm{r}}$ and $\mathcal{H}_{\mathrm{s}}$.

Finally, the last statement of the theorem follows from the following abstract lemma (see [177]). □

Lemma 5.3.2. *Let σ, γ be positive forms on $\mathcal{H}$, $\sigma, \gamma \geq 0$. Then*

$$\sigma \subseteq \sigma^{\text{cl}}, \quad \sigma \leq \gamma \implies \sigma \leq \gamma_{\text{r}}, \tag{5.17}$$

where γ_{r} is the regular component of γ in $\mathcal{H}$ defined according to (5.14).

Proof. Since $Q(\gamma) \subseteq Q(\sigma)$ and $Q(\gamma_{\text{r}}) = Q(\gamma)$, we have to prove only that $\sigma[\varphi] \leq \gamma_{\text{r}}[\varphi]$, $\varphi \in Q(\gamma) \equiv \Phi$. This inequality is equivalent to

$$\|O_\sigma \varphi\|_{\sigma+\chi} \leq \|P_{\text{r}} O_\gamma \varphi\|_{\gamma+\chi}, \quad \varphi \in \Phi, \tag{5.18}$$

where O_σ and O_γ denote the (defined on Φ) identity mappings from $\mathcal{H}$ to $\mathcal{H}_{\sigma+\chi}$ and $\mathcal{H}_{\gamma+\chi}$, respectively, and P_{r} denotes the orthogonal projection onto $\mathcal{H}_{\text{r}}$ in $\mathcal{H}_{\gamma+\chi}$. The less sharp inequality (without P_{r})

$$\|O_\sigma \varphi\|_{\sigma+\chi} \leq \|O_\gamma \varphi\|_{\gamma+\chi}, \quad \varphi \in \Phi, \tag{5.19}$$

holds due to the condition $\sigma \leq \gamma$. Let us consider the operator $j_{\sigma\gamma} : \mathcal{H}_{\gamma+\chi} \to \mathcal{H}_{\sigma+\chi}$ as the closure of the mapping $O_\gamma \varphi \to O_\sigma \varphi$, $\varphi \in \Phi$. It is contractive due to (5.19). Thus,

$$O_\sigma \varphi = j_{\sigma\gamma} O_\gamma \varphi, \quad \varphi \in \Phi, \tag{5.20}$$

$$\|j_{\sigma\gamma} \eta\|_{\sigma+\chi} \leq \|\eta\|_{\gamma+\chi}, \quad \eta \in \mathcal{H}_{\gamma+\chi}. \tag{5.21}$$

We claim that

$$J_\gamma = J_\sigma j_{\sigma\gamma}, \tag{5.22}$$

where J_γ and J_σ denote the operators extending by continuity the contractive mappings of Φ from $\mathcal{H}_{\gamma+\chi}$ and $\mathcal{H}_{\sigma+\chi}$ respectively into $\mathcal{H}$. Indeed, by (5.21) we have

$$J_\sigma j_{\sigma\gamma} O_\gamma \varphi = J_\sigma O_\sigma \varphi = \varphi = J_\gamma O_\gamma \varphi, \quad \varphi \in \Phi.$$

Therefore, (5.22) holds true for all $\eta \in \mathcal{H}_{\gamma+\chi}$. Further, since the form σ is closable, the operator J_σ is injective and hence

$$\operatorname{Ker} J_\gamma = \operatorname{Ker} j_{\sigma\gamma}.$$

In particular,

$$j_{\sigma\gamma} P_{\text{s}} O_\gamma \varphi = 0, \quad \varphi \in \Phi, \tag{5.23}$$

where P_{s} is the orthogonal projection in $\mathcal{H}_{\gamma+\chi}$ onto $\mathcal{H}_{\text{s}} \equiv \operatorname{Ker} J_\gamma$. Thus, (5.20), (5.23), and (5.21) yield

$$\|O_\sigma \varphi\|_{\sigma+\chi} = \|j_{\sigma\gamma} O_\gamma \varphi\|_{\sigma+\chi} = \|j_{\sigma\gamma} P_{\text{r}} O_\gamma \varphi\|_{\sigma+\chi} \leq \|P_{\text{r}} O_\gamma \varphi\|_{\gamma+\chi},$$

that proves (5.18). □

Note that one can establish the property $\gamma_{\rm r} \geq 0$ in a different way. Formally, it follows from $\tilde{\gamma}_{\rm r} \geq \chi$ and the equality $\gamma_{\rm r} = \tilde{\gamma}_{\rm s} - \chi$. In turn, the inequality $\tilde{\gamma}_{\rm r} \geq \chi$ follows from $\gamma \geq \gamma_{\rm s}$. Indeed,

$$\tilde{\gamma}_{\rm r}[\cdot] = (P_{\rm r}\cdot, \cdot)_+ = \| \cdot \|_+^2 - (P_{\rm s}\cdot, \cdot)_+ = (\chi + \gamma - \gamma_{\rm s})[\cdot] \geq \chi.$$

Therefore, we have to prove the inequality $\gamma \geq \gamma_{\rm s}$. To this aim we consider the forms χ, γ, $\gamma_{\rm s}$, and $\tilde{\gamma}_{\rm r}$ on $\mathcal{H}_+$. Clearly, they are positive and bounded. Their extensions by continuity to $\mathcal{H}_+$ we denote as $\chi^{{\rm cl},+}$, $\gamma^{{\rm cl},+}$, $\gamma_{\rm s}^{{\rm cl},+}$, and $\gamma_{\rm r}^{{\rm cl},+}$, respectively. It is clear also that

$$\| \cdot \|_+^2 = \gamma_{\rm s}^{{\rm cl},+}[\,\cdot\,] + \tilde{\gamma}_{\rm r}^{{\rm cl},+}[\,\cdot\,]$$

and

$$\operatorname{Ker} \chi^{{\rm cl},+} = \mathcal{H}_{\rm s} = \operatorname{Ker} \tilde{\gamma}_{\rm r}^{{\rm cl},+}.$$

Thus, we have

$$\| \cdot \|_+^2 = \chi^{{\rm cl},+}[\,\cdot\,] + \gamma^{{\rm cl},+}[\,\cdot\,].$$

This implies that

$$\|f\|_+^2 = \gamma^{{\rm cl},+}[f] = \tilde{\gamma}_{\rm r}^{{\rm cl},+}[f] + \gamma_{\rm s}^{{\rm cl},+}[f] = \gamma_{\rm s}^{{\rm cl},+}[f], \quad f \in \mathcal{H}_{\rm s}.$$

Now we can conclude that $\gamma^{{\rm cl},+} \geq \gamma_{\rm s}^{{\rm cl},+}$ because $\operatorname{Ker} \gamma_{\rm s}^{{\rm cl},+} = \mathcal{H}_{\rm r}$. In particular, it is obvious that $\gamma \geq \gamma_{\rm s}$. Hence, the property $\gamma_{\rm r} \geq 0$ is proved.

5.3.1 Properties of singular quadratic forms

Let $\gamma \geq 0$ be a positive quadratic form on a Hilbert space $\mathcal{H}$. We shall always suppose that γ has a dense domain: $Q(\gamma)^{\rm cl} = \mathcal{H}$. However, γ may not be closable. One interesting question is whether there is a nontrivial singular component $\gamma_{\rm s}$ in γ.

Let $\mathcal{H}_\gamma$ denote the Hilbert space constructed by the standard procedure, introducing on $\Phi = Q(\gamma)$ the (quasi)-inner product $(\cdot, \cdot)_\gamma = \gamma(\cdot, \cdot)$. To answer the above question, we compare the convergent sequences in the spaces $\mathcal{H}$ and $\mathcal{H}_\gamma$.

We call a vector $\varphi \in \mathcal{H}$ *regular* with respect to a quadratic form γ if for any sequence $\varphi_n \in Q(\gamma)$ satisfying conditions $\varphi_n \to \varphi$ and $\gamma[\varphi_n - \varphi_m] \to 0$ in $\mathcal{H}$, it follows that $\gamma[\varphi_n] \to a \in \mathbb{R}$. Here the number a does not depend on the choice of φ_n. We denote by $Q_{\rm reg}(\gamma)$ the set of all regular vectors of γ.

Proposition 5.3.3. *For a positive form γ on $\mathcal{H}$, the set $Q_{\rm reg}(\gamma)$ coincides with $Q(\gamma)$ if and only if this form is closed:*

$$Q(\gamma) = Q_{\rm reg}(\gamma) \Longleftrightarrow \gamma = \gamma^{\rm cl}. \tag{5.24}$$

Proof. This immediately follows from Definition 5.1.1. □

Proposition 5.3.4. *A quadratic form γ is closable if and only if the zero vector in $\mathcal{H}$ is regular:*

$$\gamma \subseteq \gamma^{\rm cl} \Longleftrightarrow 0 \in Q_{\rm reg}(\gamma). \tag{5.25}$$

Proof. The necessity is obvious. The sufficiency is established as follows. Consider a γ-convergent sequence $\varphi_n \in Q(\gamma)$. According to the definition, $\varphi_n \overset{\mathcal{H},\gamma}{\longrightarrow} \varphi$ for some $\varphi \in \mathcal{H}$, i.e., $\varphi_n \to \varphi$ and $\gamma[\varphi_n - \varphi_m] \to 0$. Since $\gamma[\varphi_n - \varphi_m] \to 0$, the sequence $\gamma[\varphi_n]$ is bounded, because

$$\begin{aligned}|\gamma[\varphi_n] - \gamma[\varphi_m]| &\le |\gamma(\varphi_n - \varphi_m, \varphi)| + |\gamma(\varphi_m, \varphi_m - \varphi_n)| \\ &\le \gamma[\varphi_n - \varphi_m] \cdot \gamma[\varphi_n] + \gamma[\varphi_m] \cdot \gamma[\varphi_m - \varphi_n].\end{aligned}$$

Hence, there exists the finite number $a = \lim_{n\to\infty} \gamma[\varphi_n] < \infty$. We put $\gamma^{\mathrm{cl}}[\varphi] := a$. However, we need to check that this value $\gamma^{\mathrm{cl}}[\varphi]$ does not depend on a choice of the γ-convergent sequence. From the condition $0 \in Q_{\mathrm{reg}}(\gamma)$ it is clear that if $\varphi_n \overset{\mathcal{H},\gamma}{\longrightarrow} \varphi$ and $\varphi'_n \overset{\mathcal{H},\gamma}{\longrightarrow} \varphi$, then

$$\lim_{n\to\infty} \gamma[\varphi_n] = \lim_{n\to\infty} \gamma[\varphi'_n] = a.$$

Moreover, it is easily seen that if $0 \in Q_{\mathrm{reg}}(\gamma)$, then the value

$$\gamma^{\mathrm{cl}}(\varphi, \psi) = \lim_{n\to\infty} \gamma(\varphi_n, \psi_n) \quad \text{with} \quad \varphi_n \overset{\mathcal{H},\gamma}{\longrightarrow} \varphi, \ \psi_n \overset{\mathcal{H},\gamma}{\longrightarrow} \psi$$

also does not depend on the choice of the γ-convergent sequences. □

In the opposite case, γ is non-closable in $\mathcal{H}$ and the zero vector is not regular with respect to this form. Then there are vectors φ such that for some couples of γ-convergent sequences $\varphi_n \overset{\mathcal{H},\gamma}{\longrightarrow} \varphi$, $\varphi'_n \overset{\mathcal{H},\gamma}{\longrightarrow} \varphi$, their limits are different:

$$a = \lim_{n\to\infty} \gamma[\varphi_n] \ne \lim_{n\to\infty} \gamma[\varphi'_n] = a'. \tag{5.26}$$

Then we say that these vectors are of singular type. Further, if in (5.26) one of the numbers a or a' is zero, then φ is said to be *singular* with respect to γ. So, we write $\varphi \in Q_{\mathrm{sing}}(\gamma)$, if there is a γ-convergent sequence $\varphi_n \overset{\mathcal{H},\gamma}{\longrightarrow} \varphi$ such that $\gamma[\varphi_n] \to 0$. Thus,

$$Q_{\mathrm{sing}}(\gamma) := \{\varphi \in \mathcal{H} \mid \exists \varphi_n \in Q(\gamma),\ \varphi_n \overset{\mathcal{H}}{\longrightarrow} \varphi, \gamma[\varphi_n] \longrightarrow 0\}.$$

It should be noted that for the value $\gamma[\varphi] \ne 0$ with $\varphi \in Q_{\mathrm{sing}}(\gamma) \cap Q(\gamma)$ and any $c > 0$ there exists a sequence $\varphi_n \in Q(\gamma)$ such that $\varphi_n \overset{\mathcal{H}}{\longrightarrow} \varphi$ and $\gamma[\varphi_n] \to c$. If the set of singular vectors for γ is dense in $\mathcal{H}$, then γ is singular.

Proposition 5.3.5. *Let $\gamma \ne 0$ be a positive densely defined form on $\mathcal{H}$. Assume that the set*

$$\Phi'_0 := Q(\gamma) \cap Q_{\mathrm{sing}}(\gamma)$$

is dense in $\mathcal{H}$. Then γ is singular.

Proof. Since Φ'_0 is dense in $\mathcal{H}$, it is easy to show that each $\varphi \in Q(\gamma)$ is singular with respect to γ. Now we have to use Definition 5.2.1. □

In the remainder to this subsection we will look at the theory of singular quadratic forms from an abstract point of view, without any connection with an underlying Hilbert space $\mathcal{H}$.

Let γ and χ be a couple of non-negative quadratic forms defined on an abstract linear space Φ. Let us denote by $\mathcal{H}_\gamma$ and $\mathcal{H}_\chi$ the Hilbert spaces constructed by the standard procedure of completion and factorization of Φ with respect to the (quasi)-inner products $(\phi, \varphi)_\gamma := \gamma(\phi, \varphi)$ and $(\phi, \varphi)_\chi := \chi(\phi, \varphi)$, respectively. We shall write $\gamma \perp \chi$, if γ is singular in $\mathcal{H}_\chi$ in the sense of Definition 5.2.1. It is easy to understand that in this abstract setting the singularity relation for forms γ and χ is reflexive (see [135]). Below we give a more precise definition of the mutual singularity relation between forms.

Definition 5.3.6. Let γ and χ be two non-negative quadratic forms on a linear space Φ. These forms are said to be mutually singular (or orthogonal), denoteed $\gamma \perp \chi$ or $\chi \perp \gamma$, if for each $\varphi \in \Phi$ there is a sequence $\varphi_n \in \Phi$ such that

$$\varphi_n \in \Sigma_\gamma \cap \Theta_\chi, \quad \varphi_n \xrightarrow{\gamma} \varphi,$$

or, equivalently, there exists a sequence $\psi_n \in \Phi$ such that

$$\psi_n \in \Theta_\gamma \cap \Sigma_\chi, \quad \psi_n \xrightarrow{\chi} \varphi.$$

Here $\varphi_n \xrightarrow{\gamma} \varphi$ means that $\gamma[\varphi_n - \varphi] \to 0$.

The next important theorem was proved in [135].

Theorem 5.3.7. *Let γ and χ be positive quadratic forms on Φ. Let $\mathcal{H}_\gamma$, $\mathcal{H}_\chi$, and $\mathcal{H}_{\gamma+\chi}$ denote the Hilbert spaces constructed by the standard procedure from Φ by means of (quasi)inner products*

$$(\phi, \varphi)_\gamma = \gamma(\phi, \varphi), \ (\phi, \varphi)_\chi = \chi(\phi, \varphi),$$

and

$$(\phi, \varphi)_{\gamma+\chi} = \gamma(\phi, \varphi) + \chi(\phi, \varphi),$$

respectively. Then

$$\gamma \perp \chi \Longleftrightarrow \mathcal{H}_{\gamma+\chi} = \mathcal{H}_\gamma \oplus \mathcal{H}_\chi. \tag{5.27}$$

This theorem is one of the basic tools for the construction of singular and super-singular perturbed operators in the method of rigged Hilbert spaces (see below Chapters 7 and 8).

Finally, we introduce a notion of regularity for quadratic forms defined on an abstract linear space Φ.

Definition 5.3.8. Let $\gamma > 0$ and $\chi > 0$ be the strictly positive quadratic forms on Φ. A form γ is said to be regular (tangential) with respect to χ, denoted $\gamma \| \chi$, if

$$\varphi_n \in \Sigma_\gamma \cap \Theta_\chi \Longrightarrow \varphi_n \xrightarrow{\gamma} 0 \quad (\varphi_n \in \Theta_\gamma).$$

Clearly, the relation $\gamma \| \chi$ implies that γ is closable in $\mathcal{H}_\chi$. However, this relation is not symmetric, i.e., in general, $\gamma \| \chi \not\Leftrightarrow \chi \| \gamma$. In addition, the given definition can be extended to the case $\gamma \geq 0$, $\chi \geq 0$ provided that $\operatorname{Ker} \chi \subseteq \operatorname{Ker} \gamma$.

5.4 Operator representation for singular forms

Similarly to Definition 5.2.1 one can introduce the concept of purely singular operator in a Hilbert space (see [96, 100, 165]).

Definition 5.4.1. A linear densely defined on a Hilbert space $\mathcal{H}$ operator S is said to be purely singular if for all $f \in \mathcal{H}$ there is a sequence $f_n \in \operatorname{Dom} S$ such that

$$f_n \longrightarrow f, \quad Sf_n \longrightarrow 0.$$

5.4.1 Singular forms and operators in the A-scale

Here we study the problem of operator representations for singular quadratic forms. It is convenient to consider this problem in scales of Hilbert spaces.

Let a linear operator S acts from $\mathcal{H}_k$ to $\mathcal{H}_{-k}, k \geq 1$ in the A-scale (see Chapter 5). We assume S has the property: $\operatorname{Ker} S \sqsubset \mathcal{H}_0$. Then, by Definition 5.4.1, S is purely singular in $\mathcal{H}_0$. We shall use operators S of this kind to provide an operator representation for singular quadratic forms.

It is well known (see, for example, [107]) that every closable, bounded from below quadratic form is associated with a self-adjoint operator. This connection can be extended to wide class of quadratic forms and linear operators, including singular ones. To this aim we shall consider this problem in a fixed A-scale of Hilbert spaces:

$$\cdots \sqsupset \mathcal{H}_{-k} \sqsupset \mathcal{H}_0 \sqsupset \mathcal{H}_k \sqsupset \cdots, \quad k \geq 0. \tag{5.28}$$

Definition 5.4.2. Let γ be a symmetric densely defined quadratic form on a Hilbert space $\mathcal{H}_0$. We say that γ belongs to the $\mathcal{H}_{-k}$-class, $k \geq 1$ of singular forms (denoted $\gamma \in \mathcal{H}_{-k}$-class) if

(1) $\operatorname{Dom} \gamma \subseteq \mathcal{H}_k$ and γ is closed in $\mathcal{H}_k$,

(2) γ is purely singular in $\mathcal{H}_{k-1}$ in the sense that $\operatorname{Ker} \gamma$ is dense in $\mathcal{H}_{k-1}$.

In particular, $\gamma \in \mathcal{H}_{-2}$-class if $\operatorname{Dom} \gamma = \mathcal{H}_2$, γ is bounded on $\mathcal{H}_2$, and $\operatorname{Ker} \gamma$ is dense in $\mathcal{H}_1$. We remark that a number publications were devoted to the study of singular perturbations given by this kind of quadratic forms (see, for example, [22, 102, 113] and references therein).

The next theorem can be considered as a generalization of the well-known first representation theorem for quadratic forms (see Theorem VI.2.1 in [107]).

Theorem 5.4.3. *Let γ be a densely defined on $\mathcal{H}_0$ quadratic form. Assume that $\gamma \in \mathcal{H}_{-k}$-class, $k \geq 1$. Then γ has the following operator representation:*

$$\gamma(\varphi, \psi) = \langle S\varphi, \psi \rangle_{-k,k}, \quad \varphi, \psi \in \operatorname{Dom} S \subseteq \mathcal{H}_k, \tag{5.29}$$

where the associated operator $S : \mathcal{H}_k \to \mathcal{H}_{-k}$ admits the factorization $S = D_{-k,k} \cdot \mathbf{s}$, where $D_{-k,k}$ is the Berezansky canonical isomorphism and $\mathbf{s}$ is a positive self-adjoint operator on $\mathcal{H}_k$ such that

$$\operatorname{Ker} \mathbf{s} = \operatorname{Ker} S = \operatorname{Ker} \gamma \sqsubset \mathcal{H}_{k-1}. \tag{5.30}$$

Proof. Since the form γ is bounded on $\mathcal{H}_k$ (see Definition 5.4.2), it has an usual operator representation: $\gamma(\varphi,\psi) = (\mathbf{s}\varphi,\psi)_k$. So, we can introduce the operator $S = D_{-k,k} \cdot \mathbf{s}$ acting from $\mathcal{H}_k$ into $\mathcal{H}_{-k}$. By using the properties of the A-scale (see Chapter 4), we get the required representation (5.29) for γ:

$$\gamma(\varphi,\psi) = (\mathbf{s}\varphi,\psi)_+ = \langle S\varphi,\psi\rangle_{-k,k}, \quad \varphi,\psi \in \operatorname{Dom}\mathbf{s} = \operatorname{Dom} S \subseteq \mathcal{H}_k.$$

The equalities in (5.30) are obviously fulfilled because γ belongs to the $\mathcal{H}_{-k}$-class. □

The next theorem gives a criterion of membership in the $\mathcal{H}_{-k}$-class.

Theorem 5.4.4 ([9, 100]). *Let γ be a bounded symmetric quadratic form on $\mathcal{H}_k$, $k \geq 1$. Define*

$$\mathcal{M}_k := \operatorname{Ker}\gamma, \quad \mathcal{N}_k = \mathcal{H}_k \ominus \mathcal{M}_k.$$

Then $\gamma \in \mathcal{H}_{-k}$-class if and only if $\mathcal{N}_{-k} \cap \mathcal{H}_{-k+1} = \{0\}$, where $\mathcal{N}_{-k} := D_{-k,k}\mathcal{N}_k$.

Proof. We only have to verify the condition (2) in Definition 5.4.2. Since due to Theorem 6.1.4

$$\mathcal{M}_k \sqsubset \mathcal{H}_{k-1} \Longleftrightarrow \mathcal{N}_{-k} \cap \mathcal{H}_{-k+1} = \{0\},$$

we can conclude that γ satisfies this condition if and only if $\mathcal{N}_{-k}\cap\mathcal{H}_{-k+1}=\{0\}$. □

Example 5.4.5 (Singular rank-one operators). Let us consider the chain of Hilbert spaces

$$\mathcal{H}_{-2}(A) \equiv \mathcal{H}_- \sqsupset \mathcal{H}_{-1} \sqsupset \mathcal{H}_0 \sqsupset \mathcal{H}_1 \sqsupset \mathcal{H}_+ \equiv \mathcal{H}_2(A) \tag{5.31}$$

from the A-scale. Fix a vector $\omega \in \mathcal{H}_-\backslash\mathcal{H}_0$. Consider the operator S acting from $\mathcal{H}_+$ into $\mathcal{H}_-$ according to the rule:

$$S\varphi = \langle\varphi,\omega\rangle_{+,-}\omega, \quad \varphi \in \mathcal{H}_+ = \operatorname{Dom} S.$$

Clearly, S is a rank-one operator on $\mathcal{H}_0$. It is singular, because the set

$$\operatorname{Ker} S = \{\varphi \in \mathcal{H}_+ \mid \langle\varphi,\omega\rangle_{+,-} = 0\}$$

is dense in $\mathcal{H}_0$. The later property follows from the fact that $\omega \notin \mathcal{H}_0$. The quadratic form associated with S has the form

$$\begin{aligned}\gamma_\omega(\varphi,\psi) &:= \langle\varphi,\omega\rangle_{+,-}\langle\omega,\psi\rangle_{-,+} = \langle S\varphi,\psi\rangle_{-,+}\\ &= \langle D_{-+}\mathbf{s}\varphi,\psi\rangle_{-,+} = (\mathbf{s}\varphi,\psi)_+,\end{aligned}$$

where we introduce another rank-one operator $\mathbf{s}$ which acts in $\mathcal{H}_+$:

$$\mathbf{s}\varphi = (\varphi,\eta_+)_+\eta_+, \quad \eta_+ := I_{+-}\omega.$$

If $\omega \in \mathcal{H}_-\backslash\mathcal{H}_{-1}$, then $\operatorname{Ker}\gamma_\omega$ is dense in $\mathcal{H}_1$ and $\gamma_\omega \in \mathcal{H}_{-2}$-class. But if $\omega \in \mathcal{H}_{-1}\backslash\mathcal{H}_0$, then by a similar argument $\gamma_\omega \in \mathcal{H}_{-1}$-class.

In the case where $\omega \in \mathcal{H}_{-k}\backslash\mathcal{H}_{-k+1}$, $k \geq 2$, the singular quadratic form

$$\gamma_\omega(\varphi,\psi) := \langle\varphi,\omega\rangle_{k,-k}\langle\omega,\psi\rangle_{-k,k}, \quad \varphi,\psi \in \mathcal{H}_k$$

belongs to the $\mathcal{H}_{-k}$-class. By Theorem 5.4.3, it has the operator representation

$$\gamma_\omega(\varphi,\psi) = \langle S\varphi,\psi\rangle_{-k,k} = \langle D_{-k,k}\mathbf{s}\varphi,\psi\rangle_{-k,k} = (\mathbf{s}\varphi,\psi)_k.$$

Here $\mathbf{s}\varphi = (\varphi,\eta_k)_k\eta_k$ with $\eta_k = I_{k,-k}\omega$. The condition $\omega \notin \mathcal{H}_{-k+1}$ implies that $\operatorname{Ker}\gamma_\omega$ is dense in $\mathcal{H}_{k-1}$, and therefore $\gamma_\omega \in \mathcal{H}_{-k}$-class.

Example 5.4.6 (Finite-rank singular operators). Let $h_i \in \mathcal{H}_0$, $i = 1,2,\ldots,n < \infty$ be orthogonal vectors. Assume that

$$\operatorname{span}\{h_i\} \cap \operatorname{Dom} A = \{0\}.$$

Then we can define on $\mathcal{H}_0$ the operator S of rank n by

$$Sf = \sum_{i=1}^{n}(Af,h_i)_0 D_{-,+}h_i = \sum_{i=1}^{n}\langle f,\omega_i\rangle_{+,-}\omega_i,$$

where $f \in \mathcal{H}_+ = \operatorname{Dom} S$, $\omega_i := D_{-+}h_i$. It is singular in $\mathcal{H}_0$ since all the vectors $\omega_i = D_{-+}h_i \in \mathcal{H}_- \setminus \mathcal{H}_0$ (their span has the same property). $\operatorname{Ker} S$ is obviously dense in $\mathcal{H}_0$. If all $h_i \in \operatorname{Dom} A^{1/2}$, then the quadratic form $\gamma[f] := \langle Sf,f\rangle_{-,+}$ belongs to the $\mathcal{H}_{-1}$-class. But if the stronger condition $\operatorname{span}\{h_i\}\cap\operatorname{Dom} A^{1/2} = \{0\}$ is fulfilled, then γ belongs to the $\mathcal{H}_{-2}$-class.

We complete this section by some definitions which will be used in the further chapters.

Let

$$\mathcal{H}_{-2} \sqsupset \mathcal{H}_{-1} \sqsupset \mathcal{H}_0 \sqsupset \mathcal{H}_1 \sqsupset \mathcal{H}_2 = \operatorname{Dom} A \tag{5.32}$$

be a part of the A-scale. Let us consider a self-adjoint operator $A = A^* \geq 1$ on $\mathcal{H} = \mathcal{H}_0$. An operator $\tilde{A} = \tilde{A}^* \neq A$ is said to be (purely) *singularly perturbed* with respect to A if the set

$$\mathfrak{D} := \{f \in \operatorname{Dom} A \cap \operatorname{Dom}\tilde{A} \mid Af = \tilde{A}f\} \tag{5.33}$$

is dense in $\mathcal{H}_0$, i.e., $\mathfrak{D} \sqsubset \mathcal{H}_0$. The family of all bounded from below operators $\tilde{A}$ singularly perturbed with respect to A we denoted by $\mathcal{P}_{\rm s}(A)$.

If $\tilde{A} \in \mathcal{P}_{\rm s}(A)$ and, in addition, $\operatorname{Dom} A^{1/2} = \operatorname{Dom}\tilde{A}^{1/2}$, then $\tilde{A}$ is said to be a *weakly singularly perturbed* operator, denoted $\tilde{A} \in \mathcal{P}_{\rm ws}(A)$.

Note that if $\operatorname{Dom} A^{1/2} = \operatorname{Dom}\tilde{A}^{1/2}$, then $\mathcal{H}_1(A) = \mathcal{H}_1(\tilde{A})$, where we recall that $\mathcal{H}_1(\tilde{A})$ denotes the completion of $\operatorname{Dom}\tilde{A}$ in the norm $\|\tilde{A}^{1/2}\cdot\|$.

If $\tilde{A} \in \mathcal{P}_{\rm s}(A)$ and the set $\mathfrak{D}$ defined by (5.33) is dense in $\mathcal{H}_1$, then $\tilde{A}$ is said to be a *strongly singularly perturbed* operator, denoted $\tilde{A} \in \mathcal{P}_{\rm ss}(A)$.

Thus, we have $\mathcal{P}_{\rm s}(A) \supset \mathcal{P}_{\rm ws}(A) \cup \mathcal{P}_{\rm ss}(A)$.

Two methods for construction of singular perturbed operators are well known in the literature. One of them is called the method of form sums and the second method is based on the theory of self-adjoint extensions of symmetric operators. The first one relates to the set $\mathcal{P}_{\mathrm{ws}}(A)$, while the second one deals with $\mathcal{P}_{\mathrm{ss}}(A)$. Both these methods are discussed in Chapter 7.

Chapter 6

Dense Subspaces in Scales of Hilbert Spaces

In this chapter we investigate the following question. Under what conditions a subset of a Hilbert space is continuously embedded into another Hilbert space? More precisely, let a couple of Hilbert spaces $\mathcal{H}$, $\mathcal{H}_+$ be such that $\mathcal{H}_+$ is a proper subset of $\mathcal{H}_0$, i.e., $\mathcal{H} \supset \mathcal{H}_+$. Moreover, suppose that $\mathcal{H}_+$ is densely and continuously embedded into $\mathcal{H}$, i.e., $\mathcal{H} \sqsupset \mathcal{H}_+$. Consider a decomposition of $\mathcal{H}_+$ into the orthogonal sum of two subspaces, $\mathcal{H}_+ = \mathcal{M}_+ \oplus \mathcal{N}_+$. Under what conditions at least one of these subspaces, for example $\mathcal{M}_+$, is dense in $\mathcal{H}$? That is, when one can write $\mathcal{H} \sqsupset \mathcal{M}_+$? This problem arises in various constructions. In particular, in the theory of self-adjoint extensions we often meet the question whether an operator produced by a formal symmetric expression on a linear set in a Hilbert space is densely defined. Similarly, in the theory of singular perturbations of self-adjoint operators, the kernel of the potential should be dense in a Hilbert space. Therefore, this fact has to be proved or to be guaranteed. Thus, finding conditions on a linear subset from the positive space under which it is densely embedded into the central Hilbert space is of considerable interest. Here we investigate this problem in the setting, in which the central and the positive spaces constitute an arbitrary pair of an A-scale of Hilbert spaces.

6.1 Densely embedding of subspace

Let $\mathcal{H}_- \sqsupset \mathcal{H}_0 \sqsupset \mathcal{H}_+$ be a rigged Hilbert space. Suppose that the positive space $\mathcal{H}_+$ is decomposed into an orthogonal sum $\mathcal{H}_+ = \mathcal{M}_+ \oplus \mathcal{N}_+$. First, we formulate a rather simple criterion for densely embedding $\mathcal{M}_+$ into the space $\mathcal{H}_0$ for the case when $\mathcal{N}_+$ is a one-dimensional subspace.

Theorem 6.1.1 ([9]). *Let $\mathcal{H}_+ = \mathcal{M}_+ \oplus \mathcal{N}_+$. Suppose that*

$$\dim \mathcal{N}_+ = 1. \tag{6.1}$$

Then the subspace $\mathcal{M}_+$ is dense in $\mathcal{H}_0$ if and only if the subspace $\mathcal{N}_- := D_{-,+}\mathcal{N}_+$ has a null intersection with $\mathcal{H}_0$:

$$\mathcal{H}_0 \sqsupset \mathcal{M}_+ \Longleftrightarrow \mathcal{N}_- \cap \mathcal{H}_0 = \{0\}, \tag{6.2}$$

where $D_{-,+} : \mathcal{H}_+ \to \mathcal{H}_-$ is the Berezansky canonical isomorphism.

Proof. Let $\mathcal{N}_- \cap \mathcal{H}_0 = \{0\}$. Since $\mathcal{N}_+$ is one-dimensional, we have

$$\mathcal{N}_+ = \{c\eta_+\}_{c\in\mathbb{C}}, \quad \eta_+ \in \mathcal{H}_+, \quad \|\eta_+\|_+ = 1. \tag{6.3}$$

Take a vector $\omega = D_{-,+}\eta_+ \in \mathcal{N}_-$. It is clear that $\omega \in \mathcal{H}_-\backslash\mathcal{H}_0$. If we suppose that the subspace $\mathcal{M}_+$ is not dense in $\mathcal{H}_0$, then a non-trivial vector $\psi \in \mathcal{H}_0$ exists such that $\psi \perp \mathcal{M}_+$, and hence

$$0 = (\psi, \mathcal{M}_+)_0 = \langle \psi, \mathcal{M}_+\rangle_{-,+} = (I_{+,-}\psi, \mathcal{M}_+)_+, \quad I_{+,-} = D^{-1}_{-,+}.$$

Since $\mathcal{N}_+$ is one-dimensional, $I_{+,-}\psi = c\eta_+$ for some $0 \neq c \in \mathbb{C}$, and then

$$\omega = D_{-,+}\eta_+ = c^{-1}\psi \in \mathcal{H}_0.$$

This contradicts the fact that $\omega \in \mathcal{H}_-\backslash\mathcal{H}_0$.

Conversely, suppose the subspace $\mathcal{M}_+$ is dense in $\mathcal{H}_0$. Then the assumption that the vector $\omega = D_{-,+}\eta_+$ belongs to $\mathcal{H}_0$ leads to a contradiction. Indeed, by this assumption and due to density $\mathcal{M}_+$ in $\mathcal{H}_0$, it follows that there exists a sequence $\varphi_n \in \mathcal{M}_+$ converging to ω in $\mathcal{H}_0$. In particular, $(\varphi_n, \omega)_0 \to \|\omega\|_0^2 \neq 0$. But since $\mathcal{N}_+ \perp \mathcal{M}_+$, this limit is zero:

$$(\varphi_n, \omega)_0 = \langle \varphi_n, \omega\rangle_{+,-} = (\varphi_n, \eta_+)_+ = 0.$$

This is a contradiction. □

Example 6.1.2. Let $\mathcal{H}_0 = L_2(\mathbb{R}^1, dx)$, and suppose the Sobolev space $W_2^1(\mathbb{R}^1)$ considered as the positive space with respect to $\mathcal{H}_0$, is decomposed into the orthogonal sum:

$$\mathcal{H}_+ = W_2^1 = \mathcal{M}_+ \oplus \mathcal{N}_+,$$

where $\mathcal{N}_+ = \{c\eta_+\}$ with $\eta_+(x) = \exp(-|x|)$. Then the subspace $\mathcal{M}_+$ is dense in $L_2(\mathbb{R}^1, dx)$, since

$$\delta_0(x) = \omega(x) = \left(1 - \frac{d^2}{dx^2}\right)\eta_+(x) \in W_2^{-1} \setminus L_2(\mathbb{R}^1, dx),$$

due to Theorem 6.1.1.

Example 6.1.3. Let $\mathcal{H}_0 = L_2(\mathbb{R}^3, dx)$, and $\omega = \delta_y \in W_2^{-2}(\mathbb{R}^3)$, $y \in \mathbb{R}^3$. Then the subspace

$$\mathcal{M}_+ = \{\varphi \in W_2^2 \mid \varphi(y) = 0\}$$

is dense in $L_2(\mathbb{R}^3, dx)$ since the subspace $\mathcal{N}_- = \{c\delta_y\}$ has a null intersection with $L_2(\mathbb{R}^3, dx)$.

Let us consider the general case when $\mathcal{N}_+$ has an arbitrary dimension

$$\dim \mathcal{N}_+ = n \leq \infty.$$

Theorem 6.1.4 (The main theorem of singular perturbation theory [9, 121, 123]). *Let $\mathcal{H}_+ = \mathcal{M}_+ \oplus \mathcal{N}_+$. The subspace $\mathcal{M}_+$ is dense in $\mathcal{H}_0$ if and only if the subspace $\mathcal{N}_- := D_{-,+}\mathcal{N}_+$ has a null intersection with $\mathcal{H}_0$:*

$$\mathcal{H}_0 \sqsupset \mathcal{M}_+ \Longleftrightarrow \mathcal{N}_- \cap \mathcal{H}_0 = \{0\}, \tag{6.4}$$

or, equivalently, the subspace $\mathcal{N}_0 := D_{0,+}\mathcal{N}_+$ has a null intersection with $\mathcal{H}_+$:

$$\mathcal{H}_0 \sqsupset \mathcal{M}_+ \Longleftrightarrow \mathcal{N}_0 \cap \mathcal{H}_+ = \{0\}. \tag{6.5}$$

Proof. Let us prove (6.4). Let $\mathcal{N}_- \cap \mathcal{H}_0 = \{0\}$ and suppose that there exists a vector $0 \neq \psi \in \mathcal{H}_0$ such that $\psi \perp \mathcal{M}_+$ in $\mathcal{H}_0$. Since $\mathcal{M}_+$ is a subspace in $\mathcal{H}_+$, then regarding on the vector ψ as an element of the space $\mathcal{H}_-$, we have

$$0 = (\psi, \mathcal{M}_+)_0 = \langle \psi, \mathcal{M}_+ \rangle_{-,+} = (I_{+,-}\psi, \mathcal{M}_+)_+.$$

This means that $I_{+,-}\psi \in \mathcal{N}_+$, and $\psi \in \mathcal{N}_-$. But this contradicts the condition $\mathcal{N}_- \cap \mathcal{H}_0 = \{0\}$.

Conversely, let the subspace $\mathcal{M}_+$ be dense in $\mathcal{H}_0$. Then the assumption that there exists a vector $0 \neq \omega \in \mathcal{N}_- \cap \mathcal{H}_0$ leads to a contradiction. Indeed, since $\mathcal{N}_- = D_{-,+}\mathcal{N}_+$,

$$\langle \omega, \mathcal{M}_+ \rangle_{-,+} = (\omega, \mathcal{M}_+)_0 = (I_{+,-}\omega, \mathcal{M}_+)_+ = 0,$$

because $I_{+,-}\omega \in \mathcal{N}_+$. This contradicts the fact $\mathcal{M}_+ \sqsubset \mathcal{H}_0$.

The equivalence of (6.4) and (6.5) follows from the first invariance principle in the A-scale (see Chapter 5), and from the fact that the operators $D_{0,+}$ and $I_{+,0}$ act from $\mathcal{H}_+$ to $\mathcal{H}_0$ and from $\mathcal{H}_0$ to $\mathcal{H}_+$, respectively. □

Example 6.1.5. Consider in $W_2^{-2}(\mathbb{R}^3)$ the subspace

$$\mathcal{N}_- = (\operatorname{span}\{\delta_y\}_{y \in E})^{\mathrm{cl},-}, \quad E \subset \mathbb{R}^3,$$

which satisfies the condition $\mathcal{N}_- \cap L_2 = \{0\}$. Then by Theorem 6.1.4, the subspace

$$\mathcal{M}_+ = \{\varphi \in W_2^2 \mid \varphi(y) = 0, \forall y \in E\}$$

is dense in L_2.

Let A be a self-adjoint operator on $\mathcal{H}_0$. One can use Theorem 6.1.4 to construct densely defined symmetric restrictions $\mathbf{A} := A \restriction \mathfrak{D}(\mathbf{A})$, with $\mathfrak{D}(\mathbf{A}) = \mathcal{M}_+$.

Proposition 6.1.6. *Let $\mathbf{A} := A \restriction \mathfrak{D}(\mathbf{A})$ be a restriction of a self-adjoint operator $A \geq 1$ in $\mathcal{H}_0$. The operator $\mathbf{A}$ is densely defined and symmetric if and only if* *the orthogonal complement to its range* $\operatorname{Ran}(\mathbf{A})$, *i.e., that the subspace* $\mathcal{N}_0 = \operatorname{Ran}(\mathbf{A})^\perp = \operatorname{Ker}\mathbf{A}^*$, *has a null intersection with* $\mathfrak{D}(A)$:

$$(\mathfrak{D}(\mathbf{A}))^{\text{cl}} = \mathcal{H}_0 \Longleftrightarrow \mathcal{N}_0 \cap \mathfrak{D}(A) = \{0\}. \tag{6.6}$$

Proof. The implications (6.6) are equivalent to (6.4) and also to (6.5). □

The next proposition follows directly from the previous results.

Proposition 6.1.7. *Let* $A \geq 1$ *be a self-adjoint operator on* $\mathcal{H}_0$ *and* $\mathfrak{D}(A) = \mathcal{H}_+$ *with the inner product* $(\cdot,\cdot)_+ = (A\cdot, A\cdot)_0$. *Then there exists a one-to-one correspondence between the set of all densely defined closed symmetric restrictions* $\mathbf{A} := A \restriction \mathfrak{D}(\mathbf{A})$ *of the operator* A *and the set of all subsets* $\mathcal{N}_0 \subset (\mathcal{H}_0 \setminus \mathcal{H}_+) \cup \{0\}$. *This correspondence is given by the expression:* $\operatorname{Ran}(\mathbf{A})^\perp = \mathcal{N}_0$, *where* $\operatorname{Ran}(\mathbf{A})$ *denotes the range of* $\mathbf{A}$.

Proof. For a fixed $A = A^* \geq 1$ on $\mathcal{H}_0$ with $\mathfrak{D}(A) = \mathcal{H}_+$, let $\mathbf{A}$ be a closed symmetric restriction of A such that $\mathfrak{D}(\mathbf{A}) \sqsubset \mathcal{H}_0$ and $\mathcal{H}_+ = \mathfrak{D}(\mathbf{A}) \oplus \mathcal{N}_+$, $\mathcal{N}_+ \neq 0$. If $\mathfrak{D}(\mathbf{A})$ is dense in $\mathcal{H}_0$, then the subspace $\operatorname{Ran}(\mathbf{A})^\perp = \mathcal{N}_0$ has a null intersection with $\mathfrak{D}(A) = \mathcal{H}_+$.

Conversely, let

$$\mathcal{N}_0 \subset (\mathcal{H}_0 \setminus \mathcal{H}_+) \cup \{0\}.$$

Then $\mathcal{H}_0 = \mathcal{M}_0 \oplus \mathcal{N}_0$ and by Theorem 6.1.4, the subspace $\mathcal{M}_+ =: A^{-1}\mathcal{M}_0$ is dense in $\mathcal{H}_0$. Therefore, the operator $\mathbf{A} =: A \restriction \mathcal{M}_+$, with $\mathfrak{D}(\mathbf{A}) = \mathcal{M}_+$ and $\operatorname{Ran}(\mathbf{A})^\perp = \mathcal{N}_0$ is dense in $\mathcal{H}_0$. □

6.2 Construction of dense subspaces

Starting with a fixed decomposition $\mathcal{H}_+ = \mathcal{M}_+ \oplus \mathcal{N}_+$ such that $\mathcal{H}_0 \sqsupset \mathcal{M}_+$, one can construct a number of new subspaces $\tilde{\mathcal{M}}_+$ in the A-scale, that are dense in $\mathcal{H}_0$. To this end one can use various methods as described below.

6.2.1 Preliminaries and notations

Let us briefly recall some relations and notations from the theory of rigged spaces and some properties of the A-scale (see Chapter 5). Let $A = A^* \geq 1$ be given on $\mathcal{H}$. Then

$$\begin{aligned} D_{0,2} &= A : \mathcal{H}_2 \longrightarrow \mathcal{H}_0; \\ I_{2,0} &= A^{-1} : \mathcal{H}_0 \longrightarrow \mathcal{H}_2; \\ D_{0,\alpha} &= A^{\alpha/2} : \mathcal{H}_\alpha \longrightarrow \mathcal{H}_0; \\ D_{-\alpha,0} &= (A^{\alpha/2})^{\text{cl}} : \mathcal{H}_0 \longrightarrow \mathcal{H}_{-\alpha}, \quad \alpha > 0; \\ (f,\varphi)_0 &= \langle f,\varphi\rangle_{-\alpha,\alpha}, \quad f \in \mathcal{H}_0, \varphi \in \mathcal{H}_\alpha,\ \alpha > 0, \end{aligned} \tag{6.7}$$

where $\langle\cdot,\cdot\rangle_{-\alpha,\alpha}$ denotes a duality pairing.

If we put $\mathcal{H}_{-\alpha} = \mathcal{H}_-$ and $\mathcal{H}_\alpha = \mathcal{H}_+$, $\alpha > 0$, then we have

$$\langle \omega, \varphi \rangle_{-,+} = (\omega, D_{-,+}\varphi)_- = (I_{+,-}\omega, \varphi)_+, \tag{6.8}$$

where $\langle \cdot, \cdot \rangle_{-,+} \equiv \langle \cdot, \cdot \rangle_{-\alpha,\alpha}$ and

$$D_{-,+} \equiv D_{-\alpha,\alpha}, \quad I_{+,-} \equiv I_{\alpha,-\alpha}, \quad \omega \in \mathcal{H}_-, \varphi \in \mathcal{H}_+.$$

6.2.2 The shift method

Let a subspace $\mathcal{M}_+$ of the positive space $\mathcal{H}_+$ be densely and continuously embedded into $\mathcal{H}_0$, i.e., $\mathcal{H}_0 \sqsupset \mathcal{M}_+$. Using the use the first invariance principle, one can easily construct a series of new subspaces $\tilde{\mathcal{M}}_+$ such that $\tilde{\mathcal{M}}_+ \sqsubset \mathcal{H}_0$.

To this end we consider a part of the A-scale which forms a rigged Hilbert space $\mathcal{H}_0 \sqsupset \mathcal{H}_+ \sqsupset \mathcal{H}_{++}$. For example, we can put $\mathcal{H}_+ = \mathcal{H}_\alpha$ with some $\alpha > 0$, and then $\mathcal{H}_{++} = \mathcal{H}_{2\alpha}$.

Theorem 6.2.1 ([1]). *Let $\mathcal{H}_+ = \mathcal{M}_+ \oplus \mathcal{N}_+$ and $\mathcal{H}_0 \sqsupset \mathcal{M}_+$. Then the subspace $\mathcal{M}_{++} := I_{++,+}\mathcal{M}_+$ is also dense in $\mathcal{H}_0$:*

$$\mathcal{H}_0 \sqsupset \mathcal{M}_+ \Longrightarrow \mathcal{H}_0 \sqsupset \mathcal{M}_{++}. \tag{6.9}$$

Moreover, $\mathcal{M}_{++}$ is dense in $\mathcal{H}_+$ and

$$\mathcal{H}_0 \sqsupset \mathcal{M}_+ \Longleftrightarrow \mathcal{H}_+ \sqsupset \mathcal{M}_{++}. \tag{6.10}$$

Proof. By Theorem 6.1.4 and the first invariance principle, from the inclusion $\mathcal{H}_0 \sqsupset \mathcal{M}_+$ it follows that $\mathcal{N}_0 \cap \mathcal{H}_+ = \{0\}$, where $\mathcal{N}_0 = D_{0,+}\mathcal{N}_+$. Indeed, we have also $\mathcal{N}_+ \cap \mathcal{H}_{++} = \{0\}$. Therefore, the subspace $\mathcal{M}_{++}$ is dense in $\mathcal{H}_+$, and also in $\mathcal{H}_0$, since $\mathcal{H}_+ \sqsubset \mathcal{H}_0$. So, we proved (6.9). Further, if $\mathcal{H}_+ \sqsupset \mathcal{M}_{++}$, then $\mathcal{N}_+ \cap \mathcal{H}_{++} = \{0\}$. Applying once more the first invariance principle, we obtain $\mathcal{N}_0 \cap \mathcal{H}_+ = \{0\}$. Hence $\mathcal{H}_0 \sqsupset \mathcal{M}_+$. Thus, (6.10) is also proved. □

The above results can be formulated in slightly different terms. We formulate them as as inddependent theorem.

Theorem 6.2.2. *Consider a part of the A-scale:*

$$\mathcal{H}_{--} \sqsupset \mathcal{H}_- \sqsupset \mathcal{H}_0 \sqsupset \mathcal{H}_+ \sqsupset \mathcal{H}_{++}.$$

For example, take

$$\mathcal{H}_\pm = \mathcal{H}_{\pm\alpha}, \quad \mathcal{H}_{++} = \mathcal{H}_{2\alpha}, \quad \mathcal{H}_{--} = \mathcal{H}_{-2\alpha}, \quad \alpha > 0.$$

Suppose additionally that $\mathcal{H}_+$ is decomposed into two subspaces: $\mathcal{H}_+ = \mathcal{M}_+ \oplus \mathcal{N}_+$. Then the assumption $\mathcal{H}_0 \sqsupset \mathcal{M}_+$ is equivalent to each of the following relations:

$$\begin{aligned}
&\text{(i)} && \mathcal{N}_0 \cap \mathcal{H}_+ = \{0\}, && \mathcal{N}_0 = D_{0,+}\mathcal{N}_+\\
&\text{(ii)} && \mathcal{N}_+ \cap \mathcal{H}_{++} = \{0\}, && \\
&\text{(iii)} && \mathcal{N}_- \cap \mathcal{H}_0 = \{0\}, && \mathcal{N}_- = D_{-,+}\mathcal{N}_+\\
&\text{(iv)} && \mathcal{H}_- \sqsupset \mathcal{M}_0, && \mathcal{M}_0 = D_{0,+}\mathcal{M}_+\\
&\text{(v)} && \mathcal{H}_+ \sqsupset \mathcal{M}_{++}, && \mathcal{M}_{++} = I_{++,+}\mathcal{M}_+.
\end{aligned}$$

Proof. The proof is carried out by means of the first invariance principle. Let $\mathcal{H}_0 \sqsupset \mathcal{M}_+$. Then by Theorem 6.1.1, $\mathcal{N}_0 \cap \mathcal{H}_+ = \{0\}$.

Now let (i) hold. We show that (ii) is valid. We suppose the contrary, i.e., there is a $\varphi \neq 0$ and $\varphi \in \mathcal{N}_+ \cap \mathcal{H}_{++}$. Then

$$D_{0,+}\varphi =: f \neq 0 \in \mathcal{N}_0 \cap \mathcal{H}_+,$$

which contradicts (i).

Suppose (ii) holds. We prove (iii). We suppose the contrary, namely, there is a $g \neq 0 \in \mathcal{N}_- \cap \mathcal{H}_0$. Then we again reach a contradiction:

$$D^{-1}_{-,+} = I_{+,-}g =: \phi \neq 0 \in \mathcal{N}_+ \cap \mathcal{H}_{++}.$$

Hence (iii) is valid.

Finally, from (iii) we obtain (iv) and (v). According to the first invariance principle and Theorem 6.1.1,

$$\mathcal{N}_0 \cap \mathcal{H}_+ = \{0\} \Longleftrightarrow \mathcal{H}_0 \sqsupset \mathcal{M}_+ \Longleftrightarrow \mathcal{H}_- \sqsupset \mathcal{M}_0,$$

and the implications

$$\mathcal{H}_- \sqsupset \mathcal{M}_0 \Longleftrightarrow \mathcal{H}_+ \sqsupset \mathcal{M}_{++} \Longleftrightarrow \mathcal{H}_0 \sqsupset \mathcal{M}_+.$$

hold. □

In the previous considerations we can replace $\mathcal{H}_{++}$ (one can think that $\mathcal{H}_{++} = \mathcal{H}_{2\alpha}$, $\alpha > 0$) by an arbitrary space $\tilde{\mathcal{H}}_+$ from the A-scale. We assume that this space is situated to the right of $\mathcal{H}_+$, i.e., $\tilde{\mathcal{H}}_+ \sqsubset \mathcal{H}_+$. For example, we can put $\tilde{\mathcal{H}}_+ = \mathcal{H}_{\alpha+\delta}$, $\delta > 0$.

Theorem 6.2.3. *If a closed subspace $\mathcal{M}_+$ of $\mathcal{H}_+$ is dense in $\mathcal{H}_0$, and a space $\tilde{\mathcal{H}}_+$ from the A-scale is dense in $\mathcal{H}_+$, then the subspace $\tilde{\mathcal{M}}_+ := \tilde{I}_{+,+}\mathcal{M}_+$ is dense in $\mathcal{H}_0$:*

$$\mathcal{H}_0 \sqsupset \mathcal{M}_+, \mathcal{H}_+ \sqsupset \tilde{\mathcal{H}}_+ \Longrightarrow \mathcal{H}_0 \sqsupset \tilde{\mathcal{M}}_+,$$

where $\tilde{I}_{+,+} : \mathcal{H}_+ \mapsto \tilde{\mathcal{H}}_+$ is the Berezansky canonical isomorphism.

Proof. Let us continue the pre-rigged couple $\mathcal{H}_0 \sqsupset \tilde{\mathcal{H}}_+$ to the whole rigged space $\tilde{\mathcal{H}}_- \sqsupset \mathcal{H}_0 \sqsupset \tilde{\mathcal{H}}_+$. By Theorem 6.1.4, the conditions $\mathcal{H}_0 \sqsupset \mathcal{M}_+$ and $\mathcal{N}_0 \cap \mathcal{H}_+ = \{0\}$ are equivalent. Therefore, $\mathcal{N}_0 \cap \tilde{\mathcal{H}}_+ = \{0\}$, since $\tilde{\mathcal{H}}_+ \sqsubset \mathcal{H}_+$. Now, again, by Theorem 6.1.4, the subspace

$$\tilde{\mathcal{M}}_+ = (\tilde{I}_{+,0}\mathcal{N}_0)^\perp \equiv \tilde{I}_{+,+}\mathcal{M}_+$$

is dense in $\mathcal{H}_0$. □

In other words, for every triplet $\mathcal{H}_0 \sqsupset \mathcal{H}_+ \sqsupset \tilde{\mathcal{H}}_+$ of the A-scale (these spaces do not necessarily form a rigged space!), the following implication holds true:

$$\mathcal{H}_+ = \mathcal{M}_+ \oplus \mathcal{N}_+, \quad \mathcal{H}_0 \sqsupset \mathcal{M}_+ \quad \Longrightarrow \quad \mathcal{H}_0 \sqsupset \tilde{\mathcal{M}}_+, \tag{6.11}$$

where $\tilde{\mathcal{M}}_+ = \tilde{I}_{+,+}\mathcal{M}_+$. The tilde operation can be treated as a finite shift to the right in the A-scale that is provided by the Berezansky canonical isomorphism $\tilde{I}_{+,+}$.

The above result is represented schematically in Figure 1.

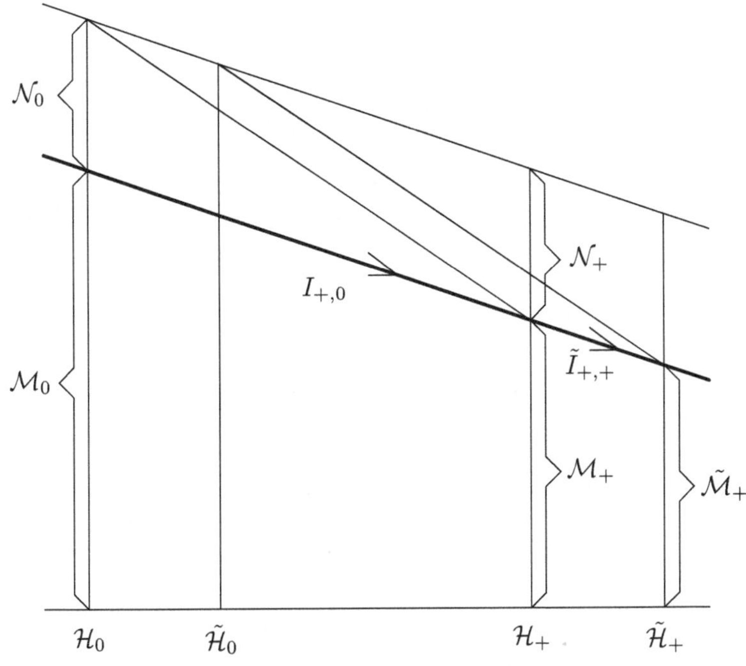

Figure 1

We note that, by moving to the right along the A-scale, the subspace $\tilde{\mathcal{M}}_+$, in general, becomes more restricted. Nevertheless, its property to be dense in the central space $\mathcal{H}_0$, is presented.

Corollary 6.2.4. *Let $\mathcal{H}_+ = \mathcal{H}_2 \equiv \mathcal{H}_2(A) = \mathcal{M}_2 \oplus \mathcal{N}_2$. If $\mathcal{H}_0 \sqsupset \mathcal{M}_2$, then $\mathcal{H}_0 \sqsupset \mathcal{M}_4$, $\mathcal{H}_2 \sqsupset \mathcal{M}_4$, where $\mathcal{M}_4 := I_{4,2}\mathcal{M}_2$.*

Corollary 6.2.5. *Let $\mathcal{H}_2 = \mathcal{M}_2 \oplus \mathcal{N}_2$ and $\mathcal{H}_0 \sqsupset \mathcal{M}_2$. Then $\mathcal{H}_0 \sqsupset \mathcal{M}_{2+\delta}$, where $\mathcal{M}_{2+\delta} := I_{2+\delta,2}\mathcal{M}_2$, $\delta > 0$.*

Corollary 6.2.6. *Let $\mathcal{H}_\alpha = \mathcal{M}_\alpha \oplus \mathcal{N}_\alpha$ for some fixed $\alpha > 0$. If $\mathcal{H}_0 \sqsupset \mathcal{M}_\alpha$, then $\mathcal{H}_0 \sqsupset \mathcal{M}_\beta$ for an arbitrary $\beta > \alpha$, where $\mathcal{M}_\beta := I_{\beta,\alpha}\mathcal{M}_\alpha$.*

Thus, the main result of this subsection, which based on the first invariance principle, can be formulated as a short theorem.

Theorem 6.2.7. *Let* $\mathcal{H}_\alpha = \mathcal{M}_\alpha \oplus \mathcal{N}_\alpha$, $\alpha > 0$. *If* $\mathcal{H}_0 \sqsupset \mathcal{M}_\alpha$, *then* $\mathcal{H}_0 \sqsupset \mathcal{M}_{\alpha+\delta}$, *where* $\mathcal{M}_{\alpha+\delta} := I_{\alpha+\delta,\alpha}\mathcal{M}_\alpha$, $0 < \delta < \infty$.

6.2.3 The intersection method

In this subsection we will construct new pairs $\mathcal{H}_0 \sqsupset \tilde{\mathcal{M}}_+$ of the A-scale, starting with a given pair $\mathcal{H}_0 \sqsupset \mathcal{M}_+$. We use the method of intersection of subspaces as a tool.

Let us consider the double rigged Hilbert space as part of the A-scale:

$$\mathcal{H}_{--} \sqsupset \mathcal{H}_- \sqsupset \mathcal{H}_0 \sqsupset \mathcal{H}_+ \sqsupset \mathcal{H}_{++}. \tag{6.12}$$

Let us suppose, as before, that the space $\mathcal{H}_+$ is decomposed into an orthogonal sum of two subspaces, so that one of them, for example $\mathcal{M}_+$, is dense in $\mathcal{H}_0$:

$$\mathcal{H}_+ = \mathcal{M}_+ \oplus \mathcal{N}_+, \quad \mathcal{H}_0 \sqsupset \mathcal{M}_+. \tag{6.13}$$

We ask under what conditions the set

$$\tilde{\mathcal{M}}_+ := \mathcal{M}_+ \cap \mathcal{H}_{++}$$

is dense in $\mathcal{H}_0$?

It is easy to see that $\tilde{\mathcal{M}}_+$ is closed in $\mathcal{H}_{++}$, i.e., it is a subspace. Indeed, if a sequence φ_n from $\tilde{\mathcal{M}}_+$ is converges in $\mathcal{H}_{++}$, $\varphi_n \to \varphi$, then it also converges in $\mathcal{H}_+$, since $\|\cdot\|_+ \lesssim \|\cdot\|_{++}$. Hence $\varphi \in \mathcal{M}_+$, since the subspace $\mathcal{M}_+$ is closed $\mathcal{H}_+$. Therefore, $\varphi \in \tilde{\mathcal{M}}_+$.

Thus, we can write

$$\mathcal{H}_{++} = \tilde{\mathcal{M}}_+ \oplus \tilde{\mathcal{N}}_+, \quad \tilde{\mathcal{N}}_+ := \tilde{\mathcal{M}}_+^\perp.$$

Theorem 6.2.8. *Let* $\mathcal{H}_0 \sqsupset \mathcal{M}_+$. *Suppose that the subspace* $\mathcal{N}_- = D_{-,+}\mathcal{N}_+$ *satisfies:*

$$\mathcal{N}_-^{\mathrm{cl},--} \cap \mathcal{H}_0 = \{0\}, \tag{6.14}$$

*where "*cl, $--$*" denotes the closure in* $\mathcal{H}_{--}$. *Then the subspace* $\tilde{\mathcal{M}}_+ = \mathcal{M}_+ \cap \mathcal{H}_{++}$ *is dense in* $\mathcal{H}_0$*:*

$$\mathcal{H}_0 \sqsupset \tilde{\mathcal{M}}_+. \tag{6.15}$$

Proof. It is easy to see that the subspace $\tilde{\mathcal{M}}_+$ can be defined as

$$\tilde{\mathcal{M}}_+ = \{\varphi \in \mathcal{H}_{++} \mid (\varphi, \psi)_+ = 0, \ \psi \in \mathcal{N}_+\}.$$

Let us consider the subspace $\tilde{\mathcal{N}}_- := D_{--,++}\tilde{\mathcal{N}}_+$ in $\mathcal{H}_{--}$ as the image of $\tilde{\mathcal{N}}_+$ under a unitary map, where $\tilde{\mathcal{N}}_+ = \tilde{\mathcal{M}}_+^{\perp}$. For $\omega = D_{-,+}\psi$, $\psi \in \mathcal{N}_+$, due to the properties of the A-scale we have

$$0 = (\varphi, \psi)_+ = \langle \varphi, \omega \rangle_{+,-} = \langle \varphi, \omega \rangle_{++,--}, \quad \varphi \in \tilde{\mathcal{M}}_+.$$

Hence, $\mathcal{N}_- \subset \tilde{\mathcal{N}}_-$ and, moreover, the subspace

$$\tilde{\mathcal{N}}_- = \{\omega \in \mathcal{H}_{--} \mid \langle \varphi, \omega \rangle_{++,--} = 0, \quad \varphi \in \tilde{\mathcal{M}}_+\}$$

coincides with the closure of $\mathcal{N}_- = D_{-,+}\mathcal{N}_+$ in $\mathcal{H}_{--}$. This can be proved by the same argument as in Lemma 6.2.10 below, or by using the Berezansky canonical isomorphism with respect to (6.17), taking into account that

$$D_{--,0}\tilde{\mathcal{N}}_0 = \tilde{\mathcal{N}}_-, \quad D_{--,0}\mathcal{N}_+{}^{\mathrm{cl},0} = \mathcal{N}_-{}^{\mathrm{cl},--}.$$

Therefore,

$$\tilde{\mathcal{N}}_- = \mathcal{N}_-^{\mathrm{cl},--}. \tag{6.16}$$

Now (6.16) and the condition (6.14), imply that $\tilde{\mathcal{N}}_- \cap \mathcal{H}_0 = \{0\}$. Hence, by Theorem 6.1.4 we conclude: $\mathcal{H}_0 \sqsupset \tilde{\mathcal{M}}_+$. □

Remark 6.2.9. The condition (6.14) is automatically satisfied if $\dim \mathcal{N}_- < \infty$ and $\mathcal{N}_- \cap \mathcal{H}_0 = \{0\}$.

Lemma 6.2.10. *Let $\mathcal{H}_+ = \mathcal{M}_+ \oplus \mathcal{N}_+$, and $\mathcal{M}_+ \sqsubset \mathcal{H}_0$. Then*

$$\mathcal{N}_+^{\mathrm{cl},0} = \tilde{\mathcal{N}}_0. \tag{6.17}$$

Hence, $\mathcal{N}_+^{\mathrm{cl},0}$ is the closure of $\mathcal{N}_+$ in $\mathcal{H}_0$, and $\tilde{\mathcal{N}}_0 = D_{0,++}\tilde{\mathcal{N}}_+$, where $\tilde{\mathcal{N}}_+ = \tilde{\mathcal{M}}_+^{\perp}$, $\tilde{\mathcal{M}}_+ = \mathcal{M}_+ \cap \mathcal{H}_{++}$.

Proof. From the definition of $\tilde{\mathcal{M}}_+$ we easily see that

$$\tilde{\mathcal{M}}_+ = \{\varphi \in \mathcal{H}_{++} \mid (\mathcal{N}_+, \varphi)_+ = \langle \mathcal{N}_+, \varphi \rangle_{0,++} = 0\}. \tag{6.18}$$

Since the triplet $\mathcal{H}_0 \sqsupset \mathcal{H}_+ \sqsupset \mathcal{H}_{++}$ forms a rigged space, we get the equality $\langle \tilde{\mathcal{N}}_0, \tilde{\mathcal{M}}_+ \rangle_{0,++} = 0$, which can be considered as the definition of the subspace $\tilde{\mathcal{N}}_0$. Further, since

$$(\mathcal{N}_+, \tilde{\mathcal{M}}_+)_+ = \langle \mathcal{N}_+, \tilde{\mathcal{M}}_+ \rangle_{0,++},$$

we conclude that $\mathcal{N}_+ \subset \tilde{\mathcal{N}}_0$, and also $\mathcal{N}_+^{\mathrm{cl},0} \subset \tilde{\mathcal{N}}_0$.

Let us suppose that the subspace $\tilde{\mathcal{N}}_0$ is larger than $\mathcal{N}_+^{\mathrm{cl},0}$. Then we can write $\tilde{\mathcal{N}}_0 = \mathcal{N}_+^{\mathrm{cl},0} \oplus \mathcal{S}_0$. We show now that $\mathcal{S}_0 = 0$. From the geometry of the A-scale it follows that:

$$0 = \langle \mathcal{N}_+^{\mathrm{cl},0} \oplus \mathcal{S}_0, \tilde{\mathcal{M}}_+ \rangle_{0,++} = (I_{++,0}[\mathcal{N}_+^{\mathrm{cl},0} \oplus \mathcal{S}_0], \tilde{\mathcal{M}}_+)_{++},$$

where $I_{++,0} = D_{0,++}^{-1} : \mathcal{H}_0 \to \mathcal{H}_{++}$. In particular,

$$0 = (I_{++,0}\mathcal{N}_+^{\mathrm{cl},0}, I_{++,0}\mathcal{S}_0 \oplus \tilde{\mathcal{M}}_+)_{++} = \langle \mathcal{N}_+^{\mathrm{cl},0}, I_{++,0}\mathcal{S}_0 \oplus \tilde{\mathcal{M}}_+ \rangle_{0,++}.$$

Since $\mathcal{N}_+ \sqsubset \mathcal{N}_+^{\mathrm{cl},0}$, we have $0 = (\mathcal{N}_+, I_{++,0}\mathcal{S}_0 \oplus \tilde{\mathcal{M}}_+)_+$, and also $I_{++,0}\mathcal{S}_0 \in \mathcal{H}_{++}$. This contradicts the fact that all vectors of $\mathcal{H}_{++}$ which are orthogonal to $\mathcal{N}_+$ necessarily belong to $\tilde{\mathcal{M}}_+$ (see (6.18)). Therefore, $I_{++,0}\mathcal{S}_0 = \{0\}$ and $\mathcal{S}_0 = \{0\}$. This also means that $\mathcal{N}_+^{\mathrm{cl},0} = \tilde{\mathcal{N}}_0$. □

Let $\tilde{\mathcal{N}}_- := D_{--,++}\tilde{\mathcal{N}}_+$ be the subspace adjoint to $\tilde{\mathcal{N}}_+$ in $\mathcal{H}_{--}$. According to the lemma proved above and the invariance of the scale (see also (6.17)) we have:

$$\tilde{\mathcal{N}}_- = \mathcal{N}_-^{\mathrm{cl},--}, \quad \mathcal{N}_- := D_{-,+}\mathcal{N}_+, \tag{6.19}$$

where "cl, −−" denotes the closure in $\mathcal{H}_{--}$. Now the condition (6.14) can be rewritten as

$$\tilde{\mathcal{N}}_- \cap \mathcal{H}_0 = \{0\}, \tag{6.20}$$

and Theorem 6.2.8 can be recast as

Theorem 6.2.11. *Let the subspace $\mathcal{M}_+$ of $\mathcal{H}_+$ be dense in $\mathcal{H}_0$, $\mathcal{H}_0 \sqsupset \mathcal{M}_+$. Assume in addition that the condition (6.20) from Lemma 6.2.10 is fulfilled. Then the subspace $\tilde{\mathcal{M}}_+ := \mathcal{M}_+ \cap \mathcal{H}_{++}$ is also dense in $\mathcal{H}_0$:*

$$\mathcal{H}_0 \sqsupset \tilde{\mathcal{M}}_+. \tag{6.21}$$

In particular, $\tilde{\mathcal{M}}_+$ is dense in $\mathcal{H}_0$ if $\mathcal{N}_+$ is a finite-dimensional subspace: $\dim \mathcal{N}_+ < \infty$.

Proof. From Lemma 6.2.10 it follows that

$$(\mathcal{N}_-)^{\mathrm{cl},--} = \tilde{\mathcal{N}}_- := \{\omega \in \mathcal{H}_{--} \mid \langle \varphi, \omega \rangle_{++,--} = 0,\ \varphi \in \tilde{\mathcal{M}}_+\}. \tag{6.22}$$

Further, since $\mathcal{M}_+$ is dense in $\mathcal{H}_0$, (6.2) and (6.20) imply that $\tilde{\mathcal{N}}_- \cap \mathcal{H}_0 = \{0\}$. Hence $\mathcal{H}_0 \sqsupset \tilde{\mathcal{M}}_+$ thanks to Theorem 6.1.4. Now, we note that condition (6.20) follows automatically from $\mathcal{H}_0 \sqsupset \mathcal{M}_+$ if

$$\dim \mathcal{N}_0 = \dim \mathcal{N}_+ < \infty.$$ □

6.2.4 Other versions of denseness conditions

It is clear that Theorem 6.2.11 remains true if the condition (6.20) is replaced by the following stronger one:

$$(\mathcal{N}_-)^{\mathrm{cl},--} \cap \mathcal{H}_- = \tilde{\mathcal{N}}_- \cap \mathcal{H}_- = \mathcal{N}_-.$$

Let us consider instead of (6.12) the larger chain

$$\mathcal{H}'_{--} \sqsupset \mathcal{H}_{--} \sqsupset \mathcal{H}_- \sqsupset \mathcal{H}_0 \sqsupset \mathcal{H}_+ \sqsupset \mathcal{H}_{++} \sqsupset \mathcal{H}'_{++}, \tag{6.23}$$

where $\mathcal{H}'_{++}$ is an arbitrary space of the A-scale, which lies to the right of $\mathcal{H}_{++}$ that is $\mathcal{H}'_{++}$ is a more "positive" space than $\mathcal{H}_{++}$.

Let us suppose that the following condition is fulfilled:

$$\tilde{\mathcal{N}}'_{-} \cap \mathcal{H}_0 = \{0\}, \tag{6.24}$$

where $\tilde{\mathcal{N}}'_{-} := (\mathcal{N}_{-})^{\mathrm{cl},'--}$ and "cl,$'--$" denotes the closure in $\mathcal{H}'_{--}$.

Theorem 6.2.12. *Under the condition* (6.24), *the closed in* $\mathcal{H}'_{++}$ *subspace*

$$\tilde{\mathcal{M}}'_{+} := \mathcal{M}_{+} \cap \mathcal{H}'_{++} \tag{6.25}$$

is dense in $\mathcal{H}_0$, *i.e.*, $\mathcal{H}_0 \sqsupset \tilde{\mathcal{M}}'_{+}$.

Proof. Essentially, it is the same as in the previous theorem. □

Let us consider now an arbitrary space $\mathcal{H}'_{+}$ instead of $\mathcal{H}'_{++}$ from the chain (6.23). We assume $\mathcal{H}'_{++}$ stands between $\mathcal{H}_{+}$ and $\mathcal{H}_{++}$.

Theorem 6.2.13. *Fix a chain of spaces*

$$\mathcal{H}_{--} \sqsupset \mathcal{H}'_{-} \sqsupset \mathcal{H}_{-} \sqsupset \mathcal{H}_0 \sqsupset \mathcal{H}_{+} \sqsupset \mathcal{H}'_{+} \sqsupset \mathcal{H}_{++},$$

and let $\tilde{\mathcal{N}}'_{-}$ *be the closure of* $\tilde{\mathcal{N}}_{-}$ *in* $\mathcal{H}'_{-}$. *If* $\tilde{\mathcal{N}}'_{-} \cap \mathcal{H}_0 = \{0\}$, *then the subspace* $\tilde{\mathcal{M}}'_{+} := \mathcal{M}_{+} \cap \mathcal{H}'_{+}$ *is closed in* $\mathcal{H}'_{+}$ *and dense in* $\mathcal{H}_0$, *i.e.*, $\mathcal{H}_0 \sqsupset \tilde{\mathcal{M}}'_{+}$.

Proof. One uses the same arguments as in the proofs of Theorems 6.2.8 and 6.2.11. □

Remark 6.2.14. It is easily seen that the condition $\tilde{\mathcal{N}}_{-} \cap \mathcal{H}_{-} = \mathcal{N}_{-}$ implies a deeper result: $\tilde{\mathcal{M}}_{+}$ is dense not only in $\mathcal{H}_0$, but also in the subspace $\mathcal{N}_{+}$ too. Noted that this condition is stronger than (6.20) of Theorem 6.2.8). The same holds true if the condition (6.24) is replaced by $\tilde{\mathcal{N}}'_{-} \cap \mathcal{H}_{-} = \mathcal{N}_{-}$.

Corollary 6.2.15. *Let an arbitrary triplet of spaces* $\mathcal{H}_0$, $\mathcal{H}_{+}$, $\mathcal{H}'_{+}$ *in the A-scale be fixed. Assume they form the rigged space* $\mathcal{H}_0 \sqsupset \mathcal{H}_{+} \sqsupset \mathcal{H}'_{+}$. *Assume also that for a decomposition* $\mathcal{H}_{+} = \mathcal{M}_{+} \oplus \mathcal{N}_{+}$ *the condition* (6.24) *is satisfied, where* $\tilde{\mathcal{N}}'_{-} := (\mathcal{N}_{-})^{\mathrm{cl},'--}$ *"cl,$'--$" denotes the closure in* $\mathcal{H}'_{-}$. *Then the implication*

$$\mathcal{H}_0 \sqsupset \mathcal{M}_{+} \Longrightarrow \mathcal{H}_0 \sqsupset \tilde{\mathcal{M}}'_{+}, \tag{6.26}$$

is valid, where $\tilde{\mathcal{M}}'_{+} = \mathcal{M}_{+} \cap \mathcal{H}'_{+}$.

Thus, if the subspace $\mathcal{M}_{+}$ of $\mathcal{H}_{+}$ is dense in $\mathcal{H}_0$, then for each other space $\mathcal{H}'_{+}$ of the A-scale such that $\mathcal{H}'_{+} \sqsubset \mathcal{H}_{+}$, the intersection $\mathcal{M}_{+} \cap \mathcal{H}'_{+}$ is densely embedded into $\mathcal{H}_0$ under the condition (6.24).

We can formulate this result in another form by using the first invariance principle.

Corollary 6.2.16. *Let $\mathcal{H}_\alpha$, $\mathcal{H}_\beta$, $\mathcal{H}_\gamma$, $\alpha < \beta < \gamma$ be an arbitrary triplet from the A-scale. Assume that the subspace $\mathcal{H}_\beta$ is decomposed into an orthogonal sum $\mathcal{H}_\beta = \mathcal{M}_\beta \oplus \mathcal{N}_\beta$ such that $\mathcal{H}_\alpha \sqsupset \mathcal{M}_\beta$. Assume also that*

$$\tilde{\mathcal{N}}_{-\beta}^{\mathrm{cl},-\alpha} \cap \mathcal{H}_{-\alpha} = \{0\}, \tag{6.27}$$

or that the stronger condition $\tilde{\mathcal{N}}_{-\beta}^{\mathrm{cl},-\alpha} \cap \mathcal{H}_{-\beta} = \mathcal{N}_{-\beta}$ is fulfilled, where $\tilde{\mathcal{N}}_{-\beta}^{\mathrm{cl},-\alpha}$ is the closure of $\mathcal{N}_{-\beta} := D_{-\beta,\beta}\mathcal{N}_+$ in $\mathcal{H}_{-\alpha}$. Then the subspace $\tilde{\mathcal{M}}_\beta := \mathcal{M}_\beta \cap \mathcal{H}_\gamma$ is dense in $\mathcal{H}_\alpha$, or respectively in $\mathcal{M}_\beta$, if the stronger condition is fulfilled.

Indeed, to show this we put $\mathcal{H}_0 = \mathcal{H}_\alpha$, $\mathcal{H}_+ = \mathcal{H}_\beta$, $\mathcal{H}'_+ = \mathcal{H}_\gamma$ and use the implication (6.26) or Remark 6.2.14.

In fact, we have a more perfect result. We formulate it for the case $\tilde{\mathcal{M}}_+ = \mathcal{M}_+ \cap \mathcal{H}_{++}$.

Theorem 6.2.17. *If the subspace $\mathcal{M}_+$ is dense in $\mathcal{H}_0$, then the set $\tilde{\mathcal{M}}_+$ is dense in $\mathcal{H}_0$ if and only if the closure of $\mathcal{N}_-$ in $\mathcal{H}_{--}$ has a null intersection with $\mathcal{H}_0$, i.e.,*

$$\mathcal{H}_0 \sqsupset \tilde{\mathcal{M}}_+ \Longleftrightarrow \mathcal{N}_-^{\mathrm{cl},--} \cap \mathcal{H}_0 = \{0\}. \tag{6.28}$$

Moreover, $\tilde{\mathcal{M}}_+$ is dense in the subspace $\mathcal{M}_+$ if and only if the intersection of $\mathcal{N}_-^{\mathrm{cl},--}$ with $\mathcal{H}_-$ coincides with $\mathcal{N}_-$, i.e.,

$$\mathcal{M}_+ \sqsupset \tilde{\mathcal{M}}_+ \Longleftrightarrow \mathcal{N}_-^{\mathrm{cl},--} \cap \mathcal{H}_- = \mathcal{N}_-. \tag{6.29}$$

Proof. The relation (6.28) is already proven. It is valid due to the main result of this subsection (see Theorem 6.2.8).

The implication (6.29) to the left side follows from the equality (6.17). Indeed, suppose that $\mathcal{M}_+ = \tilde{\mathcal{M}}_+ \oplus S_+$. Then

$$\begin{aligned} 0 = (\mathcal{N}_+, \mathcal{M}_+)_+ &= (\mathcal{N}_+, \tilde{\mathcal{M}}_+)_+ \oplus S_+ = \langle \mathcal{N}_-, \tilde{\mathcal{M}}_+ \oplus S_+\rangle_{-,+} \\ &= \langle \mathcal{N}_- \oplus D_{-,+}S_+, \tilde{\mathcal{M}}_+\rangle_{-,+} = \langle \mathcal{N}_- \oplus D_{-,+}S_+, \tilde{\mathcal{M}}_+\rangle_{--,++}. \end{aligned}$$

Hence we see that if $S_+ \neq \{0\}$, then the subspace $\tilde{\mathcal{N}}_-$ is different from $\mathcal{N}_-{}^{\mathrm{cl},--}$. But this contradicts (6.17).

Let us show the validity of the inverse implication that the condition $\mathcal{N}_-^{\mathrm{cl},--} \cap \mathcal{H}_- = \mathcal{N}_-$ is necessary for $\mathcal{M}_+ \sqsupset \tilde{\mathcal{M}}_+$. Suppose for a moment that $\mathcal{N}_-^{\mathrm{cl},--} \cap \mathcal{H}_- = \mathcal{N}_- \oplus \mathcal{S}_-$, where the subspace $\mathcal{S}_- \neq 0$. Then since $\tilde{\mathcal{N}}_- = \mathcal{N}_-^{\mathrm{cl},--}$, we have

$$0 = \langle \mathcal{N}_-^{\mathrm{cl},--}, \tilde{\mathcal{M}}_+\rangle_{--,++}.$$

It follows that

$$0 = \langle \mathcal{N}_-^{\mathrm{cl},--} \cap \mathcal{H}_-, \tilde{\mathcal{M}}_+\rangle_{-,+} = \langle \mathcal{N}_- \oplus \mathcal{S}_-, \tilde{\mathcal{M}}_+\rangle_{-,+}.$$

Therefore,

$$0 = \langle \mathcal{N}_-, I_{+,-}\mathcal{S}_- \oplus \tilde{\mathcal{M}}_+\rangle_{-,+} = (\mathcal{N}_+, I_{+,-}\mathcal{S}_- \oplus \tilde{\mathcal{M}}_+)_+.$$

This means that the subspace $\tilde{\mathcal{M}}_+$ is not dense in $\mathcal{N}_+^{\perp} = \mathcal{M}_+$. □

This result, like the previous ones, can be applied in perturbation theory for constructions of symmetric operators by the method of restrictions of a self-adjoint operator, or in the theory of self-adjoint extensions of symmetric operators.

In particular, the reformulation of the results above obtained in the case $\mathcal{H}_+ = \mathfrak{D}(A)$ leads to a construction of densely defined symmetric operators $\mathbf{A} := A \restriction \mathfrak{D}(\mathbf{A})$, with $\mathfrak{D}(\mathbf{A}) \equiv \mathcal{M}_+ \sqsubset \mathcal{H}_0$.

6.3 Dense subspaces in scales of the Sobolev spaces

The abstract results of Subsection 6.2 have applications in the constructions of dense subspaces in functional spaces, especially in Sobolev spaces. A convenient tool for this is provided by the notion of capacity for a set in the scale of Sobolev spaces.

Definition 6.3.1. The positive value

$$C_\alpha(K) \equiv \operatorname{cap}_\alpha(K) = \inf\{\|\varphi\|^2_{W_2^\alpha} \mid \varphi \in C_0^\infty, \varphi \geq 1 \text{ on } K\} \tag{6.30}$$

is called the α-capacity of a compact set $K \subset \mathbb{R}^n$.

If $\Omega \subset \mathbb{R}^n$ is an open set, then one sets

$$\operatorname{cap}_\alpha(\Omega) = \sup\{\operatorname{cap}_\alpha(K) \mid K \subset \Omega\}. \tag{6.31}$$

We remark that in (6.30) the set C_0^∞ can be replaced by the space of Schwartz test functions $\mathcal{S}(\mathbb{R}^n)$. The condition $\varphi \geq 1$ can be replaced by $\varphi(x) = 1$, $x \in K$.

Denote

$$\begin{aligned} \mathcal{M}_{-k}(\Omega) &:= \{\omega \in W_2^{-k} | \langle \omega, \varphi\rangle_{-k,k} = 0,\ \forall \varphi \in W_2^k,\ \operatorname{supp}\varphi \in \Omega\} \\ &= \{\omega \in W_2^{-k} \mid \operatorname{supp}\omega \subset \Omega^{\mathrm{c}}\}. \end{aligned} \tag{6.32}$$

Consider in W_2^k the subspace

$$\mathcal{M}_k(\Omega) = (I_{k,-k}\mathcal{M}_{-k}(\Omega))^{\mathrm{cl},k}. \tag{6.33}$$

We consider the following question: Under what conditions on $\mathcal{M}_k(\Omega)$ is this proper subspace in W_2^k ($\mathcal{M}_k(\Omega) \neq W_2^k$) dense in L_2? The next theorem gives a simple sufficient condition for a positive answer to this question.

Theorem 6.3.2. *Let $k > n/2$. Then for each set $K \subset \mathbb{R}^n$ of zero Lebesgue measure, i.e., with*

$$\lambda(K) = 0, \tag{6.34}$$

and such that

$$\mathrm{cap}_k(K) > 0, \tag{6.35}$$

$\mathcal{M}_k(\Omega)$, $\Omega = \mathbb{R} \setminus K$, is proper subspace in W_2^k and is dense in L_2.

Proof. According to the known Sobolev Theorem, for $k > n/2$ the space $W_2^k \subset C(\mathbb{R}^n)$. Therefore, for each point $y \in \Omega$ the evaluation functional $l_{\delta_y}(\varphi) := \varphi(y)$, $\varphi \in W_2^k$, is continuous on W_2^k. It follows that the set of generalized function

$$\{\omega = \delta_y \mid y \in \Omega\}$$

belong to the subspace W_2^{-k} and the subspace

$$\mathcal{N}_{-k} = (\mathrm{span}\{\omega = \delta_y \mid y \in \Omega\})^{\mathrm{cl},-k}$$

has a null intersection with L_2. Moreover, (6.32) implies that

$$\mathcal{N}_k(\Omega) = I_{k,-k}\mathcal{N}_{-k} \perp \mathcal{M}_k(\Omega).$$

So, we can use general Theorem 6.1.4. □

Let us consider the scale of Sobolev spaces

$$\cdots W_2^{-k} \sqsupset L_2(\mathbb{R}^n, dx) \sqsupset W_2^k \equiv W_2^k(\mathbb{R}^n) \cdots .$$

Let $K \subset \mathbb{R}^n$ be an arbitrary compact set. Denote its complement $K^c = \mathbb{R}^n \setminus K$ by Ω. We recall that the Sobolev space $\overset{\circ}{W}{}_2^k(\Omega)$, according to the definition, is the closure of the functions set $C_0^\infty(\Omega)$ in W_2^k. We are interested the question: under what conditions is $\overset{\circ}{W}{}_2^k(\Omega)$ dense in W_2^m, $m \leq k-1$? We provide the answer in terms of the capacity of the set K. First, we need some preparations.

Let us consider the set

$$\Phi(K) = \{\varphi \in \mathcal{S} \mid \varphi \geq 1 \text{ on } K\},$$

and denote by $\Phi^{\mathrm{cl},\alpha}(K)$ the closure of $\Phi(K)$ in W_2^α. From potential theory we know the following nontrivial result (see, for example, Theorem 2.2.7 in [29]).

Theorem 6.3.3. *Let $\alpha \geq 1$, and let for a fixed compact $K \subset \mathbb{R}^n$ its α-capacity be nonzero, i.e., $C_\alpha(K) > 0$. Then there exists a unique extreme element $\varphi_K \in \Phi^{\mathrm{cl},\alpha}(K)$ such that*

$$C_\alpha(K) = \|\varphi_K\|^2_{W_2^\alpha}.$$

Moreover, the dual space to W_2^α *contains an extreme element* $\mu^K \in W_2^{-\alpha}$, *which has the meaning of a measure (so-called* α*-capacitive measure on* K*), such that*

$$\varphi_K = G_\alpha * (G_\alpha * \mu^K),$$

where G_α *is the Bessel integral operator*

$$(G_\alpha * \mu^K)(x) = \int G_\alpha(x-y)d\mu^K(y).$$

In addition, the value of the measure μ^K *of the compact set* K *coincides with the* α*-capacity of* K*:*

$$\mu^K(K) = C_\alpha(K).$$

It is easy to see that the extreme element φ_K is orthogonal to the subspace $\overset{\circ}{W}{}_2^\alpha(\Omega)$ in W_2^α:

$$\varphi_K \perp \overset{\circ}{W}{}_2^\alpha(\Omega).$$

Indeed, let $\varphi_n \in C_0^\infty$ be a minimizing sequence in (6.30). Then it is clear that

$$(\varphi_K, \varphi)_{W_2^\alpha} = \lim_{n\to\infty} (\varphi_n, \varphi) = 0, \quad \forall \varphi \in C_0^\infty(\Omega).$$

Thus, in the case when $C_\alpha(K) > 0$, the orthogonal complement to $\overset{\circ}{W}{}_2^\alpha(\Omega)$ in W_2^α is a nontrivial subspace, which we denote by $\mathcal{N}_\alpha(K)$. In addition, it is obvious that an extreme element φ_K belongs to $\mathcal{N}_\alpha(K)$. Thus, we can write

$$W_2^\alpha = \mathcal{M}_\alpha(\Omega) \oplus \mathcal{N}_\alpha(K), \quad \mathcal{M}_\alpha(\Omega) \equiv \overset{\circ}{W}{}_2^\alpha(\Omega).$$

The dual subspace to $\mathcal{N}_\alpha(K)$ in $W_2^{-\alpha}$ we denote by $\mathcal{N}_{-\alpha}(K)$. It is connected with $\mathcal{N}_\alpha(K)$ by the Berezansky canonical isomorphism $D_{-k,k} : W_2^\alpha \mapsto W_2^{-\alpha}$:

$$\mathcal{N}_{-\alpha}(K) = D_{-k,k}\mathcal{N}_\alpha(K).$$

Proposition 6.3.4. *Under the condition* $C_\alpha(K) > 0$, *the subspace* $\mathcal{N}_{-\alpha}(K)$ *consists of generalized functions* $\omega \in W_2^{-\alpha}$ *with support in* K*:*

$$\mathcal{N}_{-\alpha}(K) = \{\omega \in W_2^{-\alpha} \mid \operatorname{supp}\omega \subseteq K\}.$$

Proof. It is based on results from [29] (see Theorem 9.1.3 and its Corollary 9.1.6 therein). So, Theorem 9.1.3, in particular, shows that $\overset{\circ}{W}{}_2^\alpha(\Omega)$ coincides with the set of functions φ such that

$$(D^\beta\varphi) \restriction K = 0, \quad 0 \le |\beta| \le \alpha - 1. \tag{6.36}$$

So, if $\operatorname{supp}\omega \subseteq K$, then it is obvious that

$$\langle \omega, \varphi \rangle_{-\alpha,\alpha} = 0, \tag{6.37}$$

for all $\varphi \in W_2^\alpha$ satisfying (6.36), since all such functions are approximated by sequences $\varphi_n \in C_0^\infty(\Omega)$. Conversely, if for $\omega \in W_2^{-\alpha}$ (6.37) holds, then $\operatorname{supp}\omega \subseteq K$. □

For a complete understanding of the above results, it is worth noting that the subspace $\mathcal{M}_\alpha(\Omega) = \overset{\circ}{W}{}_2^\alpha(\Omega)$ can be described in terms of a vector-valued operator Tr_K^α, which is defined on W_2^α by the expression

$$\mathrm{Tr}_K^\alpha \varphi = \{D^\beta \varphi \restriction K \mid \beta \le \alpha - 1\},$$

where β is an integer multi-index. Namely, by using this operator, it is easy to see that (6.37) yields the equality

$$\overset{\circ}{W}{}_2^\alpha(\Omega) = \mathrm{Ker}(\mathrm{Tr}_K^\alpha). \tag{6.38}$$

Thus, we can formulate the following important result.

Theorem 6.3.5. *Let the α-capacity of the compact $K \subset \mathbb{R}^n$ be different from zero, $C_\alpha(K) > 0$, where $\alpha \ge 1$ is an integer. Then the set $\mathcal{M}_\alpha(\Omega) = \overset{\circ}{W}{}_2^\alpha(\Omega)$, $\Omega = \mathbb{R}^n \backslash K$ is a proper subspace of W_2^α. This subspace is dense in W_2^m as a set for each integer $m \le \alpha - 1$,*

$$W_2^m \sqsupset \overset{\circ}{W}{}_2^\alpha(\Omega), \tag{6.39}$$

if and only if

$$C_m(K) = 0. \tag{6.40}$$

Proof. We observe immediately that the equality $C_m(K) = 0$ is possible only when $m \le n/2$, otherwise $C_m(\{x\}) > 0$ even for one point $x \in \mathbb{R}^n$. This follows from the Sobolev embedding theorem (see Theorem 1.2.4). In the case $C_m(K) = 0$, $m \le n/2$, the following statement (see Theorem 9.9.1 in [29]) is true. *For each fixed element $h \in W_2^m$, for an arbitrary $\varepsilon > 0$ and an arbitrary neighborhood V of a compact set K, there exists a function $\varphi \in C_0^\infty(V)$, $0 \le \varphi \le 1$, $\varphi(x) = 1$, $x \in K$ such that*

$$\|\varphi h\|_{W_2^m} < \varepsilon. \tag{6.41}$$

Let $\varepsilon_n \to 0$, $n \to \infty$. Consider a sequence of functions φ_n with the above properties. Then, due to (6.41), the sequence $\check{h}_n := \varphi_n h$ converges to 0 in W_2^m. Clearly, in the sum $h = h_n + \check{h}_n$, the sequence $h_n = (1 - \varphi_n)h \in W_2^m$ converges to h. Indeed, each element $h_n \in \overset{\circ}{W}{}_2^m(\Omega)$ according to the construction (see (6.38)). That is, $\mathrm{Tr}_K^m h_m = 0$, i.e., $h_n \in \mathrm{Ker}\,\mathrm{Tr}_K^m$. Let the function $\psi_n \in C_0^\infty(\Omega)$ be such that $\|\psi_n - h_n\|_{\overset{\circ}{W}{}_2^m} < \varepsilon_n$. Since $\psi_n \in \overset{\circ}{W}{}_\alpha^m(\Omega)$, then since $\check{h}_n \to 0$ we obtain: $\|\psi_n - h_n\|_{W_2^m} \to 0$ as $n \to \infty$. This proves that $\overset{\circ}{W}{}_2^\alpha(\Omega)$ is dense in W_2^m, namely the implication (6.40) $\Longrightarrow$ (6.39).

The inverse implication is trivial. Indeed, from (6.39) it follows that

$$\overset{\circ}{W}{}_2^m(\Omega) = W_2^m, \quad \text{since} \quad \overset{\circ}{W}{}_2^\alpha(\Omega) \subset \overset{\circ}{W}{}_2^m(\Omega) \quad \text{for } m \leq \alpha - 1.$$

Hence $C_m(K) = 0$. □

In applications to mathematical physics one often uses a simplified version of Theorem 6.3.5 (cf. with Theorem 6.3.3).

Corollary 6.3.6. *Let $\alpha > n/2$ and K be an arbitrary compact set in $\mathbb{R}^n$. Suppose that the α-capacity $C_\alpha(K) > 0$. Then the set $\overset{\circ}{W}{}_2^\alpha(\Omega)$, $\Omega = \mathbb{R}^n \setminus K$ is a proper subspace in $W_2^\alpha(\mathbb{R}^n)$. This subspace is dense in L_2 if the Lebesgue measure of the set K is equal zero, $\lambda(K) = 0$.*

6.4 A non-denseness defect

In terms of a geometry of Hilbert scales there arises naturally the question about the non-density defect of a subspace by the continuous embedding. The co-dimension of the subspace $\mathcal{M}_+$ in $\mathcal{H}_0$ under continuous embedding of $\mathcal{M}_+$ into $\mathcal{H}_0$, i.e., the dimension of the subspace $\mathcal{M}_+^\perp$ in $\mathcal{H}_0$, is called the defect of the subspace $\mathcal{M}_+$.

We denote this value by $\operatorname{def}(\mathcal{M}_+ \subset \mathcal{H}_0)$. We have the following result.

Theorem 6.4.1. *Let a scale of spaces*

$$\mathcal{H}_{--} \sqsupset \mathcal{H}_- \sqsupset \mathcal{H}_0 \sqsupset \mathcal{H}_+ \sqsupset \mathcal{H}_{++}$$

be given (one can consider that $\mathcal{H}_\pm = \mathcal{H}_{\pm k}$, $\mathcal{H}_{++} = \mathcal{H}_{2k}$ and $\mathcal{H}_{--} = \mathcal{H}_{-2k}$, with $k > 0$) and $\mathcal{H}_+ = \mathcal{M}_+ \oplus \mathcal{N}_+$. Then the condition $\operatorname{def}(\mathcal{M}_+ \subset \mathcal{H}_0) \neq 0$ is equivalent to one of the relations $\mathcal{N}_0 \cap \mathcal{H}_+ \neq \{0\}$ or $\mathcal{N}_+ \cap \mathcal{H}_{++} \neq \{0\}$, or $\mathcal{N}_- \cap \mathcal{H}_0 \neq \{0\}$.

Moreover the following relations hold:

$$(\mathcal{M}_0^{\mathrm{cl},+})^\perp = D_{-,0}(\mathcal{M}_+^{\mathrm{cl},0})^\perp, \quad (\mathcal{M}_{++}^{\mathrm{cl},+})^\perp = I_{+,0}(\mathcal{M}_+^{\mathrm{cl},0})^\perp,$$

where $(\mathcal{M}_k^{\mathrm{cl}})^\perp$ denotes orthogonal complement to $\mathcal{M}_k$ in $\mathcal{H}_l$.

Proof. It easily follows from the first invariance principle for scales of Hilbert spaces. □

The following theorem is one of the main results of the article [57].

Theorem 6.4.2. *Let $\mathcal{H}_+ = \mathcal{M}_+ \oplus \mathcal{N}_+$. Then*

$$\operatorname{def}(\mathcal{M}_+ \subset \mathcal{H}_0) = \dim(\mathcal{N}_- \cap \mathcal{H}_0), \tag{6.42}$$

where $\mathcal{N}_- = D_{-,+}\mathcal{N}_+$.

Proof. The relation (6.42) is a consequence of the equalities

$$(\mathcal{M}_+^{\mathrm{cl},0})^\perp \equiv \mathcal{H}_0 \ominus \mathcal{M}_+ = \mathcal{N}_- \cap \mathcal{H}_0. \tag{6.43}$$

Let us show that $(\mathcal{M}_+^{\mathrm{cl},0})^\perp \subset (\mathcal{N}_- \cap \mathcal{H}_0)$. Take a vector $g \in \mathcal{H}_0$ that belongs to $(\mathcal{M}_+^{\mathrm{cl},0})^\perp$. Then

$$0 = (g, \mathcal{M}_+)_0 = \langle g, \mathcal{M}_+\rangle_{-,+} = (I_{+,-}g, \mathcal{M}_+)_+,$$

where $\langle\cdot,\cdot\rangle_{-,+}$ denotes the duality pairing of the spaces $\mathcal{H}_-$ and $\mathcal{H}_+$, and $I_{+,-} := D_{-,+}^{-1} : \mathcal{H}_- \to \mathcal{H}_-$. This means that $I_{+,-}g \in \mathcal{N}_+$. Thus, $g \in \mathcal{N}_-$, since $\mathcal{N}_+ = I_{+,-}\mathcal{N}_-$. Hence, $g \in \mathcal{N}_- \cap \mathcal{H}_0$.

Let us prove the inverse inclusion $(\mathcal{N}_- \cap \mathcal{H}_0) \subset (\mathcal{M}_+^{\mathrm{cl},0})^\perp$. Let $D_{0,+} : \mathcal{H}_+ \to \mathcal{H}_0$ and $D_{-,0} : \mathcal{H}_0 \to \mathcal{H}_-$ be the usual Berezansky canonical isomorphisms. Let $g \in \mathcal{N}_- \cap \mathcal{H}_0$. Equivalently, $I_{+,0}g := \varphi \in \mathcal{N}_0 \cap \mathcal{H}_+$, where $\mathcal{N}_0 = I_{0,-}\mathcal{N}_-$ and $I_{+,0} := D_{0,+}^{-1} : \mathcal{H}_0 \to \mathcal{H}_+$, $I_{0,-} := D_{-,0}^{-1} : \mathcal{H}_- \to \mathcal{H}_0$. From this it follows that

$$0 = (\varphi, \mathcal{M}_0)_0 = \langle \varphi, \mathcal{M}_0\rangle_{+,-} = (D_{0,+}\varphi, I_{0,-}\mathcal{M}_0)_0 = (D_{0,+}\varphi, \mathcal{M}_+)_0,$$

where $\mathcal{M}_0 := D_{0,+}\mathcal{M}_+ \perp \mathcal{N}_0$ in $\mathcal{H}_0$.

Let us note that the last equality follows from the properties of the canonical isomorphism:

$$I_{0,-} \restriction \mathcal{H}_0 = I_{+,0} \Longrightarrow I_{0,-}\mathcal{M}_0 = I_{+,0}\mathcal{M}_0 = \mathcal{M}_+$$

Thus, since

$$\varphi = I_{+,0}g \in \mathcal{N}_0 \cap \mathcal{H}_+ \Longleftrightarrow g \in \mathcal{N}_- \cap \mathcal{H}_0,$$

we have that

$$0 = (D_{0,+}\varphi, \mathcal{M}_+)_0 = (D_{0,+}I_{+,0}g, \mathcal{M}_+)_0 = (g, \mathcal{M}_+)_0.$$

Hence $g \in (\mathcal{M}_+^{\mathrm{cl},0})^\perp$. Thus, if the vector $g \in \mathcal{N}_- \cap \mathcal{H}_0$, then $g \in (\mathcal{M}_+^{\mathrm{cl},0})^\perp$, and so $(\mathcal{M}_+^{\mathrm{cl},0})^\perp \sqsupset \mathcal{N}_- \cap \mathcal{H}_0$. The theorem is proved. □

Let

$$\mathcal{H}_{--} \sqsupset \mathcal{H}_- \sqsupset \mathcal{H}_0 \sqsupset \mathcal{H}_+ \sqsupset \mathcal{H}_{++}$$

and $\mathcal{H}_+ = \mathcal{M}_+ \oplus \mathcal{N}_+$, $\mathcal{H}_{++} = \tilde{\mathcal{M}}_+ \oplus \tilde{\mathcal{N}}_+$, where $\tilde{\mathcal{M}}_+ = \mathcal{M}_+ \cap \mathcal{H}_{++}$. Assume that $\mathcal{H}_0 \sqsupset \mathcal{M}_+$, and consider the following condition stronger than (6.19):

$$\tilde{\mathcal{N}}_- \cap \mathcal{H}_- = \mathcal{N}_-. \tag{6.44}$$

Then we have the following theorem.

Theorem 6.4.3. *If the condition* (6.44) *is fulfilled, then* $\tilde{\mathcal{M}}_+$ *is dense in* $\mathcal{H}_0$*, and in* $\mathcal{M}_+$ *too, namely*

$$\tilde{\mathcal{N}}_- \cap \mathcal{H}_- = \mathcal{N}_- \Longleftrightarrow \mathcal{M}_+ \sqsupset \tilde{\mathcal{M}}_+.$$

Proof. If (6.44) holds, then according to the first invariance principle for a scale we have

$$\tilde{\mathcal{N}}_- \cap \mathcal{H}_- = \mathcal{N}_- \Longleftrightarrow \tilde{\mathcal{N}}_0 \cap \mathcal{H}_+ = \mathcal{N}_+,$$

and then $\tilde{\mathcal{N}}_0 := (\mathcal{N}_+)^{\mathrm{cl},0}$. It is easy to see that $\tilde{\mathcal{N}}_0 = D_{0,++}\tilde{\mathcal{N}}_+$. Using the equality (6.43), we obtain

$$\mathcal{H}_+ \ominus \tilde{\mathcal{M}}_+ = \tilde{\mathcal{N}}_0 \cap \mathcal{H}_+ = \mathcal{N}_+,$$

which proves the necessity. The sufficiency follows from similar considerations. □

6.5 The denseness problem in terms of an A-scale

In this section we present a deeper results on the denseness problem and obtain their applications.

Theorem 6.5.1. *Let there be given positive spaces $\mathcal{H}_+ = \mathfrak{D}(A)$ and $\mathcal{H}_{++} = \mathfrak{D}(A^2)$ from the A-scale, which have an orthogonal decomposition as above, namely, $\mathcal{H}_+ = \mathcal{M}_+ \oplus \mathcal{N}_+$ and $\mathcal{H}_{++} = \tilde{\mathcal{M}}_+ \oplus \tilde{\mathcal{N}}_+$, where $\tilde{\mathcal{M}}_+ := \mathcal{M}_+ \cap \mathcal{H}_{++}$. In addition, suppose that the subspace $\mathcal{M}_+$ is dense in $\mathcal{H}_0$, i.e., $\mathcal{H}_0 \sqsupset \mathcal{M}_+$.*

Then the subspace $\tilde{\mathcal{M}}_+$ is dense in $\mathcal{M}_+$ if and only if the subspace $\mathcal{N}_+^{\mathrm{cl},0}$ (the closure $\mathcal{N}_+$ in $\mathcal{H}_0$) has a null intersection with $\mathcal{M}_+$:

$$\tilde{\mathcal{M}}_+ \sqsubset \mathcal{M}_+ \Longleftrightarrow \mathcal{N}_+^{\mathrm{cl},0} \cap \mathcal{M}_+ = \{0\}. \tag{6.45}$$

If the right-hand condition in (6.45) *is not valid and the subspace $\tilde{\mathcal{M}}_+$ is not dense in $\mathcal{M}_+$, then its non-denseness defect is given by the formula*

$$\operatorname{def}(\tilde{\mathcal{M}}_+ \subset \mathcal{M}_+) = \dim(\mathcal{N}_+^{\mathrm{cl},0} \cap \mathcal{M}_+). \tag{6.46}$$

Proof. The Proof of (6.45) follows from the equality (6.17) and the choice of the spaces $\mathcal{H}_0 \sqsupset \mathcal{M}_+ \sqsupset \mathcal{M}_{++}$ (see below Lemma 6.5.2). The second part of the theorem follows easily from the results of Subsection 6.4. □

Lemma 6.5.2. *Let a segment*

$$\mathcal{H}_{--} \sqsupset \mathcal{H}_- \sqsupset \mathcal{H}_0 \sqsupset \mathcal{H}_+ \sqsupset \mathcal{H}_{++}$$

of the A-scale of Hilbert spaces be given and let $\mathcal{H}_+ = \mathcal{M}_+ \oplus \mathcal{N}_+$ be an arbitrary decomposition of $\mathcal{H}_+$ into an orthogonal sum of subspaces. Then the set $P_{\mathcal{M}_+}\mathcal{H}_{++}$ where $P_{\mathcal{M}_+}$ is the orthogonal projection onto the subspace $\mathcal{M}_+$ in $\mathcal{H}_+$, dense in the subspace $\mathcal{M}_+$:

$$P_{\mathcal{M}_+}\mathcal{H}_{++} \sqsubset \mathcal{M}_+.$$

Let us denote $\mathcal{M}_{++} := P_{\mathcal{M}_+}\mathcal{H}_{++}$ and introduce a norm in $\mathcal{M}_{++}$ by

$$\|\varphi\|_{\mathcal{M}_{++}} := \|\psi\|_{\mathcal{H}_{++}}, \quad \varphi = P_{\mathcal{M}_+}\psi, \ \psi \in \mathcal{H}_{++}.$$

Moreover, according to the condition $\mathcal{H}_0 \sqsupset \mathcal{M}_+$, *the triplet*

$$\mathcal{H}_0 \sqsupset \mathcal{M}_+ \sqsupset \mathcal{M}_{++}$$

forms a rigged Hilbert space.

Proof. Suppose there exists $g \in \mathcal{M}_+$ such that $(g, P_{\mathcal{M}_+}\mathcal{H}_{++})_+ = 0$. Let O denote the embedding operator from $\mathcal{H}_{++}$ into $\mathcal{H}_+$. Then $\forall \varphi \in \mathcal{H}_{++}$, we have that

$$\mathcal{H}_+ \ni O\varphi = \varphi^{\mathcal{M}_+} + \varphi^{\mathcal{N}_+},$$

where

$$\varphi^{\mathcal{M}_+} := P_{\mathcal{M}_+}(O\varphi) \in \mathcal{M}_+$$

and

$$\varphi^{\mathcal{N}_+} := P_{\mathcal{N}_+}(O\varphi) \in \mathcal{N}_+$$

This means that $(g, \varphi^{\mathcal{M}_+})_+ = 0$ and $(g, \varphi^{\mathcal{N}_+})_+ = 0$, since $g \perp \mathcal{N}_+$. It follows that $(g, \varphi^{\mathcal{M}_+} + \varphi^{\mathcal{N}_+})_+ = 0$, and also $(g, O\varphi)_+ = 0$. Since the vector $\varphi \in \mathcal{H}_{++}$ is arbitrary, we get that $\mathcal{H}_{++}$ is non-dense in $\mathcal{H}_+$ which contradicts the fact that $\mathcal{H}_+ \sqsupset \mathcal{H}_{++}$. Hence $g = 0$. To prove the last assertion of the lemma we should consider the operator $\breve{A}$ (see Subsection 7.8.4, and constructions in [1, 58]). □

The next corollaries of Theorems 6.1.4, 6.4.3 and 6.5.1 are interesting in terms of the geometry of a scale of Hilbert spaces.

Corollary 6.5.3. *It holds that*

$$\mathcal{N}_+^{\mathrm{cl},0} \cap \mathcal{M}_+ = \{0\} \Longleftrightarrow \mathcal{N}_+^{\mathrm{cl},0} \cap \mathcal{H}_+ = \mathcal{N}_+. \tag{6.47}$$

Proof. In view of the properties of scales

$$D_{--,0}\mathcal{N}_+^{\mathrm{cl},0} = \mathcal{N}_-^{\mathrm{cl},--} = \tilde{\mathcal{N}}_-.$$

Recall that $D_{--,0} \mid \mathcal{H}_+ = D_{-,+}$. From this, it obviously follows that $D_{--,0}\mathcal{H}_+ = \mathcal{H}_-, D_{--,0}\mathcal{N}_+ = \mathcal{N}_-$, and then the right-hand site of (6.47) is equivalent to (6.44). Namely,

$$\mathcal{N}_+^{\mathrm{cl},0} \cap \mathcal{H}_+ = \mathcal{N}_+ \Longleftrightarrow \tilde{\mathcal{N}}_- \cap \mathcal{H}_- = \mathcal{N}_-.$$

By Theorem 6.4.3, the equality $\tilde{\mathcal{N}}_- \cap \mathcal{H}_- = \mathcal{N}_-$ is equivalent to $\tilde{\mathcal{M}}_+$ being dense in $\mathcal{M}_+$. By Theorem 6.5.1, $\mathcal{N}_+^{\mathrm{cl},0} \cap \mathcal{M}_+ = \{0\}$ is also equivalent to $\tilde{\mathcal{M}}_+$ being dense in $\mathcal{M}_+$. Hence,

$$\mathcal{N}_+^{\mathrm{cl},0} \cap \mathcal{H}_+ = \mathcal{N}_+ \Longleftrightarrow \tilde{\mathcal{N}}_- \cap \mathcal{H}_- = \mathcal{N}_- \Longleftrightarrow \mathcal{N}_+^{\mathrm{cl},0} \cap \mathcal{M}_+ = \{0\}.$$ □

Corollary 6.5.4. *Let* $\mathcal{H}_0 \sqsupset \mathcal{M}_+$. *If*

$$\mathcal{N}_-^{\mathrm{cl},--} \cap \mathcal{H}_- = \mathcal{N}_-, \tag{6.48}$$

then $\tilde{\mathcal{M}}_+$ *is dense in* $\mathcal{H}_0$.

Proof. The subspace $\tilde{\mathcal{M}}_+$ can be described as

$$\tilde{\mathcal{M}}_+ = \{\varphi \in \mathcal{H}_{++} \mid (\varphi, \psi)_+ = 0,\ \psi \in \mathcal{N}_+\}.$$

Therefore, for all $\varphi \in \tilde{\mathcal{M}}_+$, thanks to the properties of A-scales, we have:

$$0 = (\varphi, \psi)_+ = \langle \varphi, \omega \rangle_{+,-} = \langle \varphi, \omega \rangle_{++,--},$$

where $\omega = D_{-,+}\psi, \psi \in \mathcal{N}_+$. From the last statement it follows (see (6.19)) that the subspace

$$\tilde{\mathcal{N}}_- := \{\omega \in \mathcal{H}_{--} \mid \langle \varphi, \omega \rangle_{++,--} = 0,\ \varphi \in \tilde{\mathcal{M}}_+\}$$

coincides with the closure of the subspace $\mathcal{N}_- = D_{-,+}\mathcal{N}_+$ in $\mathcal{H}_{--}$. Hence, under the condition (6.48), we obtain the inclusion $\mathcal{H}_0 \sqsupset \tilde{\mathcal{M}}_+$, since $\mathcal{N}_- \cap \mathcal{H}_0 = \{0\}$. □

Chapter 7

Singular Perturbations of Self-adjoint Operators

In this chapter we give a systematic presentation of the abstract approach to the construction of singularly perturbed operators based on the method of rigged Hilbert spaces.

7.1 Orientation

The perturbation theory for self-adjoint operators together with the spectral theory occupy a central place in mathematical physics. The mathematical foundations of the theory were laid throughout the twentieth century by many scholars, physicists and mathematicians. This theory is presented in numerous publications and monographs. Among the best known we mention only a few: [7, 107, 169–172], which undoubtedly are on the shelves of researches all over the world.

Here we deal with only one specific aspect of the perturbation theory – the singular perturbation theory that is being actively developed nowadays. It attracts and challenges researchers by its nontrivial problems and the lack of universal standard methods for constructing perturbed operators.

We note here that most of results of the modern theory relate primarily to perturbations of the Laplace operator by delta-like potentials, although there are many publications devoted to the theory of singular perturbation of differential operators in various settings: [3, 17, 19–22, 24–27, 36, 52, 60–62, 69, 77, 79–81, 85, 89, 90, 94, 109, 149–151, 154, 155, 158, 160, 162–164, 167, 175, 176, 184].

In a wide sense, by singular perturbation we mean here a mathematical object (quadratic form, generalized function, expression describing some influence on a free system), which is equal to zero on a dense set in a Hilbert space. Physically, such kind of an object corresponds to a small nontrivial change of the free system. The main nontrivial problem is how to consider singular perturbations correctly.

The theory of rigged Hilbert spaces provides the most universal method for solving this problem. And the most appropriate language for describing arbitrary-degree singularities is that of quadratic forms in scales of Hilbert spaces. We will justify these statements in the next sections.

7.2 Singular perturbations

Let $A = A^* \geq m > -\infty$ be a self-adjoint bounded from below operator on a Hilbert space $\mathcal{H}$. Without loss of generality we suppose that $m = 1$, so $A \geq 1$. In what follows we assume that A is unbounded.

Here we study mainly the strictly positive self-adjoint operators $\tilde{A} > 0$ which are singularly perturbed with respect to A.

Definition 7.2.1. A self-adjoint operator $\tilde{A} = \tilde{A}^*$ $(\tilde{A} \neq A)$ is called (purely) singularly perturbed with respect to A if the linear set

$$\mathfrak{D} := \{f \in \operatorname{Dom} A \cap \operatorname{Dom} \tilde{A} \mid Af = \tilde{A}f\} \tag{7.1}$$

is dense in $\mathcal{H}$. In this case we write $\tilde{A} \in \mathcal{P}_{\mathrm{s}}(A)$.

For fixed $\tilde{A} \in \mathcal{P}_{\mathrm{s}}(A)$, the restriction of A (and $\tilde{A}$) to the set $\mathfrak{D}$ in (7.1) defines a symmetric operator

$$\mathbf{A} := A \restriction \mathfrak{D} = \tilde{A} \restriction \mathfrak{D}. \tag{7.2}$$

Clearly, the operator $\mathbf{A}$ is automatically closed, but it is not essentially self-adjoint since $\tilde{A} \neq A$. Hence, its deficiency indices are equal and nonzero:

$$n^+(\mathbf{A}) = n^-(\mathbf{A}) \neq 0.$$

Write $\tilde{A} \in \mathcal{P}_{\mathrm{s}}^n(A)$ if a number $n = n^+(\mathbf{A}) = n^-(\mathbf{A}) = n$.

An operator $\tilde{A}$ is called *strongly singularly perturbed* with respect to A if the set $\mathfrak{D}$ is dense not only in $\mathcal{H}$, but also in $\mathcal{H}_1(A) \equiv \mathfrak{D}(A^{1/2})$, equipped with the norm $\|\cdot\|_1 = \|A^{1/2}\cdot\|$. In this case we write $\tilde{A} \in \mathcal{P}_{\mathrm{ss}}(A)$ or $\tilde{A} \in \mathcal{P}_{\mathrm{ss}}^n(A)$. By $\mathcal{P}_{\mathrm{ws}}^n(A)$ we denote the family of operators which are *weakly singularly perturbed* with respect to A. For these operators the norms in $\mathcal{H}_1(A)$ and $\mathcal{H}_1(\tilde{A})$ are equivalent.

In this section we study a specific family of rank-one strongly singularly perturbed operators $\tilde{A} \in \mathcal{P}_{\mathrm{ss}}^1(A)$, although the majority of constructions and results are valid for perturbations of an arbitrary rank $n \geq 1$. We want to establish a one-to-one correspondence between the family of all such operators and a certain class of singular quadratic forms in $\mathcal{H}$.

Let us fix a positive strongly singularly perturbed operator $\tilde{A} \in \mathcal{P}_{\mathrm{ss}}(A)$, $\tilde{A} \geq 0$. In accordance with Definition 7.2.1, the difference of quadratic forms associated with the operators $\tilde{A}$ and A defines a singular form in $\mathcal{H}$:

$$\gamma[\varphi] := \gamma_{\tilde{A}}[\varphi] - \gamma_A[\varphi], \quad \varphi \in Q(\gamma) = Q(\gamma_{\tilde{A}}) \cap Q(\gamma_A). \tag{7.3}$$

It follows from Definition 7.2.1 that γ is zero on $\mathfrak{D}$. Thus, $\mathfrak{D} \subseteq \operatorname{Ker}\gamma$ and therefore, $\operatorname{Ker}\gamma$ is dense in $\mathcal{H}$. It is precisely this kind of forms that will be used to establishing connections between the operators $\tilde{A}$ and A.

We consider a segment of the A-scale

$$\mathcal{H}_{-2} \sqsupset \mathcal{H}_{-1} \sqsupset \mathcal{H} \equiv \mathcal{H}_0 \sqsupset \mathcal{H}_1 \sqsupset \mathcal{H}_2, \tag{7.4}$$

where we recall that $\mathcal{H}_2$ coincides with $\mathfrak{D}(A)$ completed with respect to the norm $\|\varphi\|_2 = \|A\varphi\|$, while $\mathcal{H}_1$ coincides with $\mathfrak{D}(A^{1/2})$ completed with respect to the norm $\|\varphi\|_1 = \|A^{1/2}\varphi\|$.

Definition 7.2.2. A symmetric densely defined and singular in $\mathcal{H}$ quadratic form $\gamma \neq 0$ belongs to the $\mathcal{H}_{-k}$-class, $k = 1, 2$, if:

(1) $\operatorname{Dom}\gamma \equiv Q(\gamma) \subseteq \mathcal{H}_k$ and γ is closed in $\mathcal{H}_k$,
(2) γ is singular in $\mathcal{H}_{k-1}$ in the sense that $\operatorname{Ker}\gamma$ is dense in $\mathcal{H}_{k-1}$, i.e., $\operatorname{Ker}\gamma = \Phi_0 \sqsubset \mathcal{H}_{k-1}$.

It is clear that this definition can be easily extended for any $k \geq 2$.

Due to the condition (1), each $\gamma \in \mathcal{H}_{-k}$ is associated with some operator S which acts in the rigged space (7.4). Namely, the next statement holds true (the proof is left to the reader).

Proposition 7.2.3. *Each form $\gamma \in \mathcal{H}_{-k}$-class, $k = 1, 2$, admits the operator representation*

$$\gamma(\varphi, \psi) = \langle S\varphi, \psi\rangle_{-k,k}, \quad \mathfrak{D}(S) \subseteq Q(\gamma) \subseteq \mathcal{H}_k, \tag{7.5}$$

where the associated operator

$$S : \mathcal{H}_k \longrightarrow \mathcal{H}_{-k}$$

can be written as the product

$$S = D_{-k,k} \cdot \mathbf{s},$$

where $\mathbf{s} = \mathbf{s}^*$ is the self-adjoint operator associated with the form γ in $\mathcal{H}_k$, i.e.,

$$\gamma(\varphi, \psi) = (\mathbf{s}\varphi, \psi)_k, \quad \varphi, \psi \in \mathfrak{D}(\mathbf{s}) \subseteq \mathcal{H}_k, \tag{7.6}$$

and $D_{-k,k} : \mathcal{H}_k \to \mathcal{H}_{-k}$ denotes the Berezansky canonical isomorphism.

Here we only note that the condition (2) from Definition 7.2.2 implies that the set

$$\Phi_0 := \operatorname{Ker} S = \operatorname{Ker}\mathbf{s} = \operatorname{Ker}\gamma, \tag{7.7}$$

is dense in $\mathcal{H}_{k-1}$ and that Φ_0 is a closed subspace of $\mathcal{H}_k$.

In what follows we often use the notation $\gamma \in \mathcal{T}_k(A)$ or $\gamma \in \mathcal{T}_k^n(A) \equiv \mathcal{T}_k^n$ for forms γ of the $\mathcal{H}_{-k}$-class, where n stands for the rank of γ. Notice that n can be defined as the dimension of the orthogonal complement to Φ_0 in $\mathcal{H}_k$, i.e., $n := \dim \Phi_0^{\perp}$ (in general, it is possible that $n = \infty$).

Example 7.2.4 (Description of singular rank-one quadratic forms). Let us show that each couple $\{\omega, \lambda\}$, which consists of a vector ω from the negative space and a number $0 \neq \lambda$, uniquely defines a positive form $\gamma_{\lambda,\omega} \in \mathcal{T}_k^1$, $k = 1, 2$.

Indeed, let $\omega \in \mathcal{H}_{-k} \setminus \mathcal{H}_{-k+1}$, $\|\omega\|_{-k} = 1$ and $0 \neq \lambda \in \mathbb{R}_+^1$ be fixed. Using ω and λ we define a rank-one bounded operator $S_{\lambda,\omega} : \mathcal{H}_k \to \mathcal{H}_{-k}$ by the formula

$$S_{\lambda,\omega}\varphi = \lambda\langle\varphi, \omega\rangle_{k,-k}\omega, \quad \varphi \in \mathcal{H}_k. \tag{7.8}$$

Obviously, the null subspace of this operator,

$$\Phi_0 = \operatorname{Ker} S_{\lambda,\omega} = \{\varphi \in \mathcal{H}_k \mid \langle\varphi, \omega\rangle_{k,-k} = 0\}, \tag{7.9}$$

is a closed subspace in $\mathcal{H}_k$. Moreover, Φ_0 is densely included in $\mathcal{H}_{k-1}$. This follows from the condition $\omega \in \mathcal{H}_{-k} \setminus \mathcal{H}_{-k+1}$ (see also Theorem 6.1.1). Further, it is easily seen that the associated quadratic form to $S_{\lambda,\omega}$,

$$\gamma_{\lambda,\omega}[\varphi] := \langle S_{\lambda,\omega}\varphi, \varphi\rangle_{-k,k} = \lambda\langle\varphi, \omega\rangle_{k,-k}\langle\omega, \varphi\rangle_{-k,k}, \tag{7.10}$$

belongs to the set $\mathcal{T}_k^1$. Indeed, this form is bounded in $\mathcal{H}_k$:

$$|\gamma_{\lambda,\omega}[\varphi]| = |\lambda||(\varphi, \eta_k)_k|^2 \leq |\lambda|\|\varphi\|_k^2, \tag{7.11}$$

where we used the equality

$$\langle\varphi, \omega\rangle_{k,-k} = (\varphi, \eta_k)_k,$$

in which the vector $\eta_k \in \mathcal{H}_k$ is defined as an isometric image of ω,

$$\eta_k = I_{k,-k}\omega, \quad \|\eta_k\|_k = \|\omega\|_{-k} = 1.$$

Thus, the condition (1) from Definition 7.2.2 is fulfilled. Since $\Phi_0 = \operatorname{Ker}\gamma_{\lambda,\omega} = \operatorname{Ker} S_{\lambda,\omega}$ is dense in $\mathcal{H}_{k-1}$, the condition (2) from Definition 7.2.2 is also fulfilled. This means that $\gamma_{\lambda,\omega}$ is singular in $\mathcal{H}_{k-1}$.

The converse is also true. Indeed, each $\gamma \in \mathcal{T}_k^1$ is associated with a self-adjoint operator $\mathbf{s}$ in $\mathcal{H}_k$ which acts as $\mathbf{s}\varphi = \lambda(\varphi, \eta_k)_k\eta_k$. Using η_k one can find $\omega = D_{-k,k}\eta_k \in \mathcal{H}_{-k} \setminus \mathcal{H}_{-k+1}$.

Thus, each rank-one quadratic form $\gamma \in \mathcal{T}_k^1$, $k = 1, 2$ is specified by some number $\lambda > 0$ and vector $\omega \in \mathcal{H}_{-k}$, i.e., it has the representation

$$\gamma_{\lambda,\omega}[\varphi] = \lambda|(\varphi, \eta_k)_k|^2. \tag{7.12}$$

In this way one can also consider the forms $\gamma_{\lambda,\omega}$ with $\lambda < 0$.

We are ready to prove

Theorem 7.2.5. *Between the family of positive strongly singularly perturbed operators $\tilde{A} \in \mathcal{P}^1_{\mathrm{ss}}(A)$ and the set of positive singular quadratic forms $\gamma_{\lambda,\omega} \in \mathcal{T}^1_2$ there exists the bijective correspondence which is described as follows.*

Given a positive form $\gamma_{\lambda,\omega} \in \mathcal{T}^1_2$, the corresponding singularly perturbed operator $\tilde{A} = A_{\lambda,\omega} \in \mathcal{P}^1_{\mathrm{ss}}(A)$ is defined by the equality

$$A_{\lambda,\omega} g = Af, \tag{7.13}$$

where

$$\mathfrak{D}(A_{\lambda,\omega}) = \{g \in \mathcal{H} \mid g = f + \lambda^{-1} c_f \eta,\ \eta = I_{0,-2}\omega,\ f \in \mathfrak{D}(A)\}, \tag{7.14}$$

and $c_f = \langle f, \omega \rangle_{2,-2}$.

Conversely, for a given perturbed operator $\tilde{A} \in \mathcal{P}^1_{\mathrm{ss}}$, the form $\gamma_{\lambda,\omega} \in \mathcal{T}^1_2$ is defined in accordance with (7.8)–(7.12), where the number $\lambda > 0$ and the vector ω are fixed by

$$\lambda^{-1} := ((\tilde{A}^{-1} - A^{-1})\eta, \eta), \tag{7.15}$$

and

$$\omega = I_{-2,0}\eta, \quad \|\eta\| = 1, \tag{7.16}$$

respectively, where $\eta \in \mathcal{N}_0$. Here, the subspace $\mathcal{N}_0 := \mathcal{M}_0^{\perp}$, with $\mathcal{M}_0 = A\mathfrak{D}$, where $\mathfrak{D}$ is defined in (7.1).

Proof. Let $\gamma_{\lambda,\omega} \in \mathcal{T}^1_2$, $\lambda > 0$, $\omega \in \mathcal{H}_{-2} \setminus \mathcal{H}_{-1}$. Then the set $\operatorname{Ker} \gamma_{\lambda,\omega} = \Phi_0$ is a closed subspace in $\mathcal{H}_2 = \mathfrak{D}(A)$. It is dense in both $\mathcal{H}_1$ and $\mathcal{H}$. By the definition of the set $\mathcal{T}^1_2$, $\dim \mathcal{N}_2 = 1$, where $\mathcal{N}_2 = \Phi_0^{\perp}$ in $\mathcal{H}_2$. That is, $\mathcal{N}_2 = \{c\eta_2 \mid c \in \mathbb{C}\}$ with $\eta_2 = I_{2,-2}\omega$. Let us put $\mathfrak{D} := \Phi_0$ and consider the symmetric operator $\mathbf{A} := A \restriction \mathfrak{D}$. Let us show that $A_{\lambda,\omega}$ defined by (7.13) and (7.14), is a self-adjoint extension of $\mathbf{A}$. Indeed, (7.14) implies that $\mathfrak{D} \subset \mathfrak{D}(A_{\lambda,\omega})$ and

$$A_{\lambda,\omega} f = Af = \mathbf{A}f, \quad f \in \mathfrak{D}.$$

We have only to check that $A_{\lambda,\omega}$ is self-adjoint. For any $g_1, g_2 \in \mathfrak{D}(A_{\lambda,\omega})$ one can write

$$\begin{aligned}
(A_{\lambda,\omega} g_1, g_2) &= (Af_1, g_2) = (Af_1, f_2) + \lambda^{-1}(Af_1, \eta)(\eta, Af_2) \\
&= (f_1, Af_2) + \lambda^{-1}((Af_1, \eta)\eta, Af_2) \\
&= (f_1, Af_2) + \lambda^{-1}(\langle f_1, \omega \rangle_{2,-2}\eta, Af_2) \\
&= (g_1, Af_2) = (g_1, A_{\lambda,\omega} g_2), \quad f_1, f_2 \in \mathfrak{D}(A),
\end{aligned}$$

where $\eta = I_{0,-1}\omega$. So, we proved that $A_{\lambda,\omega}$ is symmetric. From (7.13) by the Hellinger–Toeplitz theorem (see Chapter 1), it follows that $A_{\lambda,\omega}$ is self-adjoint since $\operatorname{Ran} A_{\lambda,\omega} = \operatorname{Ran} A = \mathcal{H}$.

Conversely, let $\tilde{A} \in \mathcal{P}^1_{\mathrm{ss}}$ be given. Then using the set $\mathfrak{D}$ defined in (7.1) we introduce the subspace $\mathcal{N}_2 = \mathfrak{D}^{\perp}$ in $\mathcal{H}_2$. Clearly, $\dim \mathcal{N}_2 = 1$. Taking a vector

$\eta_2 \in \mathcal{N}_2$, $\|\eta_2\|_2 = 1$ we define $\omega := D_{-2,2}\eta_2 \in \mathcal{H}_{-2}$. Note that η_2 is fixed up to a constant $e^{i\theta}$, $\theta \in [0, 2\pi)$. However, the quadratic form $\gamma_{\lambda,\omega}$ does not depend on this constant (see (7.12)).

By the formula (7.8) with $k = 2$, the vector ω defines a family of operators $S_{\lambda,\omega}$, parameterized by $\lambda \in \mathbb{R}^1, \lambda \neq 0$. To find the sought-for form $\gamma_{\lambda,\omega}$ we have to fix some number λ. To this end we use the equality (7.15). So, the form $\gamma_{\lambda,\omega}$ is uniquely defined by λ, ω in accordance with (7.10)–(7.12). Finally, we note that the operator A coincides with the Friedrichs extension of $\mathbf{A}$ since Φ_0 is dense in $\mathcal{H}_1$. Hence, $A_{\lambda,\omega} \leq A$. □

It is interesting to remark that one can easily generalize the proved theorem to the case $\lambda \in \mathbb{R}^1$. Then the operators $A_{\lambda,\omega}$ are in general non-positive, but they are bounded from below.

Furthermore, this theorem can be easily generalized to singular perturbations of an arbitrary rank. In any case, due to the denseness of $\mathfrak{D}$ in $\mathcal{H}_1$, the operator A will always coincide with the Friedrichs extension of $\mathbf{A}$. Therefore, for all $\tilde{A} \in \mathcal{P}^n_{\rm ss}$ the Kreĭn inequality $\tilde{A} \leq A$ holds true. In addition, if $n < \infty$, then $\operatorname{Ran} \tilde{A} = \operatorname{Ran} A = \mathcal{H}$ and therefore the inverse operators $\tilde{A}^{-1}$ and A^{-1} exist and are defined on the whole space.

7.3 The form-sum method

For the construction of singularly perturbed operators belonging to the $\mathcal{P}^n_{\rm ws}(A)$-class, it is not necessary to use the theory of self-adjoint extensions of symmetric operators. A more natural way is to use sums of quadratic forms. In such case the obtained operators are called weakly singular perturbed. There is a known simple sufficient condition under which it is possible to use the form-sum method. This condition was independently found by several authors (see [107, 169]) and is named as the KLMN-theorem (after Kato, Lions, Lax–Milgram, and Nelson).

Theorem 7.3.1. *Let γ_A be a quadratic form associated with a self-adjoint operator $A = A^* \geq 0$ on a Hilbert space $\mathcal{H}$ and let γ be a symmetric densely defined form satisfying the condition*

$$|\gamma[\varphi]| \leq a\|\varphi\|^2 + b\gamma_A[\varphi], \quad \varphi \in \mathfrak{D}(A) \subseteq Q(\gamma), \tag{7.17}$$

where $0 \leq b < 1$, $a \in \mathbb{R}^1$. Then the form-sum

$$\tilde{\gamma} = \gamma_A + \gamma, \quad Q(\tilde{\gamma}) = Q(\gamma),$$

is closable and bounded from below:

$$\tilde{\gamma} \subseteq \tilde{\gamma}^{\rm cl} \geq -a.$$

Thus, there exists the self-adjoint operator $\tilde{A} \geq -a$ such that $\gamma_{\tilde{A}} = \tilde{\gamma}^{\rm cl}$.

Proof. Rewriting the condition (7.17) in the form

$$-b\gamma_A - a\chi \leq \gamma \leq b\gamma_A + a\chi, \quad \chi[\cdot] \equiv \|\cdot\|^2, \tag{7.18}$$

we get

$$\tilde{\gamma} = \gamma_A + \gamma \geq (1-b)\gamma_A - a\chi \geq -a\chi,$$

since $1-b>0$ and $\gamma_A \geq 0$. Hence $\tilde{\gamma} \geq -a$, i.e., the form $\tilde{\gamma}$ is bounded from below. Moreover, from (7.18) it follows that

$$(1-b)\gamma_A + \chi \leq \tilde{\gamma} + (a+1)\chi \leq (1+b)\gamma_A + (2a+1)\chi.$$

Therefore, the norms

$$\|\cdot\|_{\tilde{\gamma}+(a+1)\chi}, \text{ and } \|\cdot\|_{\gamma_A+\chi}$$

are equivalent and the form $\tilde{\gamma}$ is closable. With its closure $\tilde{\gamma}^{\mathrm{cl}}$ there is associated a self-adjoint operator $\tilde{A} \geq -a$. □

Given $A = A^* \geq 0$, a symmetric form γ on $\mathcal{H}$ is called A-*bounded* if there exist numbers $b>0$ and $a \in \mathbb{R}$ such that

$$|\gamma[\varphi]| \leq a\|\varphi\|^2 + b\gamma_A[\varphi], \quad \varphi \in Q(\gamma_A) \subseteq Q(\gamma). \tag{7.19}$$

The smallest number b in (7.19) is called the A-*bound* of γ.

Example 7.3.2 (The operator $-d^2/dx^2$ with δ-potential). Let $\mathcal{H} = L_2(\mathbb{R}^1, dx)$ and $A = -\frac{d^2}{dx^2}$. We consider on $C_0^\infty(\mathbb{R}^1)$ the quadratic form

$$\gamma_\delta[\varphi] = |\varphi(0)|^2.$$

By the Sobolev embedding theorem, for each $0<b$ there exists $a \in \mathbb{R}^1$ such that

$$|\varphi(0)|^2 \leq b\gamma_A[\varphi] + a\|\varphi\|^2.$$

Then, by Theorem 7.3.1, the form-sum $\tilde{\gamma} = \gamma_A + \gamma_\delta$ is closable in $L_2(\mathbb{R}^1)$. The self-adjoint operator associated with the closure of $\tilde{\gamma}$ is usually interpreted as a singularly perturbed operator with respect to $A = -\frac{d^2}{dx^2}$. Thus, the form-sum method gives a precise rigorous meaning to the formal expression $-\frac{d^2}{dx^2} + \delta$.

7.3.1 The generalized operator sum

There we study the question of representing of $\tilde{A} \in \mathcal{P}_{\mathrm{ws}}(A)$ as a generalized operator sum, $\tilde{A} = A \tilde{+} T$, with some $T : \mathcal{H}_1 \to \mathcal{H}_{-1}$.

We recall that by Definition 7.2.1 (see also [9, 98, 103, 105, 135]), a self-adjoint operator $\tilde{A} \neq A$ is called (purely) singularly perturbed with respect to A if $\tilde{A}$ coincides with A on some linear subset $\mathfrak{D}$ which is dense in $\mathcal{H}$. The family of all such operators is denoted here by $\mathcal{P}_s(A)$. Thus, $\tilde{A} \neq A$ belongs to $\mathcal{P}_s(A)$ if and only if the set

$$\mathfrak{D} := \{f \in \mathfrak{D}(\tilde{A}) \cap \mathfrak{D}(A) \mid Af = \tilde{A}f\} \tag{7.20}$$

is dense in $\mathcal{H}$. We say that $\tilde{A} \neq A$ belongs to the class of weakly singular perturbed (with respect to A) operators and write

$$\tilde{A} \in \mathcal{P}_{\rm ws}(A), \tag{7.21}$$

if, in addition to (7.20), it holds that

$$\mathfrak{D}(\tilde{A}) \sqsubset \mathcal{H}_1. \tag{7.22}$$

This implies that the norms in the spaces $\mathcal{H}_1(A)$ and $\mathcal{H}_1(\tilde{A})$ are equivalent.

It is important that each operator $\tilde{A} \in \mathcal{P}_{\rm ws}(A)$, $\tilde{A} \neq A_\infty$, admits an additive representation as a generalized sum $\tilde{A} = A \tilde{+} T$ (see Theorem 3 in [137] and Theorem 7 in [138]). Here A_∞ denotes the Friedrichs extension of $A \restriction \mathfrak{D}$ and T acts in the A-scale, i.e., $T : \mathcal{H}_1 \to \mathcal{H}_{-1}$. For the precise formulation of this result we need some preparations.

Let $0 \neq T$ be a closed symmetric operator acting in the A-scale from $\mathcal{H}_1$ to $\mathcal{H}_{-1}$. Note that the adjoint operator T^* is defined with respect to the duality pairing $\langle \cdot, \cdot \rangle$. So, T^* acts also from $\mathcal{H}_1$ to $\mathcal{H}_{-1}$. Thus,

$$\langle T\varphi, \psi \rangle = \langle \varphi, T\psi \rangle, \quad \varphi, \psi \in \mathfrak{D}(T) \subset \mathfrak{D}(T^*) \subset \mathcal{H}_1.$$

We recall that an operator $T : \mathcal{H}_1 \to \mathcal{H}_{-1}$ is called *$\mathcal{H}$-singular* if its range $\operatorname{Ran} T$ contains at least one element which does not belong to $\mathcal{H}$. T is called *purely singular* with respect to $\mathcal{H}$, briefly, T is *purely $\mathcal{H}$-singular* (see [20, 21, 105, 135, 138]), if

$$\operatorname{Ran} T^{\mathrm{cl},-1} \cap \mathcal{H} = \{0\},$$

where cl, -1 denotes the closure in $\mathcal{H}_{-1}$. It is known (see [137], Theorem A) that T is purely $\mathcal{H}$-singular if the set

$$\operatorname{Ker} T \quad \text{is dense in } \mathcal{H}. \tag{7.23}$$

Since T is closed, the set $\mathcal{M}_1 = \operatorname{Ker} T$ is a closed subspace in $\mathcal{H}_1$. We write

$$T \in \mathcal{H}_{-1}(A)\text{-class}, \tag{7.24}$$

if the set

$$\operatorname{Ker} T \cap \mathfrak{D}(A) \quad \text{is dense in } \mathcal{M}_1. \tag{7.25}$$

It is not hard to see that the set $\operatorname{Ker} T \cap \mathfrak{D}(A)$ is a proper closed subspace in $\mathcal{H}_2$. We denote it by $\mathcal{M}_2$. Now from (7.23) and (7.25) it follows that

$$\mathcal{M}_2^{\mathrm{cl},1} = \mathcal{M}_1, \quad \mathcal{M}_1^{\mathrm{cl},0} = \mathcal{H}_0, \tag{7.26}$$

where cl, 0 (cl, 1) denotes a closure in $\mathcal{H}_0$, ($\mathcal{H}_1$). It is clear, $\mathcal{M}_2^{\mathrm{cl},0} = \mathcal{H}_0$ due to (7.26).

We will consider the operator T as a singular perturbation of A (see [9, 20, 21, 25, 105, 135]). To construct the perturbed operators $\tilde{A}$ we can use the method of generalized operator sums which, in fact, extends the well-known method of form-sums (see the previous subsection). We briefly recall this construction (see [21, 25, 42, 43, 98, 121, 135] for more detail).

Given a symmetric operator $T : \mathcal{H}_1 \to \mathcal{H}_{-1}$, we define the *generalized operator sum* $\tilde{A} = A\tilde{+}T$ as the restriction of a usual operator sum $A^{\mathrm{cl}}+T : \mathcal{H}_1 \to \mathcal{H}_{-1}$ to $\mathcal{H}_0$, where A^{cl} denotes the closure of A as an operator from $\mathcal{H}_1$ to $\mathcal{H}_{-1}$. Precisely,

$$\begin{aligned} \mathfrak{D}(\tilde{A}) &= \{\varphi \in \mathcal{H}_1 \cap \mathfrak{D}(T) \mid A^{\mathrm{cl}}\varphi + T\varphi \in \mathcal{H}\}, \\ \tilde{A}\varphi &= A^{\mathrm{cl}}\varphi + T\varphi. \end{aligned} \tag{7.27}$$

We emphasize that both components, $A^{\mathrm{cl}}\varphi$ and $T\varphi$, in the general case belong to $\mathcal{H}_{-1}$ (but not necessarily to $\mathcal{H}_0$). It is easily seen that $\tilde{A}$ in (7.27) is a Hermitian (symmetric) operator, i.e., $(\tilde{A}\varphi, \psi) = (\varphi, \tilde{A}\psi)$, $\varphi, \psi \in \mathfrak{D}(\tilde{A})$. But, in general, $\tilde{A}$ is non-densely defined in $\mathcal{H}$. Nevertheless, if $\mathfrak{D}(T) \subseteq \mathfrak{D}(A)$ and $\mathcal{R}(T) \subseteq \mathcal{H}$, then $A\tilde{+}T$ coincides with the usual operator sum $A + T$. In [9–11] it was shown (see also [17, 21, 23, 25]) that a sum of operators in the sense of quadratic forms is a particular case of the generalized operator sum.

It is obvious that for $T \in \mathcal{H}_{-1}$-class, the generalizes sum $\tilde{A} = A\tilde{+}T$ necessarily belongs to $\in \mathcal{P}_{\mathrm{ws}}(A)$. This is true, since $\mathfrak{D}$ defined in (7.20) is dense in $\mathcal{H}$ (see (7.23)–(7.26)) and therefore $\mathfrak{D}(\tilde{A}) \subset \mathcal{H}_1$. The converse statement is evidently also true.

Thus, we can formulate the above observations as

Theorem 7.3.3 ([137, 138]). *Each operator $\tilde{A} \in \mathcal{P}_{\mathrm{ws}}(A)$, $\tilde{A} \neq A_\infty$, under the condition that both couples, A, A_∞ and $\tilde{A}$, A_∞ are mutually simple with respect to* $\mathbf{A} := A \restriction \mathfrak{D} = \tilde{A} \restriction \mathfrak{D}$ *(see* [32]*), admits the generalized operator sum representation $\tilde{A} = A\tilde{+}T$, where the uniquely defined (bounded) operator $T : \mathcal{H}_1 \to \mathcal{H}_{-1}$ belongs to the $\mathcal{H}_{-1}$-class.*

For the construction of T for given $\tilde{A}$ see Section 7.4.

Note that the above definition of the generalized operator sum can be easily extended to include the case when A is replaced by other self-adjoint operator C in $\mathcal{H}_0$ such that $\mathcal{H}_1(C)$ is distinct from $\mathcal{H}_1(A) \equiv \mathcal{H}_1$. Indeed, assume that the domain $\mathfrak{D}(C) \subseteq \mathcal{H}_1$ and C is closable as a map from $\mathcal{H}_1$ to $\mathcal{H}_{-1}$. Let $C^{\mathrm{cl}} : \mathcal{H}_1 \to \mathcal{H}_{-1}$ denotes its closure. Then, similarly to the above case, we define $\tilde{C} = C\tilde{+}T$ as the restriction of the operator sum $C^{\mathrm{cl}} + T : \mathcal{H}_1 \to \mathcal{H}_{-1}$ to $\mathcal{H}$. Thus,

$$\mathfrak{D}(\tilde{C}) = \{\varphi \in \mathfrak{D}(C^{\mathrm{cl}}) \cap \mathfrak{D}(T) : C^{\mathrm{cl}}\varphi + T\varphi \in \mathcal{H}\}, \quad \tilde{C}\varphi = C^{\mathrm{cl}}\varphi + T\varphi.$$

In our construction of the generalized sum one can use the property of additivity:

$$A\tilde{+}(T_1 + T_2) = (A\tilde{+}T_1)\tilde{+}T_2 = (A\tilde{+}T_2)\tilde{+}T_1. \tag{7.28}$$

This property holds true for bounded operators T_1, T_2 which act from $\mathcal{H}_1$ to $\mathcal{H}_{-1}$ under condition that both domains $\mathfrak{D}(A\tilde{+}T_1)$, $\mathfrak{D}(A\tilde{+}T_2)$ are dense in $\mathcal{H}_1$. Then (7.28) is obviously fulfilled since in such a case the closures of $A\tilde{+}T_1$ and $A\tilde{+}T_2$ as operators from $\mathcal{H}_1$ to $\mathcal{H}_{-1}$ coincide with $A^{\rm cl}+T_1$ and $A^{\rm cl}+T_2$, respectively. Hence, by using (7.27), one can construct the generalized sum $(A\tilde{+}T_1)\tilde{+}T_2$ (or $(A\tilde{+}T_2)\tilde{+}T_1$, respectively) in the case when A is replaced by $C = A\tilde{+}T_1$ or $C = A\tilde{+}T_2$, respectively.

Now we want to find the conditions for a symmetric operator T which will ensure the essential self-adjointness of the generalized sum $A\tilde{+}T$. This problem has been studied in [10, 20, 98, 105, 137]. Here we formulate the following result on self-adjointness of $A\tilde{+}T$.

Theorem 7.3.4 ([10, 105]). *Let $T \in \mathcal{H}_{-1}$-class. Assume that one of the following conditions is satisfied:*

(a) *the operator $T_1 := (A^{\rm cl}+I)^{-1}T$ on $\mathcal{H}_1$ has purely point spectrum,*
(b) *the domain $\mathfrak{D}(A) \subset \operatorname{Ran}(T_1+I)$.*

Then the operator $\tilde{A} = A\tilde{+}T$ is essentially self-adjoint.

In particular, $\tilde{A} = A\tilde{+}T$ is self-adjoint whenever $T_1 = (A^{\rm cl}+I)^{-1}T$ is compact in $\mathcal{H}_1$.

Further we consider the issue of the additive representation of operators $\tilde{A}$ which are bounded from below, but need not belong to the family $\mathcal{P}_{\rm ws}(A)$. So, let $A \geq 0$ and $\tilde{A} = \tilde{A}^*$ be bounded from below. We say that $\tilde{A}$ belongs to the set of additively perturbed operators with respect to A and write

$$\tilde{A} \in \mathcal{P}_{\rm ad}(A) \tag{7.29}$$

if $\tilde{A}$ admits a representation as a generalized sum, $\tilde{A} = A\tilde{+}T$. Here $T : \mathcal{H}_1 \to \mathcal{H}_{-1}$ is some bounded symmetric operator which does not necessarily belong to the $\mathcal{H}_{-1}$-class.

Given $\tilde{A}$, we are interested in the question: Under what conditions $\tilde{A}$ is an additive perturbation of A? In other words, when does $\tilde{A} \in \mathcal{P}_{\rm ad}(A)$?

We introduce the $\tilde{A}$-scale of Hilbert spaces $\{\tilde{\mathcal{H}}_k\}_{k\in\mathbb{R}^1}$, where $\tilde{\mathcal{H}}_k \equiv \mathcal{H}_k(\tilde{A})$ is defined as the closure of $\mathfrak{D}(\tilde{A})$ with respect to the norm $\|f\|_{\tilde{\mathcal{H}}_k} := \|(\tilde{A}+\tilde{m})^{k/2}f\|^{1/2}$ and the constant $\tilde{m} \geq 1$ is chosen to provide the inequality

$$\|f\|^2_{\tilde{\mathcal{H}}_1} \equiv (\tilde{A}f,f) + \tilde{m}\|f\|^2 \geq \|f\|^2. \tag{7.30}$$

Thus, we have a rigged Hilbert space for each fixed $s > 0$:

$$\tilde{\mathcal{H}}_{-s} \sqsupset \mathcal{H}_0 \equiv \mathcal{H} \sqsupset \tilde{\mathcal{H}}_s,$$

where $\tilde{\mathcal{H}}_{-s} \equiv \mathcal{H}_{-s}(\tilde{A})$ coincides with the dual space to $\tilde{\mathcal{H}}_s$. Let $\tilde{D}_{-1,1} : \tilde{\mathcal{H}}_1 \to \tilde{\mathcal{H}}_{-1}$ denotes the Berezansky canonical isomorphism [42, 48]. By its construction, $\tilde{D}_{-1,1} \restriction \tilde{\mathcal{H}}_2$ coincides with $\tilde{A}+\tilde{m}I$ (for details see [42]). Thus,

$$\tilde{A}f = (\tilde{D}_{-1,1} - \tilde{m}I)f, \quad \mathfrak{D}(\tilde{A}) = \{f \in \tilde{\mathcal{H}}_1 \mid \tilde{D}_{-1,1}f \in \mathcal{H}\}. \tag{7.31}$$

In particular, the operator $\tilde{A}$ is uniquely associated with the rigged space $\tilde{\mathcal{H}}_{-1} \sqsupset \mathcal{H} \sqsupset \tilde{\mathcal{H}}_1$.

The next theorem gives a criterion membership of $\tilde{A}$ in $\mathcal{P}_{\mathrm{ad}}(A)$.

Theorem 7.3.5. *Given $A = A^* \geq 1$, let $\tilde{A} \neq A$ be a self-adjoint bounded from below operator on $\mathcal{H}$. Assume $\tilde{A} \geq \tilde{m} \geq 1$ and*

$$\mathcal{H}_1(\tilde{A}) \sqsupset \mathcal{H}_1(A) \tag{7.32}$$

in the sense of dense continuous embedding. Then $\tilde{A} \in \mathcal{P}_{\mathrm{ad}}(A)$, i.e., $\tilde{A}$ admits an additive representation as a generalized sum $\tilde{A} = A \tilde{+} T$ with a bounded self-adjoint operator $T : \mathcal{H}_1 \to \mathcal{H}_{-1}$ which satisfies the inequality

$$-\langle Tf, f\rangle \leq \langle A^{\mathrm{cl}} f, f\rangle + \lambda \|f\|^2, \quad \lambda = \tilde{m} - 1 \geq 0, \ f \in \mathcal{H}_1. \tag{7.33}$$

Conversely, if $\tilde{A} \in \mathcal{P}_{\mathrm{ad}}(A)$ and $\tilde{A} = A \tilde{+} T$, with $T : \mathcal{H}_1 \to \mathcal{H}_{-1}$ satisfying (7.33), *then* (7.32) *holds.*

Proof. Let $q_{\tilde{A}}[f]$ denote the closure of the quadratic form $\langle \tilde{A}^{\mathrm{cl}} f, f\rangle$, $f \in \mathfrak{D}(\tilde{A})$ in $\mathcal{H}$. Then the domain $Q(q_{\tilde{A}})$ coincides with $\mathcal{H}_1(\tilde{A}) \equiv \tilde{\mathcal{H}}_1$. Hence, due to (7.32), the form $q_{\tilde{A}}$ is densely defined in $\mathcal{H}_1$. Moreover, $q_{\tilde{A}}$ is continuous on $\mathcal{H}_1$. Indeed, if $f_n \to 0$ in $\mathcal{H}_1$, then by (7.32), $f_n \to 0$ in $\mathcal{H}_1(\tilde{A})$, and also in $\mathcal{H}$. Thus, due to (7.30), $q_{\tilde{A}}[f_n] \to 0$. It follows that

$$q[f] = q_{\tilde{A}}[f] - q_A[f]$$

is continuous on $\mathcal{H}_1(A)$, where $q_A[f] = \langle A^{\mathrm{cl}} f, f\rangle$.

Let us prove now the validity of the representation $\tilde{A} = A \tilde{+} T$. To this aim we consider the operator $\tilde{D}_{-1,1} : \tilde{\mathcal{H}}_1 \to \tilde{\mathcal{H}}_{-1}$. Obviously, it coincides with $\tilde{A}^{\mathrm{cl}} - \tilde{m}I$, where $\tilde{A}^{\mathrm{cl}}$ denotes the closure of $\tilde{A}$ as an embedding from $\tilde{\mathcal{H}}_1$ to $\tilde{\mathcal{H}}_{-1}$. Further, since

$$q_{\tilde{A}}[f] = q_A[f] + q[f] = \langle \tilde{A}^{\mathrm{cl}} f, f\rangle = \langle A^{\mathrm{cl}} f, f\rangle + \langle Tf, f\rangle, \quad f \in \mathcal{H}_1(A), \tag{7.34}$$

the restriction of $\tilde{D}_{-1,1} - \tilde{m}I$ to $\mathcal{H}_1(A)$ coincides with $A^{\mathrm{cl}} \tilde{+} T$. Hence, the self-adjoint in $\mathcal{H}$ operator $\tilde{A}$ is associated with the rigged Hilbert space $\tilde{\mathcal{H}}_1 \subset \mathcal{H} \subset \tilde{\mathcal{H}}_{-1}$. One can construct this operator in accordance with the standard procedure:

$$\tilde{A} = (\tilde{D}_{-1,1} - \tilde{m}I) \restriction \mathfrak{D}(\tilde{A}), \quad \mathfrak{D}(\tilde{A}) = \{f \in \tilde{\mathcal{H}}_1 : (\tilde{D}_{-1,1} - \tilde{m}I) f \in \mathcal{H}\}.$$

Surely, this operator coincides with $A \tilde{+} T$. Thus, $\tilde{A} \in \mathcal{P}_{\mathrm{ad}}(A)$. Moreover, according (7.30),

$$\|f\|^2_{\tilde{\mathcal{H}}_1} - \|f\|^2 = q_{\tilde{A}}[f] + (\tilde{m} - 1)\|f\|^2 \geq 0,$$

and therefore due to (7.34),

$$q_{\tilde{A}}[f] - q[f] + \lambda \|f\|^2 = q_A[f] + \lambda \|f\|^2 \geq -q[f]$$

with $\lambda = \tilde{m} - 1$. So, we obtain the inequality (7.33).

Conversely, let $\tilde{A} = A \tilde{+} T$ be a self-adjoint operator on $\mathcal{H}$, where $T : \mathcal{H}_1 \to \mathcal{H}_{-1}$ satisfies (7.33) with some $\lambda \geq 0$. Then it follows from (7.34) that $\tilde{A}$ is bounded from below,

$$\tilde{A} \geq \tilde{m} \geq 1, \quad \tilde{m} = \lambda + 1.$$

Thus,

$$q_{\tilde{A}}[f] + \tilde{m}\|f\|^2 \geq \|f\|^2,$$

where

$$q_{\tilde{A}}[f] := \langle A^{\text{cl}} f, f \rangle + \langle T f, f \rangle.$$

Therefore, $\tilde{\mathcal{H}}_1$ is the closure of $\mathcal{H}_1$ in the norm

$$\|f\|^2_{\tilde{\mathcal{H}}_1} = \tilde{q}[f] + \tilde{m}\|f\|^2$$

and $\mathcal{H}_1$ is densely embedded in $\tilde{\mathcal{H}}_1$. □

Corollary 7.3.6. *Let $A \geq 0$ and $\tilde{A} \geq \tilde{m}$ be the self-adjoint operators associated with the triplets $\mathcal{H}_{-1} \sqsupset \mathcal{H} \sqsupset \mathcal{H}_1$ and $\tilde{\mathcal{H}}_{-1} \sqsupset \mathcal{H} \sqsupset \tilde{\mathcal{H}}_1$, respectively. Then $\tilde{A} \in \mathcal{P}_{\text{ad}}(A)$, and hence $\tilde{A} = A \tilde{+} T$ with a bounded self-adjoint operator $T : \mathcal{H}_1 \to \mathcal{H}_{-1}$ which satisfies the inequality* (7.33), *if and only if the following chain of inclusions holds true:*

$$\mathcal{H}_{-1} \sqsupset \tilde{\mathcal{H}}_{-1} \sqsupset \mathcal{H} \sqsupset \tilde{\mathcal{H}}_1 \sqsupset \mathcal{H}_1$$

in the sense of dense continuous embeddings.

7.4 The uniqueness problem

In the general setting the problem of uniqueness for singularly perturbed operators is open. In particular, for the explicitly solvable models (see, e.g., [7]) this problem was treated by relying on physical arguments. Apparently, it is impossible to consider and construct the singularly perturbed operator in a unique way by usual methods. Here we analyze this problem in more details with respect to the singular perturbations of $\mathcal{H}_{-2}$-class.

For the first time the idea that the singularly perturbed operator can be defined in a unique way was proposed in [119, 120] by Koshmanenko. Then it was developed in detail in the work [121]. In fact, this idea preceeds the Vishik method [181]. According to the latter, every extension $\tilde{A}$ of a symmetric operator $\mathbf{A}$ on a Hilbert space $\mathcal{H}$ is fixed by some operator B which acts on the null-space $\mathcal{N}_0$ of the adjoint operator $\mathbf{A}^*$. In applications to boundary problems, the operator B is defined directly by a boundary condition written in the canonical form. Coming to the singular perturbation theory we have to define the operator B by a singular quadratic form γ which corresponds to a perturbation. However, in this way some additional problem arises. Usually, a form γ is nowhere closable in $\mathcal{H}$ (precisely, $\operatorname{Ker}\gamma$ is dense in $\mathcal{H}$). So, it is not clear how to associate with γ an operator B. For this reason, most researchers beginning with Berezin and Faddeev [53] (see also the well-known monograph [7] and the bibliography therein) considered the whole set of self-adjoint extensions

$$\mathcal{A}(\mathbf{A}) = \{\tilde{A} = \tilde{A}^* \mid \mathbf{A} \subset \tilde{A} \subset \mathbf{A}^*\}$$

of the symmetric operator $\mathbf{A} := A \restriction \operatorname{Ker}\gamma$. At least ten different ways was proposed for the construction of perturbed operator.

In particular, the approximation approach appears to be most natural and appropriate from the physical point of view. However, in [53] it was shown that in the problem of construction of the operator $-\Delta + \varepsilon\delta(x)$ whenever approximates the delta-function by smooth potentials, $\varepsilon_N V_N(x) \to \varepsilon\delta(x)$, one can obtains a non-trivial result only if the coupling constant ε_N approaches zero in a specific way. This means that the value of the coupling constant loses its meaning in the approximation approach. The same observation emerges in the general situation with arbitrary-rank singular perturbations of the $\mathcal{H}_{-2}$-class. In particular, in the works [18, 129, 137, 138] it was shown that the approximation approach is suitable for treating only a singular component of a perturbation under the assumption that it is regular in the positive space.

The perturbations of the $\mathcal{H}_{-2}$-class can be analyzed in a better way by the method of self-adjoint extensions. Different approaches were explored as well. So, in [60, 69, 81, 89], [90, 94, 95, 97, 108, 145, 146, 158], and [160, 162–164, 175, 176, 184] a series of non-trivial interesting results concerning the $\mathcal{H}_{-2}$-class of singular perturbations were obtained. Nevertheless, a complete theory in this direction is still missing.

It was clarified that the method of rigged spaces gives the most effective way to solving the above problem. Let us describe briefly an essence of this method.

In the first step one has to introduce the rigged Hilbert space $\mathcal{H}_- \sqsupset \mathcal{H} \sqsupset \mathcal{H}_+$ associated with a free (unperturbed) operator A. Here the positive space $\mathcal{H}_+$ coincides with $\operatorname{Dom} A$ in the graph-norm (it is convenient to assume that $A \geq 1$).

Let a perturbation of A be given by a singular quadratic form γ with the dense in $\mathcal{H}$ domain $Q(\gamma)$. We suppose the set $\operatorname{Dom} A \cap Q(\gamma)$ is also dense in $\mathcal{H}$. In the next step we consider γ as a form on $\mathcal{H}_+$. It often happens that γ is regular (closable) in $\mathcal{H}_+$. But its singularity property in $\mathcal{H}$ means that $\operatorname{Ker}\gamma$ is a proper subspace $\mathcal{M}_+$ in $\mathcal{H}_+$ which is dense in $\mathcal{H}$. Thus, there exists the self-adjoint operator $\mathbf{s}$ associated with the closure of γ in $\mathcal{N}_+ := \mathcal{H}_+ \ominus \mathcal{M}_+$. The restriction of the free operator to $\operatorname{Ker}\gamma$ defines the symmetric operator $\mathbf{A} := A \restriction \mathcal{M}_+$ on $\mathcal{H}$.

In the third step we have to clarify how to define by means of $\mathbf{s}$ an abstract boundary condition, which determines a new self-adjoint extension $\tilde{A}$ of $\mathbf{A}$. In accordance with the Birman–Kreĭn–Vishik theory of self-adjoint extensions of semi-bounded operators (see, e.g., [28]), each boundary condition is equivalent to a choice of a bounded operator B on the deficiency subspace $\mathcal{N}_0 = \operatorname{Ker}\mathbf{A}^*$. Here the connection between the subspaces $\mathcal{N}_0$ and $\mathcal{N}_+$ plays an important role. This connection is established by the Berezansky canonical isomorphism $D_{0,+} : \mathcal{H}_+ \to \mathcal{H}$, namely, $\mathcal{N}_0 = D_{0,+}\mathcal{N}_+$.

The next formula represents the most important step in our construction:

$$B := D_{0,+}\mathbf{s}I_{0,+}, \quad I_{0,+} = D_{0,+}^{-1}.$$

It provides the uniqueness of the singularly perturbed operator $\tilde{A} = A_B$ (for details see the text below in this section).

Thus, the key fact in the uniqueness problem is the existence of the bijective correspondence between singular quadratic forms γ, which describe the perturbation of A, and an abstract boundary conditions, B, that uniquely determines $\tilde{A}$ as self-adjoint extensions of the symmetric operator $\mathbf{A} = A \restriction \operatorname{Ker}\gamma$.

We claim that under some appropriate conditions each singular quadratic form γ contains complete information that two objects: the symmetric restriction $\mathbf{A} = A \restriction \operatorname{Ker}\gamma$ and the abstract boundary condition B needed for the construction of $\tilde{A} = A_B$.

Further we establish the above-mentioned correspondence starting with quadratic forms γ of $\mathcal{H}_{-2}$-class. In Chapter 8 we show how to generalize this correspondence to the singular perturbations of the $\mathcal{H}_{-k}$-class with $k \geq 2$, including the so-called super-singular perturbations.

Let a perturbation of an unbounded self-adjoint operator $A \geq 1$ on $\mathcal{H}$ is given by a quadratic form $\gamma \geq 0$ from the $\mathcal{H}_{-2}$-class. According to Definition 7.2.2 such a form is regular in the space $\mathcal{H}_+ \equiv \mathcal{H}_2 = \mathfrak{D}(A)$ equipped with the norm $\|\cdot\|_+ = \|A\cdot\|$. And the null-set of γ forms the subspace $\mathcal{M}_+ = \operatorname{Ker}\gamma$ in $\mathcal{H}_+$ which is dense in $\mathcal{H}_1 = \mathfrak{D}(A^{1/2})$ with the norm $\|\cdot\|_1 = \|A^{1/2}\cdot\|$, i.e.,

$$\mathcal{M}_+ \sqsubset \mathcal{H}_1.$$

Thus, the domain $\mathfrak{D}(A)$, as the Hilbert space $\mathcal{H}_+$, is decomposed into the orthogonal sum

$$\mathcal{H}_+ = \mathcal{M}_+ \oplus \mathcal{N}_+, \quad \mathcal{M}_+ = \operatorname{Ker}\gamma.$$

That is, by Proposition 7.2.3, there is a self-adjoint operator $\mathbf{s} > 0$ in $\mathcal{N}_+$ such that

$$\gamma(\cdot,\cdot) = (\mathbf{s}P_{\mathcal{N}_+}\cdot,\cdot)_+. \tag{7.35}$$

Using γ we define two objects, the dense in $\mathcal{H}$ subspace $\mathcal{M}_+ = \operatorname{Ker}\gamma$ and the operator $\mathbf{s}$ in $\mathcal{N}_+$.

Now we introduce the symmetric operator $\mathbf{A} := A \restriction \mathcal{M}_+ \equiv A \restriction \operatorname{Ker}\gamma$ in $\mathcal{H}$. The family $\mathcal{A}(\mathbf{A})$ of all self-adjoint extensions of $\mathbf{A}$ contains the uniquely defined operator $\tilde{A}$ which corresponds to the singular perturbation γ. To find it one has to use the operator $\mathbf{s}$.

In accordance with the theory of self-adjoint extensions of symmetric operators (see Chapter 3), each $\tilde{A} \in \mathcal{A}(\mathbf{A})$ is fixed by some abstract boundary condition. Since $\gamma \geq 0$, it is natural to suppose that $\tilde{A}$ is strictly positive, $\tilde{A} \in \mathcal{A}_+(\mathbf{A})$. Further, since the subspace $\mathcal{M}_+$ is dense in $\mathcal{H}_1$, the Friedrichs extension of $\mathbf{A}$ coincides with $A_\infty = A$. Consequently, $\tilde{A} \le A$. In turn, it follows that $\tilde{A}$ admits the Kreĭn formula representation (see Theorems 3.1.3, 3.1.4):

$$\tilde{A}^{-1} = A^{-1} + \tilde{B}, \quad \tilde{B} = B^{-1} P_{\mathcal{N}_0}, \tag{7.36}$$

with some $B = B^* > 0$ acting in $\mathcal{N}_0 = \operatorname{Ker} \mathbf{A}^*$. The unknown for the moment operator B in (7.36) plays the role of an abstract boundary condition for fixing of $\tilde{A}$.

Thus, to fix a certain positive extension $\tilde{A} \in \mathcal{A}_+(\mathbf{A})$ corresponding to the quadratic form γ, we have to find the operator B in $\mathcal{N}_0$. Just to this end we introduce the rigged Hilbert space

$$\mathcal{H}_- \sqsupset \mathcal{H}_0 \equiv \mathcal{H} \sqsupset \mathcal{H}_+ = \mathcal{M}_+ \oplus \mathcal{N}_+$$

and observe that the orthogonal decomposition of $\mathcal{H}_0$,

$$\mathcal{H}_0 = \mathcal{M}_0 \oplus \mathcal{N}_0, \quad \mathcal{M}_0 = \mathbf{A}\mathcal{M}_+, \quad \mathcal{N}_0 = \operatorname{Ker} \mathbf{A}^*$$

is connected directly with the similar decomposition of $\mathcal{H}_+$ by the Berezansky canonical isomorphism $D_{0,+} : \mathcal{H}_+ \to \mathcal{H}_0$. In particular,

$$\mathcal{N}_0 = D_{0,+}\mathcal{N}_+.$$

Now we define B on $\mathcal{N}_0$ as the image of $\mathbf{s}$ (see (7.35)) from $\mathcal{H}_+$:

$$B := D_{0,+}\mathbf{s}D_{0,+}^{-1}. \tag{7.37}$$

As it was noted above, the equality (7.37) is crucial for the construction of $\tilde{A}$ on γ.

Theorem 7.4.1. *For a fixed operator $A = A^* \ge 1$ on $\mathcal{H}$, the family of all positive purely singularly perturbed operators $\tilde{A} \in \mathcal{P}_{\rm ss}(A)$, $\tilde{A} > 0$ admits the parametrization $\tilde{A} = A_\gamma$ in terms of positive singular quadratic forms $\gamma \in \mathcal{H}_{-2}$-class. The bijective correspondence between $\tilde{A}$ and γ,*

$$\mathcal{P}_{\rm ss}(A) \ni \tilde{A} = A_\gamma \longleftrightarrow \gamma \in \mathcal{H}_{-2}\text{-}class,$$

is defined as follows. Given γ, the perturbed operator $\tilde{A} = A_\gamma$ is defined as the self-adjoint extension of the symmetric operator $\mathbf{A} = A \restriction \operatorname{Ker}\gamma$, $\tilde{A}^{-1} = A^{-1} + \tilde{B}$, $\tilde{B} = B^{-1}P_{\mathcal{N}_0}$, where B is associated with γ in accordance with (7.35) *and* (7.37). *Conversely, for given $\tilde{A}$ the form γ is reconstructed by the equality*

$$\gamma[\varphi] = (\tilde{A}^{-1}A\varphi, A\varphi) - \gamma_A[\varphi], \quad \varphi \in \operatorname{Dom}\gamma \subseteq \mathfrak{D}(A), \tag{7.38}$$

where

$$\operatorname{Dom}\gamma = \{\varphi \in \mathcal{H}_+ \mid A\varphi \in \operatorname{Ran}\tilde{A}\}.$$

Proof. In one direction the proof was already given. We have to show how to construct a form γ starting with the operator $\tilde{A}$ (see formulas (7.36), (7.35), and (7.37)). So, we only need to check that γ is uniquely defined by (7.38) and that it belongs to the $\mathcal{H}_{-2}$-class.

Indeed, let a perturbed operator $\tilde{A} \in \mathcal{P}_{\rm ss}(A)$, $\tilde{A} > 0$ be given. Then put $\mathcal{M}_+ = \mathfrak{D}$ (see (7.2.1)), $\mathcal{N}_+ = \mathcal{H}_+ \ominus \mathcal{M}_+$, and consider the decomposition

$$\mathcal{H}_0 = \mathcal{M}_0 \oplus \mathcal{N}_0, \quad \mathcal{M}_0 = A\mathfrak{D}, \quad \mathcal{N}_0 = A\mathcal{N}_+.$$

Since $\tilde{A}$ is a self-adjoint extension of the symmetric operator $\mathbf{A} = A \restriction \mathfrak{D} = \tilde{A} \restriction \mathfrak{D}$ (see Subsection 7.1), $\tilde{A} \in \mathcal{A}_+(\mathbf{A})$ and

$$\tilde{A}^{-1} = A_\infty^{-1} + \tilde{B} = A_\infty^{-1} + B^{-1}P_{\mathcal{N}_0},$$

where $B = B^* > 0$ on the subspace $\mathcal{N}_0$. According to Definition 7.2.1, $\tilde{A}$ and A are mutually simple with respect to $\mathbf{A}$. Moreover, $A = A_\infty$, because $\tilde{A} \in \mathcal{P}_{\rm ss}(A)$. In accordance with Theorem 3.1.4, the correspondence between the operators $\tilde{A}$ and $B = B^* > 0$ is bijective. In the space $\mathcal{H}_0$, the operator $\tilde{B}$ is associated with the quadratic form

$$\gamma_{\tilde{B}}[h] = (\tilde{A}^{-1}h, h) - (A_\infty^{-1}h, h), \quad h \in \operatorname{Ran} \tilde{A}.$$

Substituting $h = A\varphi$, $\varphi \in \mathcal{H}_+ = \mathfrak{D}(A)$, we define a quadratic form in $\mathcal{H}_+$ by

$$\gamma[\varphi] := \gamma_{\tilde{B}}[A\varphi], \quad \varphi \in \mathcal{H}_+, \quad A\varphi \in \operatorname{Ran} \tilde{A}.$$

Thanks to the equality $A = A_\infty$, this form coincides with the form defined in (7.38). Further, γ is closable in $\mathcal{H}_+$ since it is connected with $\gamma_{\tilde{B}}$ by the isometric transformation $D_{0,+}$. Thus, γ admits the operator representation. The corresponding associated operator has the form $\mathbf{s}P_{\mathcal{N}_+}$, where $\mathbf{s} = \mathbf{s}^* > 0$. Thus, $\gamma \in \mathcal{H}_{-2}$-class. □

In the last proof we used the fact that the Friedrichs extension of a positive symmetric operator coincides with A. The following theorem gives a criterion for the equality $A = A_\infty$ to hold.

Theorem 7.4.2. *Let $A \in \mathcal{A}_+(\mathbf{A})$, where $\mathbf{A} \geq 1$ is a symmetric operator in $\mathcal{H}_0$. Let*

$$\mathcal{H}_{-2} \sqsupset \mathcal{H}_{-1} \sqsupset \mathcal{H}_0 \sqsupset \mathcal{H}_1 \sqsupset \mathcal{H}_2 = \operatorname{Dom} A,$$

be a segment of the scale associated with A. Let us assume that $\mathcal{H}_2 = \operatorname{Dom} \mathbf{A} \oplus \mathcal{N}_2$ with $\mathcal{N}_2 \neq \{0\}$. Then A coincides with the Friedrichs extension of $\mathbf{A}$ if and only if the domain of $\mathbf{A}$ is dense in $\mathcal{H}_1$:

$$A = A_\infty \Longleftrightarrow \operatorname{Dom} \mathbf{A} \sqsubset \mathcal{H}_1. \tag{7.39}$$

Proof. Let $A = A_\infty$. We assume for a moment that $\operatorname{Dom} \mathbf{A} =: \mathcal{M}_2$ has a defect in $\mathcal{H}_1$, i.e., $\mathcal{H}_1 = \mathcal{M}_2^{\mathrm{cl},1} \oplus \mathcal{X}$ with $\mathcal{X} \neq \{0\}$, where cl, 1 denotes the closure in $\mathcal{H}_1$. We consider the quadratic form associated with $\mathbf{A}$,

$$\gamma_{\mathbf{A}}(\varphi, \psi) = (\mathbf{A}\varphi, \psi)_0, \quad \varphi, \psi \in \mathcal{M}_2.$$

The equality $A\varphi = \mathbf{A}\varphi,\ \varphi \in \mathcal{M}_2$ implies that

$$\gamma_{\mathbf{A}}(\varphi, \psi) = (\mathbf{A}\varphi, \psi)_0 = (A\varphi, \psi)_0 = (\varphi, \psi)_1,$$

where we recall that $(\cdot,\cdot)_1 = (A\cdot,\cdot)_0$. This shows that the domain of $\gamma_{\mathbf{A}}^{\mathrm{cl}}$ coincides with $\mathcal{M}_2^{\mathrm{cl},1}$, write $\mathcal{M}_2^{\mathrm{cl},1} = \operatorname{Dom} \gamma_{\mathbf{A}}^{\mathrm{cl}}$. By our assumption that $\mathcal{X} \neq \{0\}$, we get $\mathcal{H}_2 = \operatorname{Dom} A \not\subset \operatorname{Dom} \gamma_{\mathbf{A}}^{\mathrm{cl}}$. However, it is a well-known fact (see, e.g., [55, 82, 148]) that the domain of the Friedrichs extension of a symmetric operator is a dense set in $\operatorname{Dom} \gamma_{\mathbf{A}}^{\mathrm{cl}}$. Thus, we reached a contradiction with our starting assumption. So, $\mathcal{X} = \{0\}$ and $\mathcal{M}_2 = \operatorname{Dom} \mathbf{A} \sqsubset \mathcal{H}_1$.

Conversely, let $\operatorname{Dom} \mathbf{A} \sqsubset \mathcal{H}_1$. Then it is obvious that

$$\operatorname{Dom} \gamma_{\mathbf{A}}^{\mathrm{cl}} = \mathcal{H}_1.$$

This means that $\gamma_{A_\infty} = \gamma_A$ and therefore $A = A_\infty$. □

As an immediate corollary of the above theorem we get

Theorem 7.4.3. *Let A and $\tilde{A}$ be a couple of positive self-adjoint extensions of some symmetric operator $\mathbf{A} \geq 1$. Then*

$$\tilde{A} \in \mathcal{P}_{\mathrm{ss}}(A) \Longrightarrow A = A_\infty. \tag{7.40}$$

Therefore, for each $\tilde{A} \in \mathcal{P}_{\mathrm{ss}}(A)$, $\tilde{A} \neq A$, the space $\tilde{\mathcal{H}}_1 = \mathcal{H}_1(\tilde{A})$ necessarily contains vectors which do not belong to $\mathcal{H}_1$. The next theorem gives a more precise description of the structure of the space $\tilde{\mathcal{H}}_1$ corresponding to $\tilde{A} \in \mathcal{P}_{\mathrm{ss}}(A)$.

Theorem 7.4.4. *Given $A = A^* \geq 1$, consider the rigged space*

$$\mathcal{H}_{-2} \sqsupset \mathcal{H}_{-1} \sqsupset \mathcal{H}_0 \sqsupset \mathcal{H}_1 \sqsupset \mathcal{H}_2 = \operatorname{Dom} A$$

associated with A. We assume that $\tilde{A} \in \mathcal{P}_{\mathrm{s}}(A)$, $\tilde{A} = \tilde{A}^* \geq 1$. Let $\tilde{\mathcal{H}}_1$ denote the domain $\mathfrak{D}(\tilde{A})$ equipped with the norm $\|\cdot\|_{\tilde{1}} := \|\tilde{A}^{1/2}\cdot\|_0$. Then $\tilde{A}$ is a strongly singularly perturbed operator, $\tilde{A} \in \mathcal{P}_{\mathrm{ss}}(A)$, if and only if $\tilde{\mathcal{H}}_1$ contains $\mathcal{H}_1$ as a proper subspace. That is,

$$\tilde{A} \in \mathcal{P}_{\mathrm{ss}}(A) \Longleftrightarrow \tilde{\mathcal{H}}_1 = \mathcal{H}_1 \oplus \tilde{\mathcal{H}}_B,$$

where the subspace $\tilde{\mathcal{H}}_B$ is constructed in a certain way by means of the operator B (see Theorem 3.1.3) which acts in $\mathcal{N}_0$ and determines the extension $\tilde{A}$.

Proof. Let $\tilde{A} \in \mathcal{P}_{\rm ss}(A)$. Then there exists a dense in $\mathcal{H}_1$ set

$$\mathfrak{D} := \{f \in \operatorname{Dom} A \cap \operatorname{Dom} \tilde{A} \mid Af = \tilde{A}f\}.$$

Since $\tilde{A} \geq 1$, one has that

$$\tilde{A}^{-1} = A^{-1} + \tilde{B}, \quad \tilde{B} = B^{-1} P_{\mathcal{N}_0},$$

where $B > 0$ is a bounded and symmetric operator on $\mathcal{N}_0$. Thus,

$$\operatorname{Ker} \tilde{B} = \mathcal{M}_0 = A\mathfrak{D} = \tilde{A}\mathfrak{D}.$$

Let us recall that the domain of $\tilde{A}$ admits the description in terms of B:

$$\operatorname{Dom} \tilde{A} = \{g \in \mathcal{H}_0 \mid g = f + \tilde{B}Af, \ f \in \mathcal{H}_2 = \operatorname{Dom} A\}, \quad Af = \tilde{A}g.$$

Now we rewrite the inner product in the space $\tilde{\mathcal{H}}_1$ as

$$\begin{aligned}(g_1, g_2)_{\tilde{1}} &= (\tilde{A}g_1, g_2)_0 = (Af_1, g_2)_0 = (Af_1, f_2 + \tilde{B}Af_2)_0 \\ &= (Af_1, f_2)_0 + (Af_1, \tilde{B}Af_2)_0 \\ &= (f_1, f_2)_1 + (Af_1, B^{-1}P_{\mathcal{N}_0}Af_2)_0.\end{aligned}$$

It is easily seen that the quadratic form

$$\gamma(f_1, f_2) := (Af_1, B^{-1}P_{\mathcal{N}_0}Af_2)_0, \quad f_1, f_2 \in \operatorname{Dom} A = \mathcal{H}_2, \tag{7.41}$$

is well defined on $\mathcal{H}_2$ and singular in $\mathcal{H}_1$. This follows from the fact that this form is equal to zero for all

$$f \in \mathcal{M}_2 = \mathfrak{D} = A^{-1}\mathcal{M}_0.$$

Hence, by Theorem 5.3.7, the closure of $\mathcal{H}_2$ with respect to the norm $\|f\|_{\tilde{1}} = (\|f\|_1^2 + \gamma[f])^{1/2}$ gives the space $\mathcal{H}_1 \oplus \tilde{\mathcal{H}}_B$, where, in turn, the subspace $\tilde{\mathcal{H}}_B$ is the completion of the manifold

$$\{\eta \in \mathcal{N}_0 \mid \eta = \eta_f, \ \eta_f = P_{\mathcal{N}_0}Af, \ f \in \mathcal{H}_2\}$$

with respect to the norm $\|\eta_f\|_B := (B^{-1}f, f)^{1/2}$. Thus, $\tilde{\mathcal{H}}_1 = \mathcal{H}_1 \oplus \tilde{\mathcal{H}}_B$.

Conversely, let for a positive operator $\tilde{A} \in \mathcal{P}_{\rm s}(A)$ the corresponding space $\tilde{\mathcal{H}}_1 = \mathcal{H}_1 \oplus \tilde{\mathcal{H}}_B$. Let us show that $\tilde{A} \in \mathcal{P}_{\rm ss}(A)$. From the description of the domain $\mathfrak{D}(\tilde{A})$ it follows that the inner product in $\tilde{\mathcal{H}}_1$ admits the representation as a sum of two positive forms:

$$(g_1, g_2)_{\tilde{1}} = (Af_1, f_2)_0 + (B^{-1}P_{\mathcal{N}_0}Af_1, Af_2)_0 = \chi_1(f_1, f_2) + \gamma(f_1, f_2).$$

By Theorem 5.3.7, the orthogonal decomposition $\tilde{\mathcal{H}}_1 = \mathcal{H}_1 \oplus \mathcal{H}_B$ means the mutual singularity of the forms, $\chi_1 \perp \gamma$. Further, the restriction of the form γ to subspace $\mathcal{N}_2 = \mathfrak{D}^{\perp}$ is strongly positive in $\mathcal{H}_2$. Therefore, $\chi_1 \perp \gamma$ implies the denseness of $\mathfrak{D}$ in $\tilde{\mathcal{H}}_1$. This proves that $\tilde{A} \in \mathcal{P}_{\rm ss}(A)$. $\square$

Example 7.4.5. Let $A = A^* \geq 1$ be a self-adjoint operator on $\mathcal{H}_0$ and $\mathcal{H}_+ = \mathfrak{D}(A)$ with the norm $\|\cdot\|_+ = \|A\cdot\|_0$. We consider the perturbation of the operator A given by a system of abstract boundary conditions:

$$\{\omega_i(\varphi) = \langle\varphi, \omega_i\rangle_{+,-} = 0,\ i = 1, 2, \dots, n \leq \infty,\ \omega_i \in \mathcal{H}_-,\ \varphi \in \mathfrak{D}(A)\}.$$

Introduce the Hermitian operator $\mathbf{A} := A \restriction \mathfrak{D}(\mathbf{A})$ with $\mathfrak{D}(\mathbf{A}) = \{\varphi \in \mathfrak{D}(A) \mid \omega_i(\varphi) = 0\}$. We are interested in what condition on vectors ω_i ensures that $\mathbf{A}$ is densely defined. The answer follows from Theorem 6.1.1. The set $\mathfrak{D}(\mathbf{A})$ is dense in $\mathcal{H}_0$ if and only if $\operatorname{def}(\mathfrak{D}(\mathbf{A}) \subset \mathcal{H}_0) = 0$. It is equivalent to the condition that $(\operatorname{span}\{\omega_i\})^{\mathrm{cl}} \cap \mathcal{H}_0 = \{0\}$. Assume that this condition is not fulfilled, i.e., the operator $\mathbf{A}$ is not densely defined. Then the defect of non-denseness for $\mathfrak{D}(\mathbf{A})$ in $\mathcal{H}_0$ can have an arbitrary value. It depends on the properties of subspaces $\mathcal{N}_- :=\operatorname{span}\{\omega_i\}$. Namely,

$$\dim \mathfrak{D}(\mathbf{A})^\perp = \operatorname{def}(\mathfrak{D}(\mathbf{A}) \subset \mathcal{H}_0) = \dim(\mathcal{N}_- \cap \mathcal{H}_0).$$

7.5 Rigged spaces and singular perturbations

In this section we analyze the structural properties of the rigged Hilbert spaces generated by singularly perturbed operators (for more details see [2, 58, 59, 143]).

Let $A = A^* \geq 1$ be an unbounded self-adjoint operator on a Hilbert space $\mathcal{H}$. As above, by $\{\mathcal{H}_k(A)\}_{k\in\mathbb{R}^1}$ we denote the scale of Hilbert spaces associated with A. We need the following auxiliary result.

Theorem 7.5.1. *Let γ be a bounded symmetric quadratic form on $\mathcal{H}_k$ with a fixed $k > 1$. Let the space $\mathcal{H}_k$ be decomposed onto an orthogonal sum, $\mathcal{H}_k = \mathcal{M}_k \oplus \mathcal{N}_k$, where*

$$\mathcal{M}_k = \operatorname{Ker}\gamma, \quad \mathcal{N}_k = \mathcal{H}_k \ominus \mathcal{M}_k.$$

Assume that $\mathcal{N}_k \neq \{0\}$ and define $\mathcal{N}_{-k} := D_{-k,k}\mathcal{N}_k$, where $D_{-k,k} : \mathcal{H}_k \to \mathcal{H}_{-k}$ is the Berezansky canonical isomorphism. Then the subspace $\mathcal{M}_k$ is dense in $\mathcal{H}_{k-1}$, if and only if the following condition is satisfied:

$$\mathcal{N}_{-k} \cap \mathcal{H}_{-k+1} = \{0\}. \tag{7.42}$$

Proof. By part (iii) of Theorem 6.2.2,

$$\mathcal{M}_k \sqsubset \mathcal{H}_{k-1} \Longleftrightarrow \mathcal{N}_{-k} \cap \mathcal{H}_{-k+1} = \{0\}.$$ □

We recall that the form γ belongs to the $\mathcal{H}_{-k}$-class if (7.42) holds.

Consider now a singularly perturbed operator $\tilde{A} \geq m > -\infty$ defined by the method of self-adjoint extensions starting with A and some quadratic form γ of $\mathcal{H}_{-k}$-class, $k = 1, 2$ (see Section 7.3). Here we will show that operator $\tilde{A}$ arises also in another way. Namely, we claim that $\tilde{A}$ can be defined as an operator associated with a new rigged Hilbert space constructed directly from A and γ, without the use of the method of self-adjoint extensions.

Below we prove this claim.

Let

$$\mathcal{H}_- \sqsupset \mathcal{H}_0 \sqsupset \mathcal{H}_+ \tag{7.43}$$

be a part of the A-scale associated with the given fixed operator $A \geq 1$ on $\mathcal{H}_0$. We recall that the positive space $\mathcal{H}_+$ in (7.43) coincides with $\operatorname{Dom} A$ equipped with the norm $\|\varphi\|_+ := \|A\varphi\|_0$.

Let us consider a singularly perturbed operator $\tilde{A} \in \mathcal{P}_{\mathrm{s}}(A)$. Without loss of generality we can assume that $\tilde{A} \geq 1$. Indeed, if $-\infty < m = \inf \sigma(\tilde{A}) \leq 0$, then we take $\tilde{A} + (1-m)\mathbf{1}$ in the role of $\tilde{A}$, where $\mathbf{1}$ denotes the identity operator. And if $0 < m < 1$, then we consider $m^{-1}\tilde{A}$ instead of $\tilde{A}$.

With $\tilde{A}$ we associate the rigged Hilbert space

$$\tilde{\mathcal{H}}_- \sqsupset \mathcal{H}_0 \sqsupset \tilde{\mathcal{H}}_+, \tag{7.44}$$

constructed in the standard way (see Chapter 4 and [42, 44]). We want to analyze the structure of the spaces $\tilde{\mathcal{H}}_+$ and $\tilde{\mathcal{H}}_-$ taking into account that $\tilde{A}$ is singularly perturbed with respect to A.

First, note that the positive space $\tilde{\mathcal{H}}_+$ in (7.44) coincides with $\operatorname{Dom} \tilde{A}$ as a set because $\tilde{A} \geq 1$ and the positive inner product can be defined as

$$(f, g)_+^{\sim} = (\tilde{A}f, \tilde{A}g)_0, \quad f, g \in \operatorname{Dom} \tilde{A}.$$

Further, since $\tilde{A} \in \mathcal{P}_{\mathrm{s}}(A)$, there exists a dense in $\mathcal{H}_0$ linear set $\mathfrak{D} \subset \operatorname{Dom} A \cap \operatorname{Dom} \tilde{A}$ such that

$$(f, g)_+ = (f, g)_+^{\sim}, \quad f, g \in \mathfrak{D}. \tag{7.45}$$

Moreover, by Definition 7.2.1, the set $\mathfrak{D}$ is a closed subspace in both $\mathcal{H}_+$ and $\tilde{\mathcal{H}}_+$. Hence, one can write

$$\mathcal{H}_+ = \mathcal{M}_+ \oplus \mathcal{N}_+, \quad \tilde{\mathcal{H}}_+ = \tilde{\mathcal{M}}_+ \oplus \tilde{\mathcal{N}}_+, \tag{7.46}$$

where

$$\mathcal{M}_+ = \tilde{\mathcal{M}}_+ = \mathfrak{D}. \tag{7.47}$$

Now (7.46) and (7.47) imply

$$\mathcal{H}_0 = \mathcal{M}_0 \oplus \mathcal{N}_0, \tag{7.48}$$

$$\mathcal{M}_0 = A\mathcal{M}_+ = \tilde{A}\mathcal{M}_+, \quad \mathcal{N}_0 = A\mathcal{N}_+ = \tilde{A}\tilde{\mathcal{N}}_+.$$

And moreover, similar orthogonal decompositions are valid for the spaces with negative norms.

Proposition 7.5.2. *Given two rigged Hilbert spaces,* (7.43) *and* (7.44), *we assume that the positive spaces,* $\mathcal{H}_+$ *and* $\tilde{\mathcal{H}}_+$ *are decomposed according to* (7.46) *and* (7.47). *Then the negative spaces* $\mathcal{H}_-$ *and* $\tilde{\mathcal{H}}_-$ *admit the similar orthogonal decompositions*

$$\mathcal{H}_- = \mathcal{M}_- \oplus \mathcal{N}_-, \quad \tilde{\mathcal{H}}_- = \tilde{\mathcal{M}}_- \oplus \tilde{\mathcal{N}}_-, \tag{7.49}$$

with

$$\mathcal{M}_- \approx \tilde{\mathcal{M}}_-, \tag{7.50}$$

where $\approx$ *means that*

$$\tilde{\mathcal{M}}_- = \tilde{D}_{-,+} I_{+,-} \mathcal{M}_-,$$

where $I_{+,-} = D^{-1}_{-,+}$, $\tilde{D}_{-,+}$ *denote the Berezansky canonical isomorphisms in* (7.43) *and* (7.44), *respectively. In addition,*

$$\mathcal{N}_- \cap \mathcal{H}_0 = \{0\} = \tilde{\mathcal{N}}_- \cap \mathcal{H}_0. \tag{7.51}$$

Proof. Applying the mappings $D_{-,+}$ and $\tilde{D}_{-,+}$ to the corresponding decompositions in (7.46), we obtain (7.49) with

$$\tilde{\mathcal{M}}_- = \tilde{D}_{-,+} \tilde{\mathcal{M}}_+, \quad \tilde{\mathcal{N}}_- = \tilde{D}_{-,+} \tilde{\mathcal{N}}_+.$$

In particular, for $\omega = D_{-,+}\varphi$ and $\tilde{\omega} = \tilde{D}_{-,+}\varphi$ with $\varphi \in \mathfrak{D}$, due to (7.47) we have

$$\langle \omega, \psi \rangle_{-,+} = (\varphi, \psi)_+ = (\varphi, \psi)^{\sim}_+ = \langle \tilde{\omega}, \psi \rangle^{\sim}_{-,+}, \quad \psi \in \mathfrak{D}. \tag{7.52}$$

It is obvious that

$$\|\omega\|_- = \|D_{-,+}\varphi\|_- = \|\varphi\|_+ = \|\tilde{D}_{-,+}\varphi\|^{\sim}_- = \|\tilde{\omega}\|^{\sim}_-,$$

because $D_{-,+}$ and $\tilde{D}_{-,+}$ are isometric. Moreover,

$$l_\omega(\eta) = \langle \omega, \eta \rangle_{-,+} = 0 = \langle \tilde{\omega}, \tilde{\eta} \rangle^{\sim}_{-,+} = l_{\tilde{\omega}}(\tilde{\eta}), \quad \eta \in \mathcal{N}_+, \ \tilde{\eta} \in \tilde{\mathcal{N}}_+, \tag{7.53}$$

where l_ω and $l_{\tilde{\omega}}$ denote the linear functionals on $\mathcal{H}_+$ and $\tilde{\mathcal{H}}_+$ which are generated by ω and $\tilde{\omega}$, respectively. In addition, $\mathfrak{D} = \mathcal{M}_+ = \tilde{\mathcal{M}}_+$. Now, by the polarization identity we easily obtain

$$(\omega_1, \omega_2)_- = (\tilde{\omega}_1, \tilde{\omega}_2)^{\sim}_-, \quad \forall \omega_1, \omega_2 \in \mathcal{M}_-,$$

where we used that $\tilde{\omega}_i = \tilde{D}_{-,+} I_{+,-} \omega_i$, $i = 1, 2$. Therefore, the mapping

$$D_{-,+}\varphi = \omega \longmapsto \tilde{\omega} = \tilde{D}_{-,+}\varphi, \quad \varphi \in \mathfrak{D}$$

is isometric and thus (7.50) is proved. Finally, (7.51) follows directly from the denseness of $\mathfrak{D}$ in $\mathcal{H}_0$ (see Theorem 7.5.1). □

Since we supposed that $\tilde{A} \geq 1$, there exists the bounded inverse operator $\tilde{A}^{-1}$. By the Kreĭn formula (see Chapter 3),

$$\tilde{A}^{-1} = A^{-1} + \tilde{B}, \tag{7.54}$$

where $\tilde{B} = B^{-1}P_{\mathcal{N}_0}$. Here B is a bounded positive operator on $\mathcal{N}_0$. We recall that B can be constructed based on the quadratic form γ in such a way that $\operatorname{Ker}\tilde{B} = \mathcal{M}_0$, where $\mathcal{M}_0 := A\mathfrak{D}$. We recall also that the domain of $\tilde{A}$ has an explicit description in terms of the operator $\tilde{B}$:

$$\operatorname{Dom}\tilde{A} = \{g \in \mathcal{H}_0 \mid g = f + \tilde{B}Af,\ f \in \mathcal{H}_+ = \operatorname{Dom}A\}. \tag{7.55}$$

Another characterization of $\operatorname{Dom}\tilde{A}$ is established in the following proposition.

Proposition 7.5.3. *For every $\tilde{A} \in \mathcal{P}_{\mathrm{s}}(A)$, $\tilde{A} \geq 1$, the space $\tilde{\mathcal{H}}_+ = \operatorname{Dom}\tilde{A}$ admits the orthogonal decomposition:*

$$\tilde{\mathcal{H}}_+ = \tilde{\mathcal{M}}_+ \oplus \tilde{\mathcal{N}}_+ = \mathcal{M}_+ \oplus \tilde{\mathcal{N}}_+, \tag{7.56}$$

where

$$\tilde{\mathcal{M}}_+ = \mathcal{M}_+ = \mathfrak{D} \sqsubset \mathcal{H}_0,$$

and the subspace $\tilde{\mathcal{N}}_+$ is connected with $\mathcal{N}_+$ similarly to (7.55)*:*

$$\tilde{\mathcal{N}}_+ = \{\theta_+ \in \mathcal{H}_0 \mid \theta_+ = \eta_+ + \tilde{B}A\eta_+,\ \ \eta_+ \in \mathcal{N}_+\}. \tag{7.57}$$

In addition,

$$\tilde{A}\theta_+ = A\eta_+.$$

Proof. The representation of $\tilde{\mathcal{H}}_+$ in the form (7.56) holds due to (7.46) and (7.47). Further, since

$$\mathcal{H}_+ = \mathcal{M}_+ \oplus \mathcal{N}_+, \quad \mathcal{M}_+ = \mathfrak{D},$$

one can write for each $f \in \mathcal{H}_+$

$$f = \varphi_+ \oplus \eta_+, \quad \varphi_+ = P_{\mathcal{M}_+}f, \quad \eta_+ = P_{\mathcal{N}_+}f,$$

where $P_{\mathcal{L}}$, $\mathcal{L} = \mathcal{M}_+$, and $\mathcal{N}_+$, stand for the respective orthogonal projections. Now using (7.55) and (7.56), for every $g \in \operatorname{Dom}\tilde{A} = \tilde{\mathcal{H}}_+$ we obtain:

$$g = \varphi_+ + \eta_+ + \tilde{B}A(\varphi_+ + \eta_+) = \varphi_+ + \theta_+, \quad \theta_+ := \eta_+ + \tilde{B}A\eta_+.$$

Here $\tilde{B}A\varphi_+ = 0$ since $A\varphi_+ \in \operatorname{Ker}\tilde{B} = \mathcal{M}_0$. Thus, (7.57) is proved. Finally, $A\eta_+ = \tilde{A}\theta_+$ follows directly from $Af = \tilde{A}g$. □

We are now in position to formulate a result that is important for further considerations.

Theorem 7.5.4. *For every $\tilde{A} \in \mathcal{P}_s(A)$, $\tilde{A} \geq 1$, the inner product in the negative space $\tilde{\mathcal{H}}_-$ from* (7.44) *admits a form-sum representation:*

$$(\cdot,\cdot)_-^{\sim} = (\cdot,\cdot)_- + \tau(\cdot,\cdot), \tag{7.58}$$

where

$$\tau(\cdot,\cdot) := (A^{-1}\cdot, \tilde{B}\cdot)_0 + (\tilde{B}\cdot, A^{-1}\cdot)_0 + (\tilde{B}\cdot, \tilde{B}\cdot)_0. \tag{7.59}$$

is a singular quadratic form on $\mathcal{H}_-$.

Proof. By definition, the space $\tilde{\mathcal{H}}_-$ is the completion of $\mathcal{H}_0$ with respect to the inner product

$$(h_1, h_2)_-^{\sim} := (\tilde{A}^{-1}h_1, \tilde{A}^{-1}h_2)_0, \quad h_1, h_2 \in \mathcal{H}_0.$$

By the Kreĭn formula (7.54) for $\tilde{A}^{-1}$, we obtain

$$(h_1, h_2)_-^{\sim} = (A^{-1}h_1, A^{-1}h_2)_0 + \tau(h_1, h_2),$$

where $\tau(\cdot,\cdot)$ is defined by (7.59). It is obvious that this form is symmetric and non-positive. From (7.59) it easily follows that

$$\operatorname{Ker}\tau = \operatorname{Ker}\tilde{B} = \mathcal{M}_0.$$

We recall that $\mathcal{M}_0 = A\mathfrak{D}$. Hence, the inner products in $\tilde{\mathcal{H}}_-$ and $\mathcal{H}_-$ coincide after their restrictions onto $\mathcal{M}_0$:

$$(\cdot,\cdot)_- \restriction \mathcal{M}_0 = (\cdot,\cdot)_-^{\sim} \restriction \mathcal{M}_0. \tag{7.60}$$

In particular, it follows that τ is singular in $\mathcal{H}_-$ since the set $\operatorname{Ker}\tau = \mathcal{M}_0$ is dense in $\mathcal{H}_-$. □

We remark that (7.60) implies $\mathcal{M}_- \approx \tilde{\mathcal{M}}_-$, where $\mathcal{M}_-$ and $\tilde{\mathcal{M}}_-$ are the completions of $\mathcal{M}_0$ in the norms $\|\cdot\|_-$ and $\|\cdot\|_-^{\sim}$, respectively. In fact, these subspaces coincide as sets, although they are included in different spaces, $\mathcal{H}_-$ and $\tilde{\mathcal{H}}_-$, respectively.

It is well known (see, for example, [135]) that for each weakly singularly perturbed operator $\tilde{A} \in \mathcal{P}_{ws}(A)$, the space $\tilde{\mathcal{H}}_1$ can be constructed by the form-sum method. In other words, the inner product $(\cdot,\cdot)_1^{\sim}$ in $\tilde{\mathcal{H}}_1$ can be defined as $(\cdot,\cdot)_1 + \gamma(\cdot,\cdot)$, where a singular quadratic form γ belongs to the $\mathcal{H}_{-1}$-class. The previous theorem shows that for the case with a more singular γ, when it belongs to the $\mathcal{H}_{-2}$-class, the space $\tilde{\mathcal{H}}_-$ can be constructed by the form-sum method too. Moreover, the singularly perturbed operator $\tilde{A}$, which originally is defined by the self-adjoint extensions method, can be recovered as the operator associated with the chain (7.44). These observations lead to

Theorem 7.5.5. *For every operator $\tilde{A} \in \mathcal{P}_s(A)$, $\tilde{A} \geq 1$, its inverse $\tilde{A}^{-1}$ is uniquely associated, in the sense of the second representation theorem (see [107]), with the positive quadratic form*

$$\chi_-^{\sim}(h_1, h_2) = (Th_1, Th_2)_0, \quad T \equiv \tilde{A}^{-1}, \ h_1, h_2 \in \mathcal{H}_0,$$

which is defined as the inner product in $\mathcal{H}_-^{\sim}$.

Proof. Given $\tilde{A} \in \mathcal{P}_s(A)$, we define the form τ by (7.59). To this end we use the positive operator $\tilde{B}$ acting in $\mathcal{H}_0$. Then the form $\chi_-^{\sim}$, as the inner product in $\mathcal{H}_-$, admits the form-sum representation (see Theorem 7.5.4)

$$\chi_-^{\sim}[\cdot] = (\cdot, \cdot)_- + \tau(\cdot, \cdot).$$

It is clear that τ is singular in $\mathcal{H}_-$, since $\operatorname{Ker} \tilde{B} = \mathcal{M}_0$ is dense in this space. Since $\chi_-[\cdot] = \|\cdot\|_-^2$, the form $\chi_-^{\sim}$ is positive. However it defines the negative inner product since $\chi_-^{\sim} \leq \chi_0$, where $\chi_0[\cdot] = (\cdot, \cdot)_0$. Therefore, the second representation theorem for quadratic forms yields $\chi_-^{\sim}(\cdot, \cdot) = (T\cdot, T\cdot)_0$, where $T = \tilde{A}^{-1}$. □

Example 7.5.6 (Treating rank-one singular perturbations by means of the rigged space method)**.** We will apply the rigged Hilbert space method to construct the perturbed operator in the case of singular rank-one perturbations. Let us consider the singularly perturbed operator $\tilde{A}$ given by the formal expression $\tilde{A} = A + \lambda\gamma_\omega$, $\lambda > 0$, where the rank-one quadratic form γ_ω is defined by

$$\gamma_\omega(\cdot, \cdot) = \langle \cdot, \omega \rangle \langle \omega, \cdot \rangle, \quad \omega \in \mathcal{H}_- \backslash \mathcal{H}_0, \ \|\omega\|_- = 1.$$

Clearly, γ_ω is singular in $\mathcal{H}_0$, since $\omega \notin \mathcal{H}_0$ and $\operatorname{Ker} \gamma_\omega$ is dense in $\mathcal{H}_0$. Precisely, the operator $\tilde{A} \in \mathcal{P}_s(A)$ is uniquely defined by the Kreĭn formula:

$$\tilde{A}^{-1} = A^{-1} + \beta P_\eta = A^{-1} + \beta(\cdot, \eta)_0 \eta, \tag{7.61}$$

where

$$\eta = A^{-1}\omega, \quad \beta = 1/\lambda \in \mathbb{R}.$$

According to the above general construction, we have

$$\tilde{A}g = Af, \quad g \in \operatorname{Dom} \tilde{A}, \ f \in \operatorname{Dom} A, \tag{7.62}$$

where

$$\operatorname{Dom} \tilde{A} = \{g \in \mathcal{H} \mid g = f + \beta c_f \eta, \ f \in \operatorname{Dom} A\}, \tag{7.63}$$

and

$$c_f := (Af, \eta)_0 = \langle f, \omega \rangle.$$

Since $\tilde{A}^{-1}$ is the sum of two bounded positive operators, $\tilde{A}$ is strictly positive, $\tilde{A} \geq a > 0$. Without loss of generality we will suppose that $\tilde{A} \geq 1$.

Let us define the perturbed Hilbert space $\tilde{\mathcal{H}}_1$ as the completion of $\operatorname{Dom}\tilde{A}$ with respect to the inner product

$$(g_1, g_2)_{\tilde{1}} := (\tilde{A}g_1, g_2)_0, \quad g_1, g_2 \in \operatorname{Dom}\tilde{A}.$$

Using (7.62) and (7.63) we can rewrite this inner product as

$$\begin{aligned}(g_1, g_2)_{\tilde{1}} &= (Af_1, g_2)_0 = (Af_1, f_2)_0 + \beta(Af_1, \eta)_0(\eta, Af_2)_0 \\ &= (f_1, f_2)_1 + \beta\langle f_1, \omega\rangle\langle\omega, f_2\rangle = (f_1, f_2)_1 + \beta\gamma_\omega(f_1, f_2).\end{aligned}$$

If we assume that $\omega \in \mathcal{H}_{-2}\setminus\mathcal{H}_{-1}$, then the form γ_ω is singular in $\mathcal{H}_1$ and belongs to the $\mathcal{H}_{-2}$-class. In this case the space $\tilde{\mathcal{H}}_1$ admits the orthogonal decomposition:

$$\tilde{\mathcal{H}}_1 = \mathcal{H}_1 \oplus \tilde{\mathcal{N}}_1. \tag{7.64}$$

The subspace $\tilde{\mathcal{N}}_1$ in (7.64) is one-dimensional. It can be constructed by means of the form γ_ω by the standard procedure. We recall that the singularity of γ_ω in $\mathcal{H}_1$ follows from the denseness of the set $\operatorname{Ker}\gamma_\omega$ in $\mathcal{H}_1$. It should be noted that all vectors $f \in \operatorname{Dom}A$ with $c_f \neq 0$ do not belong to $\operatorname{Dom}\tilde{A}$, since $\eta \notin \operatorname{Dom}A$ (see (7.63)).

One can define the space $\tilde{\mathcal{H}}_{-1}$ adjoint to $\tilde{\mathcal{H}}_1$ as the completion of $\mathcal{H}_0$ with respect to the inner product

$$\begin{aligned}(h, l)_{\tilde{-1}} &:= (\tilde{A}^{-1}h, l)_0 = (A^{-1}h, l)_0 + \beta(h, \eta)_0(\eta, l)_0 \\ &= (h, l)_{-1} + \beta\langle A^{-1}h, \omega\rangle\langle\omega, A^{-1}l\rangle, \quad h, l \in \mathcal{H}_0.\end{aligned}$$

Hence we have

$$(\cdot,\cdot)_{\tilde{-1}} = (\cdot,\cdot)_{-1} + \beta\gamma_\eta(\cdot,\cdot) = (\cdot,\cdot)_{-1} + \beta\gamma_\omega(A^{-1}\cdot, A^{-1}\cdot),$$

where $\gamma_\eta(\cdot,\cdot) := (\cdot,\eta)(\eta,\cdot)$ with $\eta = A^{-1}\omega$. It is obvious the quadratic form γ_η is singular in $\mathcal{H}_{-1}$, because the set

$$\operatorname{Ker}\gamma_\eta = \mathcal{M}_0 := \{h \in \mathcal{H}_0 \mid (h, \eta)_0 = 0\}$$

is dense in $\tilde{\mathcal{H}}_{-1}$ due to $\omega \notin \mathcal{H}_{-1}$. By analogy with (7.64), the space $\tilde{\mathcal{H}}_{-1}$ admits the orthogonal decomposition:

$$\tilde{\mathcal{H}}_{-1} = \mathcal{H}_{-1} \oplus \tilde{\mathcal{N}}_{-1}, \tag{7.65}$$

where $\tilde{\mathcal{N}}_{-1}$ is the one-dimensional subspace constructed by means of the form γ_η. Thus, $\tilde{\mathcal{N}}_{-1} = \{c\eta \mid c \in \mathbb{C}\}$.

Further, the space $\tilde{\mathcal{H}}_+ \equiv \tilde{\mathcal{H}}_2$ coincides with $\operatorname{Dom}\tilde{A}$, which is equipped with the inner product:

$$(g_1, g_2)_{\tilde{+}} = (\tilde{A}g_1, \tilde{A}g_2)_0 = (Af_1, Af_2)_0 = (f_1, f_2)_+, \tag{7.66}$$

where $f_1, f_2 \in \operatorname{Dom} A$. These vectors are connected with $g_1, g_2 \in \operatorname{Dom} \tilde{A}$ via (7.63). In particular, if f_1, f_2 are orthogonal to ω in the sense of the duality pairing, then $g_1 = f_1$, $g_2 = f_2$, and they belong to $\mathcal{M}_+ = \operatorname{Ker} \gamma_\omega \equiv \mathfrak{D}$. Moreover, (7.66) implies the equality of the inner products restricted on $\mathfrak{D}$:

$$(\cdot,\cdot)_+^{\sim} \restriction \mathfrak{D} = (\cdot,\cdot)_+ \restriction \mathfrak{D}.$$

Therefore, $\tilde{\mathcal{M}}_+$ coincides with $\mathcal{M}_+$ despite of the fact that these subspaces belong to distinct Hilbert spaces. Thus, we can write

$$\tilde{\mathcal{H}}_+ = \mathcal{M}_+ \oplus \tilde{\mathcal{N}}_+, \tag{7.67}$$

where $\tilde{\mathcal{N}}_+$ is a one-dimensional subspace. Its structure is described in Proposition 7.5.3.

Finally, we assert that the space $\tilde{\mathcal{H}}_-$ adjoint to $\tilde{\mathcal{H}}_+$ can be defined as the completion of $\mathcal{H}_0$ with respect to the inner product

$$(h_1, h_2)_-^{\sim} := (\tilde{A}^{-1} h_1, \tilde{A}^{-1} h_2)_0, \quad h_1, h_2 \in \mathcal{H}_0.$$

Indeed, by the Kreĭn formula (7.61), we have

$$\begin{aligned}(h_1, h_2)_-^{\sim} &= (A^{-1} h_1 + \beta(h_1, \eta)_0 \eta, A^{-1} h_2 + \beta(h_2, \eta)_0 \eta)_0 \\ &= (A^{-1} h_1, A^{-1} h_2)_0 + \tau_\omega(h_1, h_2) = (h_1, h_2)_- + \tau_\omega(h_1, h_2),\end{aligned}$$

where the Hermitian quadratic form τ_ω admits the representation

$$\begin{aligned}\tau_\omega(\cdot,\cdot) &= \beta(A^{-1}\cdot, \eta)_0 (\eta, \cdot)_0 + \beta(\cdot, \eta)_0 (\eta, A^{-1}\cdot)_0 + \beta^2(\cdot, \eta)_0 (\eta, \cdot)_0 \\ &= \beta(\cdot, \eta_+)_0 (\eta, \cdot)_0 + \beta(\cdot, \eta)_0 (\eta_+, \cdot)_0 + \beta^2(\cdot, \eta)_0 (\eta, \cdot)_0,\end{aligned} \tag{7.68}$$

where $\eta_+ := A^{-1}\eta$ and we took into account that $\|\eta\|_0^2 = 1$. The quadratic form τ_ω is singular in $\mathcal{H}_-$ since the vector $\omega \notin \mathcal{H}_-$. Thus,

$$(\cdot,\cdot)_-^{\sim} = (\cdot,\cdot)_- + \tau_\omega(\cdot,\cdot). \tag{7.69}$$

It is easy to see that τ_ω is non-positive. For this reason it is impossible to decompose the space $\tilde{\mathcal{H}}_-$ into an orthogonal sum of the type $\mathcal{H}_- \oplus \tilde{\mathcal{N}}_-$. Nevertheless, we can write:

$$\tilde{\mathcal{H}}_- = \tilde{\mathcal{M}}_- \oplus \tilde{\mathcal{N}}_-, \quad \tilde{\mathcal{M}}_- \approx \mathcal{M}_-, \tag{7.70}$$

where $\tilde{\mathcal{M}}_- \approx \mathcal{M}_-$ means that these subspaces coincide, yet belong to distinct spaces. In fact, $\tilde{\mathcal{M}}_- = \tilde{D}_{-,0} I_{0,-} \mathcal{M}_-$ (cf. with (7.50)). Indeed, each vector $\tilde{\omega}$ from $\tilde{\mathcal{M}}_-$, as the adjoint (dual) to some vector $\tilde{A}^{-1}\mu = \varphi \in \tilde{\mathcal{M}}_+$, $\mu \in \mathcal{M}_0$, can be identified with $\omega \in \mathcal{M}_-$, which is in turn adjoint to

$$A^{-1}\mu = \varphi \in \mathcal{M}_+ = \tilde{\mathcal{M}}_+.$$

Of course, by this construction, the subspace $\tilde{\mathcal{N}}_-$ is the image of $\mathcal{N}_-$ under a unitary operator, although $\tilde{A}^{-1}\eta \neq A^{-1}\eta$, $\eta \in \mathcal{N}_0$.

Summarizing our analysis, we conclude that an arbitrary rank-one singular perturbation given by a quadratic form $\gamma_\omega \in \mathcal{H}_{-2}$-class can be treated in the rigged Hilbert space by setting the generalized form-sum method. Moreover, this can be done in two different ways:

(**1**) One can define the corresponding perturbed operator $\tilde{A}$ as the one associated with a new rigged triple $\tilde{\mathcal{H}}_{-1} \sqsupset \mathcal{H}_0 \sqsupset \tilde{\mathcal{H}}_1$, where the inner products in $\tilde{\mathcal{H}}_{-1}$ and $\tilde{\mathcal{H}}_1$ have the form-sum representations:

$$(\cdot,\cdot)_{-1}^{\sim} = (\cdot,\cdot)_{-1} + \beta\gamma_\eta(\cdot,\cdot) = (\cdot,\cdot)_{-1} + \beta\gamma_\omega(A^{-1}\cdot, A^{-1}\cdot),$$

$$(g_1, g_2)_1^{\sim} = (f_1, f_2)_1 + \beta\gamma_\omega(f_1, f_2), \quad g_i = f_i + c_{f_i}\eta,\ i = 1, 2.$$

(**2**) In the second way we begin with the form $\tilde{\chi}_-(\cdot,\cdot) = (\cdot,\cdot)_-^{\sim} \equiv (\cdot,\cdot)_{-2}^{\sim}$. It produces a singular form-sum perturbation of the negative inner product $(\cdot,\cdot)_-$. Then, the operator $\tilde{A}^{-1}$ is defined by means of $\tilde{\chi}_-(\cdot,\cdot)$ via the second representation theorem for quadratic forms (see [107]).

7.6 The singularity phenomenon

We start with an example of Hilbert space with two non-equivalent norms.

Example 7.6.1 (Functional spaces with non-equivalent norms). Let us consider the rigging of $L_2(\mathbb{R}^1, dx)$ by the Sobolev spaces:

$$W_2^{-2} \sqsupset L_2 \sqsupset W_2^2, \quad W_2^2 = \operatorname{Dom}\left(\mathbf{1} - \frac{d^2}{dx^2}\right)$$

and construct the operator S_δ in L_2 with the domain $\operatorname{Dom} S_\delta = W_2^2$:

$$S_\delta f := f(0)\eta, \quad f \in W_2^2, \quad \eta = I_{0,-2}\delta \in L_2.$$

Here $\delta(x)$ is the Dirac δ-function and $I_{0,-2}$ denotes the integral operator defined as the closure of the isometric mapping

$$\left(\mathbf{1} - \frac{d^2}{dx^2}\right)^{-1} : W_2^{-2} \longrightarrow L_2.$$

This mapping has an explicit representation in terms of Bessel functions (see [182, 183]). Note that the set

$$\mathfrak{D} := \{\varphi \in W_2^2 \mid \langle\delta, \varphi\rangle = \varphi(0) = 0\} \tag{7.71}$$

is dense in L_2 and

$$\operatorname{Ker} S_\delta = \mathfrak{D}.$$

Therefore, the operator S_δ is singular in L_2 (see Section 5.2).

Now consider the operator $T = (\mathbf{1} + S_\delta) : W_2^2 \longrightarrow L_2$ in L_2 which acts according to the rule

$$Tf = f + f(0)\eta, \quad f \in \operatorname{Dom} T = W_2^2.$$

Note that T coincides with the identity operator on the set $\mathfrak{D}$. However T differs from the continuous extension to W_2^2 of this identity operator. Indeed, for functions $f \in W_2^2$ such that $f(0) \neq 0$, $Tf \neq f$.

Now we introduce a new inner product $(\cdot,\cdot)^\sim$ on W_2^2, different from the usual one $(\cdot,\cdot)$ in L_2:

$$(f_1, f_2)^\sim := (Tf_1, Tf_2) = (g_1, g_2), \quad g_i = Tf_i, \ \ i = 1, 2.$$

It is easy to check that $(\cdot,\cdot)^\sim$ is an inner product, but not quasi-inner.

It is also clear that the norm $\|\cdot\|^\sim$ corresponding to this new inner product coincides on $\mathfrak{D}$ with the usual norm in L_2:

$$\|\varphi\|^\sim = \|\varphi\|, \quad \varphi \in \mathfrak{D}. \tag{7.72}$$

However

$$\|f\|^\sim \neq \|f\| \ \text{ for } \ f \in W_2^2 \setminus \mathfrak{D}, \ f(0) \neq 0. \tag{7.73}$$

Let $\tilde{L}_2$ denote the completion of W_2^2 with respect to the norm $\|\varphi\|^\sim$. Obviously, the norm $\|\cdot\|^\sim$ is not equivalent to the usual norm in L_2. Indeed, any sequence $f_n \in W_2^2$ which converges to zero in L_2, does not converge to zero in $\tilde{L}_2$ if $f_n(0) = c \neq 0$ for all n.

Proposition 7.6.2. *The space L_2 is included in $\tilde{L}_2$:*

$$L_2 \subset \tilde{L}_2. \tag{7.74}$$

Proof. It immediately follows from the equality (7.72) since $\mathfrak{D}$ is dense in L_2. □

Proposition 7.6.3.

$$\tilde{L}_2 \ominus L_2 = \{0\}. \tag{7.75}$$

Proof. It follows from Lemma 7.6.4 below. Indeed, let us assume that there exists a vector $0 \neq\in \tilde{L}_2 \ominus L_2$. Then due to (7.76) and since $h \perp L_2$, there exists a sequence $\varphi_n \in \mathfrak{D}$ which converges to h in $\tilde{L}_2$ and converges to zero in L_2. But this is impossible, because $\|\varphi_n\|_{\tilde{L}_2} \to \|h\|_{\tilde{L}_2}$ implies that $\|\varphi_n\|_{L_2}$ does not go to zero since $\|\varphi_n\|_{\tilde{L}_2} = \|\varphi_n\|_{L_2}$ (see (7.72)). □

Lemma 7.6.4. *The set $\mathfrak{D}$ is dense in both L_2 and $\tilde{L}_2$:*

$$\mathfrak{D} \sqsubset L_2, \quad \mathfrak{D} \sqsubset \tilde{L}_2. \tag{7.76}$$

Proof. Recall that $\tilde{L}_2$ is defined as the completion of W_2^2 with respect to the norm $\|f\|^\sim = \|Tf\|$, $f \in W_2^2$. To prove (7.76) it is sufficient to show that for any $f \in W_2^2$ there exists a sequence $\psi_n \in \mathfrak{D}$ that converges to $f \in \tilde{L}_2$ in the norm $\|\cdot\|^\sim$. Put $g = Tf \in L_2$ and choose an arbitrary sequence $\psi_n \in \mathfrak{D}$ that converges to g in L_2, i.e., $\|\psi_n - g\| \to 0$. Such a sequence exists since $\mathfrak{D}$ is dense in L_2. We claim that every such sequence converges to f in $\tilde{L}_2$, i.e., $\|\psi_n - f\|^\sim \to 0$. Indeed,

$$\|\psi_n - f\|^\sim = \|T(\psi_n - f)\| = \|\psi_n - Tf\| = \|\psi_n - g\| \longrightarrow 0,$$

where we used the equality $T\psi_n = \psi_n$, which is true due to $\psi_n \in \mathfrak{D}$. □

Thus, we proved that $\tilde{L}_2$ and L_2 are identical as functional spaces, thought their norms are not equivalent. So,

$$\|f\| \neq \|f\|^\sim = \|g\|, \quad g = f + f(0)\eta,$$

if $f(0) \neq 0$. In other words, we proved that $\tilde{L}_2$ and L_2 are identical as the vector spaces consisting of the equivalence classes of sequences φ_n from $\mathfrak{D}$ converging in the norm $\|\cdot\|_{\tilde{L}_2} = \|\cdot\|_{L_2}$ (see (7.72)). At the same time these spaces are different, if they are constructed by using non-equivalent norms on W_2^2.

Of course, there is no contradiction with the standard fact of functional analysis that every Hilbert space can be completely reconstructed from a dense subset. In the example discussed above, the space $\tilde{L}_2$, originally constructed from W_2^2, admits a renewal on the dense set $\mathfrak{D}$. The space L_2 can be restored from $\mathfrak{D}$ just in the same way. It is important that at first one has to fix the space and then choose a dense subset in it. So, the space L_2, as a completion of W_2^2, slightly differs from L_2, which is obtained as the completion of the set $\mathfrak{D} \subset W_2^2$. Indeed, the equivalence classes in these spaces are different. So, in the last case, each function $f \in W_2^2$ with $f(0) \neq 0$ is absent in $\mathfrak{D}$. It arises in L_2 only as the limit vector under the completion procedure for the set $\mathfrak{D}$. One can separate each such function f from its equivalence class and assign to it another norm $\|f\|^\sim \neq \|f\|$. This splitting procedure for equivalence classes can be performed by the use a non-continuous extension of the densely defined identity mapping. The operator $T = \mathbf{1} + S_\delta$ provides just this kind of extension on W_2^2. By the use of this operator, each function $f \in W_2^2$, $f(0) \neq 0$, is separated from its equivalence class in the usual space L_2 and then another norm $\|f\|^\sim = \|Tf\|$ is assigned to it. Of course, this kind of function belongs to a new equivalence class of convergent sequences from $\mathfrak{D}$ in $\tilde{L}_2$.

Let us consider another example.

Example 7.6.5 (Singular extension of the identity mapping). Let $\mathbf{A} \geq 1$ be an unbounded symmetric operator on $\mathcal{H}$ with a domain $\mathfrak{D}(\mathbf{A}) = \mathfrak{D}$. And let $A, \tilde{A} \geq a > 0$, be a couple of self-adjoint extensions of $\mathbf{A}$ with domains $\mathfrak{D}(A)$ and $\mathfrak{D}(\tilde{A})$, respectively. Since both operators A and $\tilde{A}$ are positive definite, their inverses

A^{-1} and $\tilde{A}^{-1}$ are bounded and defined on the whole space $\mathcal{H}$. We introduce the operator

$$W := \tilde{A}^{-1}A : \mathfrak{D}(A) \longrightarrow \mathfrak{D}(\tilde{A}).$$

We see that the restriction of W to the dense in $\mathcal{H}$ set $\mathfrak{D}(\mathbf{A})$ acts as the identity map:

$$W\varphi = \varphi, \quad \varphi \in \mathfrak{D}. \tag{7.77}$$

We introduce now a new inner product $(\cdot,\cdot)^{\sim}$ on $\mathfrak{D}(A)$ by

$$(f_1, f_2)^{\sim} := (Wf_1, Wf_2) = (g_1, g_2), \quad g_i = Wf_i, \ f_i \in \mathfrak{D}(A), \ i = 1, 2.$$

Since $A \neq \tilde{A}$, $(\cdot,\cdot)^{\sim}$ differs from the inner product $(\cdot,\cdot)$ in $\mathcal{H}$:

$$(f_1, f_2)^{\sim} \neq (f_1, f_2), \quad \text{if} \quad f_1, f_2 \in \mathfrak{D}(A) \setminus \mathfrak{D}.$$

Let $\tilde{\mathcal{H}}$ denote the new Hilbert space constructed in the standard way from $\mathfrak{D}(A)$ with the inner product $(\cdot,\cdot)^{\sim}$.

It is easily seen that $\tilde{\mathcal{H}}$ contains $\mathcal{H}$ as a subset, $\mathcal{H} \subset \tilde{\mathcal{H}}$. This is true since the set $\mathfrak{D}$ is dense in $\mathcal{H}$ and the inner product $(\cdot,\cdot)^{\sim}$ coincides with $(\cdot,\cdot)$ on vectors from $\mathfrak{D}$:

$$(\varphi, \psi)^{\sim} = (\varphi, \psi), \quad \varphi, \psi \in \mathfrak{D}.$$

Hence, $\mathcal{H}$ arises in $\tilde{\mathcal{H}}$ as a closure of $\mathfrak{D}$. However, at the same time the set $\mathfrak{D}$ is dense in $\tilde{\mathcal{H}}$. To prove this it is sufficient to show that for each $f \in \mathfrak{D}(A)$ there exist a sequence $\psi_n \in \mathfrak{D}$ convergent to f in $\tilde{\mathcal{H}}$:

$$\|\psi_n - f\|^{\sim} \longrightarrow 0, \quad n \longrightarrow \infty.$$

For example, one can take any sequence ψ_n from $\mathfrak{D}$ which converges to the vector $Wf = g \in \mathfrak{D}(\tilde{A})$ in $\mathcal{H}$:

$$\|\psi_n - g\| = \|\psi_n - Wf\| \longrightarrow 0.$$

Then it is obvious that

$$\|\psi_n - f\|^{\sim} = \|W(\psi_n - f)\| = \|\psi_n - Wf)\| \longrightarrow 0,$$

where we used the fact that $W\psi_n = \psi_n$. This means that one can get each vector from $\tilde{\mathcal{H}}$, which is obtained by the completion of $\mathfrak{D}(A)$ with respect to the norm $\|\cdot\|^{\sim}$, as a limit of some sequence ψ_n from $\mathfrak{D}$. It is important that all such sequences ψ_n are convergent in $\mathcal{H}$. Furthermore, the corresponding equivalence classes from $\mathfrak{D}$ in $\mathcal{H}$ and $\tilde{\mathcal{H}}$ coinside. Therefore, $\mathcal{H}$ and $\tilde{\mathcal{H}}$ are identical as vector spaces of equivalence classes of sequences from $\mathfrak{D}$. However, the norms of vectors $f \in \mathfrak{D}(A) \setminus \mathfrak{D}$ in $\mathcal{H}$ and $\tilde{\mathcal{H}}$ are different:

$$\|f\|^{\sim} = \|g\| = \|\tilde{A}^{-1}Af\| \neq \|f\|, \quad f \in \mathfrak{D}(A) \setminus \mathfrak{D}.$$

Therefore, the metrics in these spaces are not equivalent.

The observed effect follows from the fact that the operator $W = \tilde{A}^{-1}A$ is unbounded and even non-closable. This operator represents a singular extension of the identity map (7.77) which is densely defined on $\mathfrak{D}$.

We will explain the above-observed non-trivial effect once more. One needs to understand how each vector $f \in \mathfrak{D}(A)\backslash\mathfrak{D}$ is separated from its equivalence class under the embedding $\mathcal{H}$ into $\tilde{\mathcal{H}}$. In fact, all sequences $\varphi_n \in \mathfrak{D}$ convergent to f in $\mathcal{H}$, once considered in the space $\tilde{\mathcal{H}}$, will converge to another vector $\tilde{f}$ with the norm $\|\tilde{f}\|^\sim = \|f\|$. Of course, f also belongs to the space $\tilde{\mathcal{H}}$, but here its norm is different, $\|f\|^\sim \neq \|f\|$. In turn, the vector $\tilde{f}$ belongs to $\mathcal{H}$, and $\|\tilde{f}\| \neq \|f\|$. In general, all equivalence classes in $\mathcal{H}$ arising from convergent sequences of elements from $\mathfrak{D}$ when we pass in the same equivalence classes in the space $\tilde{\mathcal{H}}$. However, go over from the metric $\mathcal{H}$ to the metric $\tilde{\mathcal{H}}$, some kind of splitting occurs. Each $f \in \mathfrak{D}(A) \setminus \mathfrak{D}$, as a limit element of some equivalence class in $\mathcal{H}$, is connected to another equivalence class, which is composed of the sequences $\psi_n \in \mathfrak{D}$ convergent in $\mathcal{H}$ to the vector Wf. Note again that W is a singular extension of the identical in $\mathcal{H}$ mapping $\mathbf{A}^{-1}\mathbf{A}$ restricted onto the set $\mathfrak{D}$. The singularity property for W follows from the fact that the set $\mathrm{Ker}(\mathbf{1}-W) = \mathfrak{D}$ is dense in both spaces, $\mathcal{H}$ and $\tilde{\mathcal{H}}$.

Let us recast the above observations in the general setting.

Given a couple of linear vector spaces $\mathcal{S}, \mathcal{G}$ satisfying $\mathcal{S} \subset \mathcal{G}$, and two inner products $(\cdot,\cdot)$ and $(\cdot,\cdot)^\sim$ on $\mathcal{G}$, let $\mathcal{H}$ and $\tilde{\mathcal{H}}$ denote the corresponding Hilbert spaces constructed from $\mathcal{G}$ in the standard way. We assume the following two conditions are fulfilled:

(1) the above inner products coincide on $\mathcal{S}$:

$$(\cdot,\cdot) \upharpoonright \mathcal{S} = (\cdot,\cdot)^\sim \upharpoonright \mathcal{S},$$

(2) the linear space $\mathcal{S}$ is dense in both spaces $\mathcal{H}$ and $\tilde{\mathcal{H}}$:

$$\mathcal{S} \sqsubset \mathcal{H}, \quad \mathcal{S} \sqsubset \tilde{\mathcal{H}}.$$

From **(2)** it follows that both spaces $\mathcal{H}$ and $\tilde{\mathcal{H}}$ can be constructed as completions of S with respect to the inner products $(\cdot,\cdot)$ and $(\cdot,\cdot)^\sim$, respectively. Then, under the condition **(1)**, one may naively think that the spaces $\mathcal{H}$ and $\tilde{\mathcal{H}}$ coincide. However, this is an erroneous conclusion. The examples considered above lead to explanation. The spaces $\mathcal{H}$, $\tilde{\mathcal{H}}$ have the same supply of vectors, but in general $\mathcal{H} \neq \tilde{\mathcal{H}}$, in the following sense. For vectors $g \in \mathcal{G} \setminus \mathcal{S}$ their norms in $\mathcal{H}$ and $\tilde{\mathcal{H}}$ are different: $\|g\| \neq \|g\|^\sim$. So, in the general case the quadratic form

$$\tau[\cdot] := (\cdot,\cdot)^\sim - (\cdot,\cdot), \quad Q(\tau) = \mathcal{G}$$

is non-trivial, i.e., it is not equal to zero. It is easily seen that this form is singular in both $\mathcal{H}$ and $\tilde{\mathcal{H}}$. The set of null-vectors for τ is dense in each of these spaces:

$$\mathrm{Ker}\,\tau = \mathcal{S}, \quad \mathcal{S} \sqsubset \mathcal{H}, \quad \mathcal{S} \sqsubset \tilde{\mathcal{H}}.$$

Note that by our constructions, the form τ is Hermitian, but non-positive. Indeed, if we assume that $\tau \geq 0$, then by its singularity property, each of the spaces $\tilde{\mathcal{H}}$, $\mathcal{H}$ should have a structure of the orthogonal sum: $\tilde{\mathcal{H}} = \mathcal{H} \oplus \mathcal{H}_\tau$, $\mathcal{H} = \tilde{\mathcal{H}} \oplus \mathcal{H}_\tau$ (see Chapter 5). However, this plainly is a contradiction.

We call the situation described above by conditions (**1**), (**2**) the *singularity phenomenon.*

This phenomenon has already occurred implicitly in the preceding subsections. Indeed, for each $\tilde{A} \in \mathcal{P}_{\mathrm{s}}(A)$, $\tilde{A} \geq 1$, the negative space $\tilde{\mathcal{H}}_-$ contains the linear set $\mathcal{M}_0 = A\mathfrak{D} = \tilde{A}\mathfrak{D}$ (the manifold $\mathfrak{D}$ is defined in (7.1)) which has two properties:

(**1**) the inner products in $\tilde{\mathcal{H}}_-$ and $\mathcal{H}_-$, after their restrictions to $\mathcal{M}_0$, are equal:

$$(\cdot,\cdot)_-^\sim \restriction \mathcal{M}_0 = (\cdot,\cdot)_- \restriction \mathcal{M}_0,$$

(**2**) $\mathcal{M}_0$ is dense in both, $\tilde{\mathcal{H}}_-$ and $\mathcal{H}_-$:

$$\mathcal{M}_0 \sqsubset \mathcal{H}, \quad \mathcal{M}_0 \sqsubset \tilde{\mathcal{H}}.$$

Thus, in the role of $\mathcal{G}$ one has to take the space $\mathcal{H}_0$ as the starting set for constructing $\mathcal{H}_-$ and $\tilde{\mathcal{H}}_-$.

We recall that $\tilde{\mathcal{H}}_-$ is constructed as the completion of $\mathcal{H}_0$ with respect to the inner product $(\cdot,\cdot)_-^\sim = (\tilde{A}^{-1}\cdot, \tilde{A}^{-1}\cdot)_0$. Here $\tilde{A}^{-1}$ is defined by the Kreĭn formula (7.54) based on a positive operator $\tilde{B}$ which differs from zero only on the subspace $\mathcal{N}_0 := \mathcal{H}_0 \ominus \mathcal{M}_0$. Thus, (**1**) with $\mathcal{S} = \mathcal{M}_0$ is in fact fulfilled due to (7.60). The condition (**2**) is also fulfilled. Indeed, that $\mathcal{M}_0$ is dense in $\mathcal{H}_-$ and $\tilde{\mathcal{H}}_-$ follows from the fact that $\mathcal{N}_0 \cap \mathcal{H}_+ = \mathcal{N}_0 \cap \tilde{\mathcal{H}}_+ = \{0\}$ (see Chapter 6). Nevertheless, we will prove the fact that $\mathcal{M}_0$ is dense in $\tilde{\mathcal{H}}_-$ independently.

Lemma 7.6.6. *Let $\tilde{A} \in \mathcal{P}_{\mathrm{s}}(A)$, $\tilde{A} \geq 1$, and let $\tilde{\mathcal{H}}_-$ be the completion of $\mathcal{H}_0$ with respect to the inner product* (7.58), *where the quadratic form τ is defined in accordance with* (7.59). *Then the subspace $\mathcal{M}_0 := \operatorname{Ker}\tau$ is dense in both, $\mathcal{H}_-$ and $\tilde{\mathcal{H}}_-$:*

$$\mathcal{M}_0 \sqsubset \mathcal{H}_-, \quad \mathcal{M}_0 \sqsubset \tilde{\mathcal{H}}_-. \tag{7.78}$$

Proof. That $\mathcal{M}_0$ is dense in $\mathcal{H}_-$ follows from the geometry of the rigged spaces (see Chapter 4). So, we only have to prove $\mathcal{M}_0 \sqsubset \tilde{\mathcal{H}}_-$. Let $h \in \mathcal{H}_0 = \operatorname{Ran}\tilde{A}$. Then $h = \tilde{A}g$ with some $g \in \operatorname{Dom}\tilde{A}$. Since $\mathfrak{D} = \mathcal{M}_+ := A^{-1}\mathcal{M}_0$ is dense in $\mathcal{H}_0$, there exists a sequence $\varphi_n \in \mathcal{M}_+$ such that $\|\varphi_n - g\|_0 \to 0$. Let us put $f_n := A\varphi_n = \tilde{A}\varphi_n$. It is obvious that $f_n \in \mathcal{M}_0$. We claim that the sequence f_n converges to the vector h in $\tilde{\mathcal{H}}_-$. Indeed, since $\tilde{A}^{-1}A\varphi_n = \varphi_n$, we have

$$\|h - f_n\|_-^\sim = \|\tilde{A}^{-1}(h - f_n)\|_0 = \|\tilde{A}^{-1}(\tilde{A}g - A\varphi_n)\|_0 = \|g - \varphi_n\|_0 \to 0. \quad \square$$

As a simple corollary, we conclude that the quadratic form τ is singular not only in $\mathcal{H}_-$, but also in the space $\tilde{\mathcal{H}}_-$. Indeed, by our construction $\operatorname{Ker}\tau = \mathcal{M}_0$, which is dense in $\tilde{\mathcal{H}}_-$.

7.6.1 Effects of the singularity phenomenon

The singularity phenomenon discussed above has different unexpected effects. In particular, they manifest in the construction of rigged Hilbert spaces.

In this subsection we show that an abstract rigged space

$$\mathcal{H}_- \sqsupset \mathcal{H}_0 \sqsupset \mathcal{H}_+ \tag{7.79}$$

admits an extension to a new rigged space

$$\tilde{\mathcal{H}}_- \sqsupset \mathcal{H}_0 \sqsupset \tilde{\mathcal{H}}_+ \tag{7.80}$$

such that the orthogonal decompositions

$$\tilde{\mathcal{H}}_+ = \mathcal{H}_+ \oplus \tilde{\mathcal{N}}_+ \text{ and } \tilde{\mathcal{H}}_- = \mathcal{H}_- \oplus \tilde{\mathcal{N}}_-$$

occur simultaneously. This seems impossible, since $\mathcal{H}_+$ is a part of $\tilde{\mathcal{H}}_+$ and hence the space $(\tilde{\mathcal{H}}_+)_- = \tilde{\mathcal{H}}_-$ dual to $\tilde{\mathcal{H}}_+$ should be contained in $\mathcal{H}_-$. However, we claim that the embedding $\mathcal{H}_+ \subset \tilde{\mathcal{H}}_+$ can be fulfilled simultaneously with $\mathcal{H}_- \subset \tilde{\mathcal{H}}_-$. Precisely, we claim that the chain of Hilbert spaces

$$\mathcal{H}_- \oplus \tilde{\mathcal{N}}_- = \tilde{\mathcal{H}}_- \sqsupset \mathcal{H}_0 \sqsupset \tilde{\mathcal{H}}_+ = \mathcal{H}_+ \oplus \tilde{\mathcal{N}}_+, \tag{7.81}$$

with $\tilde{\mathcal{N}}_+ \neq \{0\} \neq \tilde{\mathcal{N}}_-$, admits a non-contradictory to (7.79) interpretation under suitable additional requirements.

Let us describe in detail the construction of the chain (7.81) starting with a given rigged space of the form (7.79).

Let us extend the triple (7.79) to five spaces

$$\mathcal{H}_{--} \sqsupset \mathcal{H}_- \sqsupset \mathcal{H}_0 \sqsupset \mathcal{H}_+ \sqsupset \mathcal{H}_{++} \tag{7.82}$$

in such a way that the triple $\mathcal{H}_0 \sqsupset \mathcal{H}_+ \sqsupset \mathcal{H}_{++}$ forms a rigged Hilbert space. Then decompose $\mathcal{H}_{++}$ into the orthogonal sum of two non-trivial subspaces,

$$\mathcal{H}_{++} = \mathcal{M}_{++} \oplus \mathcal{N}_{++},$$

and require that $\mathcal{M}_{++}$ be dense in $\mathcal{H}_+$:

$$\mathcal{H}_+ \sqsupset \mathcal{M}_{++}. \tag{7.83}$$

Further, we consider two subspaces

$$\mathcal{M}_0 = D_{0,++}\mathcal{M}_{++}, \quad \mathcal{M}_+ = D_{+,++}\mathcal{M}_{++}, \tag{7.84}$$

in $\mathcal{H}_0$ and $\mathcal{H}_+$, respectively. Here $D_{0,++} : \mathcal{H}_{++} \to \mathcal{H}_0$ and $D_{+,++} : \mathcal{H}_{++} \to \mathcal{H}_+$ denote the Berezansky canonical isomorphisms. It is clear, that $\mathcal{M}_0$ is dense in $\mathcal{H}_-$ and $\mathcal{M}_+$ is dense in $\mathcal{H}_0$:

$$\mathcal{H}_- \sqsupset \mathcal{M}_0, \quad \mathcal{H}_0 \sqsupset \mathcal{M}_+. \tag{7.85}$$

Now, let us define on $\mathcal{H}_{++}$ the non-negative quadratic form

$$\gamma(\varphi, \psi) := (P_{\mathcal{N}_0} D_{0,++}\varphi, D_{0,++}\psi)_0, \quad \varphi, \psi \in Q(\gamma) = \mathcal{H}_{++},$$

where $P_{\mathcal{N}_0}$ stands for the orthogonal projection onto the subspace $\mathcal{N}_0 = \mathcal{H}_0 \ominus \mathcal{M}_0$. It is easy to see that γ is zero on $\mathcal{M}_{++}$. Hence,

$$\operatorname{Ker} \gamma = \mathcal{M}_{++}. \tag{7.86}$$

Indeed, if $\varphi \in \mathcal{M}_{++}$, then

$$P_{\mathcal{N}_0} D_{0,++}\varphi = D_{0,++} P_{\mathcal{N}_{++}}\varphi = 0,$$

where $P_{\mathcal{N}_{++}}$ denotes the orthogonal projection onto $\mathcal{N}_{++} = \mathcal{H}_{++} \ominus \mathcal{M}_{++}$. We recall that $\mathcal{N}_0 = D_{0,++}\mathcal{N}_{++}$. Therefore,

$$\gamma[\varphi] = 0, \quad \varphi \in \mathcal{M}_{++}.$$

This means that γ is singular in $\mathcal{H}_+$ since the set $\mathcal{M}_{++}$ is dense in $\mathcal{M}_+$.

Further, we introduce the space $\tilde{\mathcal{H}}_+$ as the completion of the set

$$\tilde{\mathfrak{D}} := \{g \in \mathcal{H}_0 \mid g = \varphi + \beta P_{\mathcal{N}_0} D_{0,++}\varphi, \ \varphi \in \mathcal{H}_{++}\}$$

with respect to the norm $\|g\|_+^{\sim}$ generated by the inner product

$$(g_1, g_2)_+^{\sim} := (\varphi_1, \varphi_2)_+ + \beta\gamma(\varphi_1, \varphi_2), \quad \varphi_1, \varphi_2 \in \mathcal{H}_{++}, \ \beta > 0. \tag{7.87}$$

Note that

$$(\varphi_1, \varphi_2)_+ = (D_{0,++}\varphi_1, \varphi_2)_0.$$

Now we assert that due to the conditions (7.83) and (7.86), the space $\tilde{\mathcal{H}}_+$ admits the orthogonal decomposition

$$\tilde{\mathcal{H}}_+ = \mathcal{H}_+ \oplus \tilde{\mathcal{N}}_+. \tag{7.88}$$

Indeed, $\mathcal{H}_+$ arises in $\tilde{\mathcal{H}}_+$ as a proper subspace due to the completion procedure with respect to the norm $\|\varphi\|_+^{\sim}$. Here we take into account that $\mathcal{M}_{++} \subset \tilde{\mathfrak{D}}$ and that γ equals to zero for all $g = \varphi \in \mathcal{M}_{++}$. So, $\|g\|_+^{\sim} = \|\varphi\|_+$ for all such vectors.

Moreover, it is not hard to show that under the condition

$$0 < \beta \leq 1 - \sup \|\eta\|_-, \tag{7.89}$$

where the supremum is taken over all $\eta \in \mathcal{N}_0$, $\|\eta\|_0 = 1$, the norm $\|g\|_+^{\sim}$ satisfies the inequality

$$\|g\|_0 \leq \|g\|_+^{\sim},$$

and defines the positive Hilbert space $\tilde{\mathcal{H}}_+$ (see Section 4) such that the embedding operator from $\tilde{\mathcal{H}}_+$ into $\mathcal{H}_0$ is injective. By this, the pair $\mathcal{H}_0$ and $\tilde{\mathcal{H}}_+$ produces the pre-rigged space

$$\mathcal{H}_0 \sqsupset \tilde{\mathcal{H}}_+, \tag{7.90}$$

which can be extended to the triple (7.80) in the standard way.

The next non-trivial result we formulate as an abstract theorem.

Theorem 7.6.7. *Let the positive space $\mathcal{H}_{++}$ from (7.82) be decomposed into the orthogonal sum $\mathcal{H}_{++} = \mathcal{M}_{++} \oplus \mathcal{N}_{++}$ in such a way that the subspace $\mathcal{M}_{++}$ is dense in $\mathcal{H}_+$, i.e., $\mathcal{M}_{++} \sqsubset \mathcal{H}_+$. Let $\tilde{\mathcal{H}}_+$ be constructed as the orthogonal extension of the space $\mathcal{H}_+$ (see (7.88)). In fact $\tilde{\mathcal{H}}_+$ is a completion of the set*

$$\tilde{\mathfrak{D}} := \{g \in \mathcal{H}_0 \mid g = \varphi + \beta D_{0,++} P_{\mathcal{N}_{++}} \varphi,\ \varphi \in \mathcal{H}_{++}\}$$

with respect to the inner product (7.87). Then the space $\tilde{\mathcal{H}}_-$, adjoint to $\tilde{\mathcal{H}}_+$ with respect to $\mathcal{H}_0$, extends the pre-rigging pair (7.90) to the rigged space (7.80). Furthermore, the space $\tilde{\mathcal{H}}_-$ admits an orthogonal decomposition which contains the space $\mathcal{H}_-$ from (7.79) as a proper subspace:

$$\tilde{\mathcal{H}}_- = \mathcal{H}_- \oplus \tilde{\mathcal{N}}_-. \tag{7.91}$$

Proof. By the definition of the negative space, $\tilde{\mathcal{H}}_-$ is a completion of $\mathcal{H}_0$ with respect to the negative norm

$$\|h\|_-^{\sim} := \sup_{\|g\|_+^{\sim}=1} |(h,g)_0|, \quad g \in \tilde{\mathcal{H}}_+. \tag{7.92}$$

Let us show that $\mathcal{H}_-$ exists in $\tilde{\mathcal{H}}_-$ as a subspace. To this end we note that $\mathcal{H}_-$ can be obtained from $\mathcal{M}_0$ (see (7.85)) by the completion procedure. In addition, we use the fact that vectors g in (7.92) can be taken from any subset of $\tilde{\mathcal{H}}_+$ which is dense in $\mathcal{H}_0$. In particular, vectors $g = \varphi$ can be taken from the set $\mathcal{M}_{++}$, since it is dense in $\mathcal{H}_+$ and in $\mathcal{H}_0$ also. We recall that the norms of these vectors in $\tilde{\mathcal{H}}_+$ and $\mathcal{H}_+$ are equal:

$$\|g\|_+^{\sim} = \|\varphi\|_+, \quad g = \varphi \in \mathcal{M}_{++}.$$

This follows from the equality $\|g\|_+^{\sim} = \|\varphi\|_+ + \gamma[\varphi]$, since $\gamma[\varphi] = 0$ for $\varphi \in \mathcal{M}_{++}$. This proves that for $h \in \mathcal{M}_0$:

$$\|h\|_-^{\sim} := \sup_{\|g\|_+^{\sim}=1} |(h,g)_0| = \sup_{\|\varphi\|_+=1} |(h,\varphi)_0| = \|h\|_-, \quad \varphi \in \mathcal{M}_{++}.$$

Therefore,

$$\|h\|_-^{\sim} = \|h\|_-, \quad h \in \mathcal{M}_0. \tag{7.93}$$

Now it is clear that since $\mathcal{M}_0$ is dense in $\mathcal{H}_-$, the closure of $\mathcal{M}_0$ forms the space $\mathcal{H}_-$ as a subspace in $\tilde{\mathcal{H}}_-$. Hence, (7.91) is proved. That completes the proof. □

Below we give another proof of this theorem.

To show (7.91) we consider in $\mathcal{H}_0$ the operator $\tilde{D}$ defined on

$$\operatorname{Dom} \tilde{D} = \tilde{\mathfrak{D}} = \{g \in \mathcal{H}_0 \mid g = \varphi + \beta P_{\mathcal{N}_0} D_{0,++} \varphi,\ \varphi \in \mathcal{H}_{++}\},$$

by the formula

$$\tilde{D} g = D_{0,++} \varphi \in \mathcal{H}_0.$$

It is not hard to see that $\tilde{D}$ is self-adjoint in $\mathcal{H}$ and satisfies the condition $\tilde{D} \geq 1$ due to (7.89). Moreover, the quadratic form generated on $\tilde{\mathfrak{D}}$ by the formula

$$\begin{aligned}(\tilde{D}g_1, g_2)_0 &= (D_{0,++}\varphi_1, \varphi_2 + \beta P_{\mathcal{N}_0} D_{0,++}\varphi_2)_0 \\ &= (D_{0,++}\varphi_1, \varphi_2) + \beta(D_{0,++}\varphi_1, P_{\mathcal{N}_0} D_{0,++}\varphi_2),\end{aligned}$$

coincides with the inner product in $\tilde{\mathcal{H}}_+$. This means that the operator $\tilde{D}$ is associated with the triple (7.80). By Proposition 4.9, the negative inner product in $\tilde{\mathcal{H}}_-$ can be written in terms of the inverse operator to $\tilde{D}$:

$$(h_1, h_2)_-^{\sim} = (\tilde{D}^{-1}h_1, h_2)_0.$$

Thus, for $h_1, h_2 \in \mathcal{H}_0$ we have

$$\begin{aligned}(h_1, h_2)_-^{\sim} &= (\tilde{D}^{-1}h_1, h_2)_0 = (g_1, h_2)_0 \\ &= (\varphi_1 + \beta P_{\mathcal{N}_0} D_{0,++}\varphi_1, h_2)_0 \\ &= (D_{0,++}^{-1}h_1, h_2)_0 + \beta(P_{\mathcal{N}_0}h_1, h_2)_0.\end{aligned}$$

From this it follows that

$$\|h\|_-^{\sim} = \|h\|_-, \quad h \in \mathcal{M}_0. \tag{7.94}$$

Hence, the completion of $\mathcal{M}_0$ as a set from the space $\tilde{\mathcal{H}}_-$ yields the space $\mathcal{H}_-$. So, (7.91) is proved in a different way.

Note that under the embedding in $\tilde{\mathcal{H}}_-$ each vector $h \in \mathcal{H}_0 \backslash \mathcal{M}_0$ splits into two parts, its projection on the space $\mathcal{H}_-$ and a non-zero component in the subspace $\tilde{\mathcal{N}}_-$. Its projection on $\mathcal{H}_-$ has the same norm in $\tilde{\mathcal{H}}_-$ as the negative norm of h in $\mathcal{H}_-$ from the rigged space (7.79):

$$\|P_{\mathcal{H}_-}h\|_{\tilde{\mathcal{H}}_-} = \|h\|_{\mathcal{H}_-}.$$

Thus, we can say that a similar "splitting" occurs for all equivalence classes of vectors $h \in \mathcal{H}_0 \setminus \mathcal{M}_0$. So, the sequences from $\mathcal{M}_{++}$ convergent to h in $\mathcal{H}_0$ yield in $\tilde{\mathcal{H}}_-$ the vector $P_{\mathcal{H}_1}h$. In particular, there exist sequences $\varphi_n \in \mathcal{M}_{++}$ which have non-zero limits in the norm $\|\varphi_n\|_\gamma = \gamma[\varphi_n]$ and converge to zero in $\mathcal{H}_-$. The corresponding vectors g_n converge to the projection of h on $\tilde{\mathcal{N}}_-$.

7.7 $\tilde{A}$-scales generated by singular quadratic forms

In this section we study in detail the structure of the new rigged space of the form (7.44) and its connection with quadratic forms $\gamma \in \mathcal{H}_{-2}(A)$-class, considered as singular perturbations of the operator A.

Let us begin with the triple (7.43) which is the rigged Hilbert space associated with the unperturbed (free) operator $A = A^* \geq 1$ on $\mathcal{H}_0$. Consider the five spaces chain

$$\mathcal{H}_- \equiv \mathcal{H}_{-2} \sqsupset \mathcal{H}_{-1} \sqsupset \mathcal{H}_0 \sqsupset \mathcal{H}_1 \sqsupset \mathcal{H}_2 \equiv \mathcal{H}_+ (= \mathrm{Dom}\, A) \tag{7.95}$$

as a part of the A-scale. We recall (for details see [42]) that any triple of the type (7.43), as well as the whole A-scale can be reconstructed from a pair of spaces: $\mathcal{H}_0 \sqsupset \mathcal{H}_k$ or $\mathcal{H}_{-k} \sqsupset \mathcal{H}_0$, $k > 0$.

Given a positive quadratic form γ of the $\mathcal{H}_{-2}$-class, we define a new inner product on $\mathcal{H}_0$ by

$$(h_1, h_2)_{-1}^{\sim} := (A^{-1}h_1, h_2)_0 + \gamma(A^{-1}h_1, A^{-1}h_2), \quad h_1, h_2 \in \mathcal{H}_0. \tag{7.96}$$

It is well defined since the operator A^{-1} maps $\mathcal{H}_0$ into $\mathcal{H}_+$ and hence the vectors $A^{-1}h_1, A^{-1}h_2 \in \mathcal{H}_+ = \operatorname{Dom}\gamma$. Let $\tilde{\mathcal{H}}_{-1}$ denote the Hilbert space corresponding to the inner product (7.96) that is, $\tilde{\mathcal{H}}_{-1}$ is the completion of $\mathcal{H}_0$ with respect to the norm

$$\|\cdot\|_{-1}^{\sim} := (\|A^{-1/2}\cdot\|_0^2 + \gamma[A^{-1}\cdot])^{1/2}. \tag{7.97}$$

Now let us assume that γ satisfies the inequality

$$\|\cdot\|_{-1}^{\sim} \leq \|\cdot\|_0. \tag{7.98}$$

Then

$$\tilde{\mathcal{H}}_{-1} \sqsupset \mathcal{H}_0. \tag{7.99}$$

Moreover, this couple admits an extension to the rigged space

$$\tilde{\mathcal{H}}_{-1} \sqsupset \mathcal{H}_0 \sqsupset \tilde{\mathcal{H}}_1. \tag{7.100}$$

In addition, by using (7.100), one can introduce in $\mathcal{H}_0$ a new operator associated with this rigging:

$$\tilde{A} := \tilde{D}_{-1,1} \restriction \{f \in \tilde{\mathcal{H}}_1 \mid \tilde{D}_{-1,1},\ f \in \mathcal{H}_0\}, \tag{7.101}$$

where $\tilde{D}_{-1,1} : \tilde{\mathcal{H}}_1 \to \tilde{\mathcal{H}}_{-1}$ is the Berezansky canonical isomorphism. It is clear that $\tilde{A} \geq 1$ since due to (7.98),

$$\|\cdot\|_0 \leq \|\cdot\|_1^{\sim} = \|\tilde{A}\cdot\|_0. \tag{7.102}$$

Further, having $\tilde{A}$, one can introduce a chain which is similar to (7.95) and consists of five spaces:

$$\tilde{\mathcal{H}}_- \equiv \tilde{\mathcal{H}}_{-2} \sqsupset \tilde{\mathcal{H}}_{-1} \sqsupset \mathcal{H}_0 \sqsupset \tilde{\mathcal{H}}_1 \sqsupset \tilde{\mathcal{H}}_2 \equiv \tilde{\mathcal{H}}_+ (= \operatorname{Dom}\tilde{A}). \tag{7.103}$$

Proposition 7.7.1. *Suppose the quadratic form $\gamma \in \mathcal{H}_{-2}$-class satisfies the condition:*

$$-\|f\|_1^2 \leq \gamma[f] \leq \|f\|_2^2 - \|f\|_1^2, \quad f \in \mathcal{H}_2 = \operatorname{Dom} A. \tag{7.104}$$

Then the operator $\tilde{A}$ associated with the rigging (7.103) *is strongly singularly perturbed, i.e., $\tilde{A} \in \mathcal{P}_{\rm ss}(A)$.*

Proof. From (7.104) we have

$$-(Af,f)_0 \le \gamma[f] \le \|Af\|_0^2 - (Af,f)_0, \quad f \in \mathcal{H}_+.$$

This inequality is equivalent to

$$-(A^{-1}h,h)_0 \le \gamma[A^{-1}h] \le \|h\|_0^2 - (A^{-1}h,h)_0, \quad h \in \mathcal{H}_0,$$

since each vector f can be written as $A^{-1}h$ for some $h \in \mathcal{H}_0$. In other terms,

$$-\|h\|_{-1}^2 \le \gamma[A^{-1}h] \le \|h\|_0^2 - \|h\|_{-1}^2,$$

which is equivalent to

$$0 \le \gamma[A^{-1}h] + \|h\|_{-1}^2 \le \|h\|_0^2.$$

Hence, the inequality (7.98) holds true. Now, proceeding as described above (see (7.101)), we can construct the operator $\tilde{A} \ge 1$. So, we only need to check that $\tilde{A} \in \mathcal{P}_{\rm ss}(A)$. To this end we notice that $\gamma[A^{-1}h] = 0$ for all $h \in \mathcal{M}_0$, where $\mathcal{M}_0 := \operatorname{Ker}\gamma$. Thus, $\tilde{A}^{-1}h = A^{-1}h$, $h \in \mathcal{M}_0$ and $\tilde{A}f = Af$, $f \in \operatorname{Ker}\gamma := \mathfrak{D}$. Therefore, $\tilde{A} \in \mathcal{P}_{\rm ss}(A)$ since $\operatorname{Ker}\gamma \sqsubset \mathcal{H}_1$. □

We note that the chain (7.103) can be constructed using the operator $S : \mathcal{H}_+ \to \mathcal{H}_-$ associated with the form γ (see (5.29)). Indeed, let $S = A^{\rm cl}A\mathbf{s}$, where $\mathbf{s}$ is a bounded positive operator in $\mathcal{H}_+$ such that $(\mathbf{s}\cdot,\cdot)_+ = \gamma[\cdot]$. Let us introduce the operator $T : \mathcal{H}_0 \to \mathcal{H}_-$ acting by the rule

$$Th = (\mathbf{1} + SA^{-1})h = h + SA^{-1}h, \quad h \in \mathcal{H}_0,$$

where $\mathbf{1}$ denotes the identity operator. Now we use T to define a new inner product on $\mathcal{H}_0$ by

$$(h,l)_-^{\sim} := (Th,Tl)_-, \quad h,l \in \mathcal{H}_0. \tag{7.105}$$

Now $\tilde{\mathcal{H}}_-$ arises as the completion of $\mathcal{H}_0$ with respect to the norm $\|h\|_-^{\sim}$. We obtain the pre-rigged couple $\tilde{\mathcal{H}}_- \sqsupset \mathcal{H}_0$ if the condition

$$\|h\|_-^{\sim} \le \|h\|_0, \quad h \in \mathcal{H}_0 \tag{7.106}$$

is satisfied. Now by the standard procedure we construct $\tilde{\mathcal{H}}_+$ and extend the pre-rigged space to a rigged space. Finally, we can define $\tilde{A}$ as an operator associated with the triple $\tilde{\mathcal{H}}_- \sqsupset \mathcal{H}_0 \sqsupset \tilde{\mathcal{H}}_+$.

Proposition 7.7.2. *Let $\mathbf{s} = \mathbf{s}^*$ be a positive bounded operator on $\mathcal{H}_+$. We assume that*

$$-(Af,f)_0 \le (\mathbf{s}f,f)_+ \le \|f\|_+^2 - (Af,f)_0, \quad f \in \mathcal{H}_+, \tag{7.107}$$

and

$$\operatorname{Ker}\mathbf{s} = \mathcal{M}_+ \sqsubset \mathcal{H}_1.$$

Let $\tilde{\mathcal{H}}_- \sqsupset \mathcal{H}_0 \sqsupset \tilde{\mathcal{H}}_+$ be the rigged Hilbert space constructed using the operators T and $S = A^{\rm cl}A\mathbf{s}$ in the way described above. Then the operator $\tilde{A}$ associated with this rigging belongs to the family $\mathcal{P}_{\rm ss}(A)$ and moreover, $\tilde{A} \ge 1$.

Proof. From (7.105) it follows that for $\tilde{A}$ the following representation holds true:

$$\tilde{A}^{-1} := A^{-1}T \equiv A^{-1} + A\mathbf{s}A^{-1} = A^{-1} + \tilde{B}.$$

Here $\tilde{B} = A\mathbf{s}A^{-1}$. We observe that $\operatorname{Ker}\tilde{B} = \mathcal{M}_0 := A\mathcal{M}_+$. In addition,

$$\tilde{A} \restriction \mathfrak{D} = A \restriction \mathfrak{D}, \quad \mathfrak{D} \equiv \mathcal{M}_+.$$

Thus, $\tilde{A} \in \mathcal{P}_{\rm ss}(A)$ since the set $\mathfrak{D}$ is dense in $\mathcal{H}_1$. Finally, (7.107) implies (7.106), which is equivalent to $\tilde{A} \geq 1$. □

Definition 7.7.3. We say that the two chains of spaces defined in (7.95) and (7.103), respectively, are singularly similar (shortly, s-*similar*), if for the set $\mathfrak{D} := \mathcal{H}_+ \cap \tilde{\mathcal{H}}_+$ the following conditions are fulfilled:

$$\mathfrak{D} \sqsubset \mathcal{H}_1 \tag{7.108}$$

and

$$\|f\|_1^{\sim} = \|f\|_1, \quad f \in \mathfrak{D}. \tag{7.109}$$

From the previous constructions we get the following important result.

Theorem 7.7.4. *The chains* (7.95) *and* (7.103) *are* s-*similar if and only if the operator* $\tilde{A} \geq 1$ *associated with* (7.103) *belongs to the family* $\mathcal{P}_{\rm ss}(A)$.

Proof. By (7.109), we have

$$(\tilde{A}f, l)_0 = (Af, l)_0, \quad f, l \in \mathfrak{D}.$$

So, we can introduce the symmetric operator

$$\mathbf{A} := \tilde{A} \restriction \mathfrak{D} = A \restriction \mathfrak{D}$$

since $\mathfrak{D}$ is dense in both $\mathcal{H}_1$ and $\mathcal{H}_0$. Thus, A and $\tilde{A}$ are distinct self-adjoint extensions of $\mathbf{A}$. That is, $\tilde{A} \in \mathcal{P}_{\rm ss}(A)$ again due to the density of $\mathfrak{D}$ in $\mathcal{H}_1$. □

We remark that making use of the spectral theorem, one can replace the condition (7.109) by $\|f\|_+^{\sim} = \|f\|_+$, $f \in \mathfrak{D}$.

The next theorem gives the main result of this subsection.

Theorem 7.7.5. *Let an operator* $A = A^* \geq 1$ *be fixed. Then there exist one-to-one correspondences between three sets of objects: the set of singularly perturbed operators* $\tilde{A} \in \mathcal{P}_{\rm ss}(A)$, $\tilde{A} \geq 1$, *the set of quadratic forms* $\gamma \in \mathcal{H}_{-2}$-*class satisfying the condition* (7.104), *and the set of chains of the type* (7.103) *which are* s-*similar to* (7.95). *These correspondences are affected by the following formulas*

$$\gamma[f] = (\tilde{A}^{-1}h, h)_0 - (Af, f)_0, \quad h = Af, \quad f \in \mathcal{H}_+, \tag{7.110}$$

$$(h, l)_{-1}^{\sim} = (\tilde{A}^{-1}h, l)_0 = (h, l)_{-1} + \gamma(A^{-1}h, A^{-1}l), \quad h, l \in \mathcal{H}_0. \tag{7.111}$$

Proof. Given $\tilde{A} \in \mathcal{P}_{\rm ss}(A)$, $\tilde{A} \geq 1$, the quadratic form $\gamma \in \mathcal{H}_{-2}$-class is uniquely defined by (7.110). This form satisfies the condition (7.104) due to $\tilde{A} \geq 1$. Using γ one can introduce the space $\tilde{\mathcal{H}}_{-1}$ as the completion of $\mathcal{H}_0$ with respect to the norm

$$(h,l)^{\sim}_{-1} := (h,l)_{-1} + \gamma(A^{-1}h, A^{-1}l), \quad h,l \in \mathcal{H}_0.$$

Then, since the spaces $\tilde{\mathcal{H}}_{-1}$ and $\mathcal{H}_0$ form a pre-rigged couple, $\tilde{\mathcal{H}}_{-1} \sqsupset \mathcal{H}_0$, one can construct the chain of spaces (7.103). According to Theorem 7.7.4, this yields the chain which is s-similar to (7.95). Finally, starting with the chain of spaces (7.103), we can reconstruct $\tilde{A}$ in the standard way as the operator associated to this chain. □

Clearly, the same result holds true in the more general case when $\tilde{A}$ is not necessarily strictly positive but only bounded from below. Indeed, let

$$\tilde{A} \in \mathcal{P}_{\rm ss}(A), \quad \tilde{A} \geq m, \quad m := \inf \sigma(\tilde{A}) < 1.$$

Then the form γ can be defined by the formula (7.110) with the operators $\tilde{A}$ and A are replaced by $\tilde{A}_a = \tilde{A} + a$ and $A_a = A + a$, respectively, where $a = 1 - m > 0$:

$$\gamma[f] = (\tilde{A}_a^{-1}h, h)_0 - (A_a f, f)_0, \quad h = Af, \ f \in \mathcal{H}_+.$$

It is obvious that $\gamma \in \mathcal{H}_{-2}$-class and satisfies the inequalities

$$-(A_a f, f)_0 \leq \gamma[f] \leq (A_a f, A_a f)_0 - (A_a f, f)_0, \quad f \in \mathcal{H}_+.$$

Starting with γ one can introduce the space $\tilde{\mathcal{H}}_{-1}$ as the completion of $\mathcal{H}_0$ with respect to the norm corresponding to the inner product

$$(\cdot,\cdot)^{\sim}_{-1} := (A_a^{-1}\cdot, \cdot)_0 + \gamma[A_a^{-1}\cdot].$$

Then, using the pre-rigged couple $\tilde{\mathcal{H}}_{-1} \sqsupset \mathcal{H}_0$, we construct in standard way a chain of spaces of the form (7.103). In fact, this chain is s-similar to the chain (7.95) which is constructed based on the operator A_a. Finally, one can reconstruct the operator $\tilde{A}_a$ associated with the latter chain and return to $\tilde{A} = \tilde{A}_a - a$. It is clear that in the above implications one can start with any object: the operator $\tilde{A} \in \mathcal{P}_{\rm ss}(A)$, the quadratic form $\gamma \in \mathcal{H}_{-2}$-class, or, finally, a chain (7.103) that is s-similar to (7.95).

7.7.1 Singular rank-one perturbations of higher orders

In this subsection we show how the rigged space method can be applied to the theory of singular perturbation of higher orders, i.e., to the theory of super-singular perturbations (for more details, see Chapter 8 and [59]). Note that the method presented here differs from the one proposed in [150], where an orthogonal extension procedure is used.

Definition 7.7.6. A bounded from below self-adjoint operator $\tilde{A}$ is said to be super-singularly perturbed with respect to A, if $\tilde{A} \neq A$ and at least for one $\lambda < m_{\tilde{A}}$ there exist $k > 2$ such that the linear set

$$\mathfrak{D}_k := \{f \in \operatorname{Dom}(A-\lambda)^k \cap \operatorname{Dom}(\tilde{A}-\lambda)^k \mid (A-\lambda)^k f = (\tilde{A}-\lambda)^k f\} \quad (7.112)$$

is dense in $\mathcal{H}_{k-1}$ ($m_{\tilde{A}}$ denotes a lower bound of $\tilde{A}$). In such a case we write $\tilde{A} \in \mathcal{P}_{\mathrm{s},k}(A)$.

Let us consider at first the simplest case of the rank-one positive super-singular perturbations.

Let

$$\mathcal{H}_- \equiv \mathcal{H}_{-k} \sqsupset \mathcal{H}_{-k/2} \sqsupset \mathcal{H}_0 \sqsupset \mathcal{H}_{k/2} \sqsupset \mathcal{H}_k \equiv \mathcal{H}_+, \ k > 2, \quad (7.113)$$

be the A-scale of Hilbert spaces associated with $A = A^* \geq 1$. Let us fix $\omega \in \mathcal{H}_{-k} \backslash \mathcal{H}_{-k+1}$, $k > 2$, $\|\omega\|_{-k} = 1$, and consider the quadratic form

$$\gamma_\omega(\varphi, \psi) := \langle \varphi, \omega \rangle_{k,-k} \langle \omega, \psi \rangle_{-k,k}, \quad \varphi, \psi \in \mathcal{H}_k. \quad (7.114)$$

It is obvious it belongs to the $\mathcal{H}_{-k}$-class since due to $\omega \notin \mathcal{H}_{-k+1}$ the set

$$\operatorname{Ker} \gamma_\omega = \mathcal{M}_+ \equiv \mathcal{M}_k := \{\varphi \in \mathcal{H}_k \mid \langle \varphi, \omega \rangle_{k,-k} = 0\}$$

is dense in $\mathcal{H}_{k-1}$. For example, if $k = 3$, then the set $\mathcal{M}_k$ is dense in $\mathcal{H}_2 = \operatorname{Dom} A$. In this case A is essentially self-adjoint on $\mathcal{M}_k$. So, the operator A is not perturbed by γ_ω on this domain. In such a case it is impossible to construct the perturbed operator $\tilde{A}$ by using the form-sum method or the method of self-adjoint extensions. Nevertheless we will show that a non-trivial $\tilde{A}$ can be defined on $\mathcal{H}_0$ by the method of rigged spaces.

To this end we will construct, based on A and γ_ω, a new scale of Hilbert spaces

$$\tilde{\mathcal{H}}_- \equiv \tilde{\mathcal{H}}_{-k} \sqsupset \tilde{\mathcal{H}}_{-k/2} \sqsupset \mathcal{H}_0 \sqsupset \tilde{\mathcal{H}}_{k/2} \sqsupset \tilde{\mathcal{H}}_k \equiv \tilde{\mathcal{H}}_+, \ k > 2, \quad (7.115)$$

and then we define $\tilde{A}$ as the operator associated with this scale. We recall that the chain (7.115) can be obtained starting with any couple of spaces which forms a pre-rigging $\mathcal{H}_0 \sqsupset \tilde{\mathcal{H}}_j$ or $\tilde{\mathcal{H}}_{-j} \sqsupset \mathcal{H}_0$, $j > 0$.

So, put $\tilde{\mathcal{H}}_{-j} = \tilde{\mathcal{H}}_{-k/2}$, where $\tilde{\mathcal{H}}_{-k/2}$ is uniquely determined by A and γ_ω as the completion of $\mathcal{H}_0$ with respect to the inner product

$$(h_1, h_2)^{\sim}_{-k/2} := (A^{-k/2} h_1, h_2)_0 + \beta \gamma_\omega (A^{-k/2} h_1, A^{-k/2} h_2), \quad (7.116)$$

with $h_1, h_2 \in \mathcal{H}_0$, $\beta > 0$. We recall that $A^{-k/2}$ acts as an isometric mapping from $\mathcal{H}_0$ to $\mathcal{H}_k$. Therefore, the form $(\cdot,\cdot)^{\sim}_{-k/2}$ is bounded on $\mathcal{H}_0$ and the inequality

$$\|\cdot\|_{-k/2} \leq \alpha \|\cdot\|^{\sim}_{-k/2} \leq \|\cdot\|_0 \quad (7.117)$$

holds with some $\alpha > 0$. Since the multiplication by $\alpha > 0$ does not change essentially the space metric, from (7.117) it follows that the embedding $\tilde{\mathcal{H}}_{-k/2} \sqsupset \mathcal{H}_0$ is dense and continuous. Now one can extend this pre-rigged couple to the scale (7.115), and use the latter to define the operator $\tilde{A}^{k/2}$ as follows,

$$\tilde{A}^{k/2} := \tilde{D}_{0,k} = \tilde{D}_{-k/2,k/2} \upharpoonright \{\varphi \in \mathcal{H}_{k/2} \mid \tilde{D}_{-k/2,k/2}\varphi \in \mathcal{H}_0\},$$

where $\tilde{D}_{0,k}$ and $\tilde{D}_{-k/2,k/2}$ denote the Berezansky canonical isomorphisms in the scale (7.115). It is clear that the operator $\tilde{A}^{k/2}$ is strongly singularly perturbed with respect to $A^{k/2}$, i.e., $\tilde{A}^{k/2} \in \mathcal{P}_{\text{ss}}(A^{k/2})$. Finally, applying the spectral theorem one can define the operator $\tilde{A} := (\tilde{A}^{k/2})^{2/k} \equiv \tilde{D}_{0,2}$. It is this operator that we call the super-singularly perturbed operator which uniquely corresponds to the rank-one form $\gamma_\omega \in \mathcal{H}_{-k}$-class, $k > 2$ with $\omega \in \mathcal{H}_{-k} \backslash \mathcal{H}_{-k+1}$.

Theorem 7.7.7. *Let two positive self-adjoint operators A and $\tilde{A}$ on $\mathcal{H}_0$ generate two chains of Hilbert spaces:* (7.113) *and* (7.115), *respectively. We assume that for some $k > 2$ the difference between inner products in $\tilde{\mathcal{H}}_{-k/2}$ and $\mathcal{H}_{-k/2}$ defines a positive rank-one quadratic form on $\mathcal{H}_0$:*

$$\beta\gamma_\omega(\cdot,\cdot) := (\cdot,\cdot)^{\sim}_{-k/2} - (\cdot,\cdot)_{-k/2}, \quad \omega \in \mathcal{H}_{-k} \setminus \mathcal{H}_{-k+1},\ \beta > 0,$$

where the constant β arises from the condition $\|\omega\|_{-k} = 1$. Then the form γ_ω admits an interpretation as a super-singular perturbation of the $\mathcal{H}_{-k}$-class with respect to A. From this form and the operator A one can uniquely reconstruct the triple $\tilde{\mathcal{H}}_{-k/2} \sqsupset \mathcal{H}_0 \sqsupset \tilde{\mathcal{H}}_{k/2}$ and the associated super-singularly perturbed operator $\tilde{A}^{k/2} \in \mathcal{P}_{\text{ss}}(A^{k/2})$.

Proof. It follows from the previous constructions and arguments in the proof of Theorem 7.2.5. □

7.7.2 On s-similarity of Hilbert scales

Let $A, \tilde{A} \geq 1$ be a couple of self-adjoint operators on $\mathcal{H}_0$ and

$$\mathcal{H}_{-k} \sqsupset \mathcal{H}_0 \sqsupset \mathcal{H}_k, \tag{7.118}$$

$$\tilde{\mathcal{H}}_{-k} \sqsupset \mathcal{H}_0 \sqsupset \tilde{\mathcal{H}}_k, \quad k > 0, \tag{7.119}$$

be the scales of Hilbert spaces associated with A and $\tilde{A}$.

We say that the scales (7.118), (7.119) are s-*similar in the generalized sense* and write $\{\mathcal{H}_k\} \sim \{\tilde{\mathcal{H}}_k\}$, if there exists $k \geq 1$ such that the set

$$\mathfrak{D}_k := \mathcal{H}_{2k} \cap \tilde{\mathcal{H}}_{2k} \tag{7.120}$$

is dense in $\mathcal{H}_k$,

$$\mathcal{H}_k \sqsupset \mathfrak{D}_k, \tag{7.121}$$

and

$$\|\varphi\|_k = \|\varphi\|_{\tilde{k}}, \quad \varphi \in \mathfrak{D}_k. \tag{7.122}$$

The conditions (7.120) and (7.121) imply the following properties: the spaces $\mathcal{H}_{2k}$, $\tilde{\mathcal{H}}_{2k}$ admit the orthogonal decompositions,

$$\mathcal{H}_{2k} = \mathcal{M}_{2k} \oplus \mathcal{N}_{2k}, \quad \tilde{\mathcal{H}}_{2k} = \tilde{\mathcal{M}}_{2k} \oplus \tilde{\mathcal{N}}_{2k}, \tag{7.123}$$

where the subspaces $\mathcal{M}_{2k}$ and $\tilde{\mathcal{M}}_{2k}$ coincide and are dense in $\mathcal{H}_k$:

$$\mathcal{M}_{2k} = \tilde{\mathcal{M}}_{2k} \equiv \mathfrak{D}_k \sqsubset \mathcal{H}_k. \tag{7.124}$$

Thus, if $\tilde{A}^k \in \mathcal{P}_{\text{ss}}(A^k)$, then there exists a symmetric operator $\mathbf{A}^k$ such that the set $\mathfrak{D}_k = \mathfrak{D}(\mathbf{A}^k)$ satisfies the conditions (7.120) and (7.121). This proves in one direction the following theorem.

Theorem 7.7.8. *The scales of Hilbert spaces* (7.118) *and* (7.119) *are* s-*similar in the generalized sense,* $\{\mathcal{H}_k\} \sim \{\tilde{\mathcal{H}}_k\}$, *if and only if for some* $k \geq 1$ *the operator* $\tilde{A}^k$ *is singularly perturbed with respect to* A^k, *i.e.,* $\tilde{A}^k \in \mathcal{P}_{\text{ss}}(A^k)$.

Proof. By the construction of the scales (7.118) and (7.119), the inner products in $\mathcal{H}_k$ and $\tilde{\mathcal{H}}_k$ are defined by the quadratic forms

$$\chi_k(\varphi, \psi) := (A^k\varphi, \psi)_0 = (\varphi, \psi)_k$$

and

$$\tilde{\chi}_k(\varphi, \psi) := (\tilde{A}^k\varphi, \psi)_0 = (\varphi, \psi)_{\tilde{k}},$$

respectively. Thus, due to (7.122) we have:

$$(\varphi, \psi)_k = (A^k\varphi, \psi)_0 = (\varphi, \psi)_{\tilde{k}} = (\tilde{A}^k\varphi, \psi)_0, \quad \varphi, \psi \in \mathfrak{D}_k.$$

Further, since the set $\mathfrak{D}_k$ is dense in $\mathcal{H}_k$ (see (7.121)), the restrictions of A^k and $\tilde{A}^k$ to $\mathfrak{D}_k$ coincide:

$$A^k \restriction \mathfrak{D}_k = \tilde{A}^k \restriction \mathfrak{D}_k. \tag{7.125}$$

Therefore, these restrictions yield the same densely defined symmetric operator $\mathbf{A}^k$ in $\mathcal{H}_0$. It is closed because the set $\mathfrak{D}_k = \mathcal{M}_{2k} = \tilde{\mathcal{M}}_{2k}$ yields a closed subspace in both $\mathcal{H}_{2k}$ and $\tilde{\mathcal{H}}_{2k}$. It is clear that both A^k and $\tilde{A}^k$ are self-adjoint extensions of $\mathbf{A}^k$. We emphasize that the Friedrichs extension of $\mathbf{A}^k$ coincides with A^k because the set $\mathfrak{D}_k$ is dense in $\mathcal{H}_k$. Then, by the definition of the perturbed operator, all other bounded from below extensions of $\mathbf{A}^k$ belong to $\mathcal{P}_{\text{ss}}(A^k)$. Thus, $\tilde{A}^k \in \mathcal{P}_{\text{ss}}(A^k)$. The converse statement was already obtained. $\square$

Note that due to (7.124),

$$\|\varphi\|_{2k} = \|\varphi\|_{\widetilde{2k}}, \quad \varphi \in \mathfrak{D}_k. \tag{7.126}$$

However, in the general case, the set $\mathfrak{D}_k$ does not belong to $\mathcal{H}_{4k} \cap \tilde{\mathcal{H}}_{4k}$. Therefore, from (7.126) it does not follow that $\tilde{A}^{2k}$ is a singularly perturbed operator.

Let us remark that according to Theorem 7.7.5 (see (7.110) and (7.111)), the singular quadratic form defined by

$$\gamma(A^{-k}\cdot, A^{-k}\cdot) := (\tilde{A}^{-k}\cdot, \cdot)_0 - (A^{-k}\cdot, \cdot)_0,$$

belongs to the $\mathcal{H}_{-2}$-class with respect to the operator A^k since the set $\mathfrak{D}_k$ is dense in $\mathcal{H}_k$. However, in the case $\tilde{A}^{-k} \in \mathcal{P}_{\rm ws}(A^k)$, the spaces $\tilde{\mathcal{H}}_k$ and $\mathcal{H}_k$ coincide as subsets in $\mathcal{H}_0$, although they have different, yet equivalent norms. Thus, the form $\gamma[\varphi] := (\tilde{A}^k\varphi, \varphi)_0 - (A^k\varphi, \varphi)_0$ is bounded on $\mathcal{H}_k$ and belongs to the $\mathcal{H}_{-1}$-class with respect to A^k.

7.8 The operator associated with a dense subspace

In this subsection we discuss an additional original way of constructing a singularly perturbed operator.

Let $A \geq 1$ be an unbounded self-adjoint operator on the space $\mathcal{H}_0$, which is rigged, $\mathcal{H}_- \sqsupset \mathcal{H}_0 \sqsupset \mathcal{H}_+$, in such a way that the domain $\mathfrak{D}(A) = \mathcal{H}_+$ in the graph-norm. We assume that $\mathcal{H}_+$ is decomposed into an orthogonal sum $\mathcal{H}_+ = \mathcal{M}_+ \oplus \mathcal{N}_+$ such that the subspace $\mathcal{M}_+$ is dense in $\mathcal{H}_0$. In what follows we will construct and study a specific singularly perturbed operator $\breve{A}$ which is associated with the rigged space $\breve{\mathcal{H}}_- \sqsupset \mathcal{H}_0 \sqsupset \breve{\mathcal{H}}_+$, where $\breve{\mathcal{H}}_+ = \mathcal{M}_+ = \mathfrak{D}(\breve{A})$. We are interested in what way $\breve{A}$ is connected with A.

7.8.1 The setting of the problem

Let us consider an unbounded self-adjoint operator $A = A^* \geq 1$ with domain $\mathfrak{D}(A)$ on a Hilbert space $\mathcal{H}_0$. According to [42, 44], every such A is associated with a rigged Hilbert space

$$\mathcal{H}_- \sqsupset \mathcal{H}_0 \sqsupset \mathcal{H}_+, \quad \mathcal{H}_+ = \mathfrak{D}(A).$$

Here $\sqsupset$ stands for dense and continuous embedding. The space $\mathcal{H}_+$ coincides with $\mathfrak{D}(A)$ in the graph-norm, and $\mathcal{H}_-$ is the dual space to $\mathcal{H}_+$. $\mathcal{H}_-$ can be obtained from $\mathcal{H}_0$ by introducing the new norm $\|f\|_- := \|A^{-1}f\|$, $f \in \mathcal{H}_0$. We assume that a singular perturbation of A is given by an operator $T : \mathcal{H}_+ \to \mathcal{H}_-$ such that the set $\mathcal{M}_+ := \operatorname{Ker} T$ is dense in $\mathcal{H}_0$. According to the method of self-adjoint extensions (see, e.g., [7, 167]), the singularly perturbed operator $\tilde{A}$ corresponding to the formal sum $A \tilde{+} T$ can be defined as a self-adjoint extension of the symmetric operator $\mathbf{A} := A|\mathcal{M}_+$, i.e., $\tilde{A} \in \mathcal{A}(\mathbf{A})$. In what follows we will present an additional method for constructing a singularly perturbed operator.

First we explain the idea of this new method. Beginning with the orthogonal decomposition $\mathcal{H}_+ = \mathcal{M}_+ \oplus \mathcal{N}_+$ and taking into account that $\mathcal{M}_+ = \operatorname{Ker} T$ is dense in $\mathcal{H}_0$, we introduce the new rigged space, $\breve{\mathcal{H}}_- \sqsupset \mathcal{H}_0 \sqsupset \breve{\mathcal{H}}_+$, putting $\breve{\mathcal{H}}_+ \equiv \mathcal{M}_+$. Then we define a specific singularly perturbed operator $\breve{A}$ as the uniquely determined operator associated with this new rigging of $\mathcal{H}_0$. Thus, $\breve{A}$ is fixed by the condition: $\mathfrak{D}(\breve{A}) = \mathcal{M}_+$ in the graph-norm.

We emphasize that the construction just described leads to the extension of the usual family of singularly perturbed operators. This statement is justified by the fact that $\breve{A}$ is not a self-adjoint extension of the symmetric operator $\mathbf{A} = A \restriction \mathcal{M}_+$ because their domains coincide: $\mathfrak{D}(\breve{A}) = \mathfrak{D}(\mathbf{A}) = \mathcal{M}_+$. One can expect that the spectral properties of the operators $\tilde{A} \in \mathcal{P}_s(A)$ and $\breve{A}$ will be essentially different. Hence, in applications to mathematical physics, the choice of $\breve{A}$ as a singularly perturbed operator is related most likely to the physical conditions of absolutely hard core or completely non-transparent screen. Namely, these conditions fix the set $\mathcal{M}_+$ as the domain of the corresponding Hamiltonian. That is the physical conditions mentioned above restrict the starting domain of the unperturbed operator to the set $\mathcal{M}_+ = \mathfrak{D}(\mathbf{A})$ without any additional information for getting its extension. We claim that $\breve{A}$ is a self-adjoint operator with $\mathcal{M}_+ = \mathfrak{D}(\breve{A}) \subset \mathfrak{D}(A)$.

To develop this idea we need in some preparations. To this end we will partly repeat the principal points of the theory of rigged spaces and then present the new constructions. We formulate the main result of the subsection in Theorem 7.8.4. In particular, we provide the detailed definition of the operator $\breve{A}$ and establish the connections between operators A and $\breve{A}$.

7.8.2 Once more on rigged spaces

Let us recall that, by definition, a triple of Hilbert spaces

$$\mathcal{H}_- \sqsupset \mathcal{H}_0 \sqsupset \mathcal{H}_+ \tag{7.127}$$

form the rigged Hilbert space, if the following conditions are fulfilled: both embedding are dense and continuous, which is denoted by $\sqsupset$; the norms in the spaces $\mathcal{H}_-$, $\mathcal{H}_0$ and $\mathcal{H}_+$ satisfy the inequality

$$\|\cdot\|_- \leq \|\cdot\|_0 \leq \|\cdot\|_+; \tag{7.128}$$

and finally, the spaces $\mathcal{H}_-$ and $\mathcal{H}_+$ are dual to one another with respect to $\mathcal{H}_0$.

The last condition means that the linear functional $l_\varphi(f) := (f, \varphi)_0$, $f \in \mathcal{H}_0$ defined by a vector $\varphi \in \mathcal{H}_+$, can be extended by continuity to the whole space $\mathcal{H}_-$. Then, one can define the so-called positive norm $\|\varphi\|_+$ by the formula

$$\|\varphi\|_+ = \sup_{\|f\|_-=1} |(f, \varphi)_0|, \quad f \in \mathcal{H}_0.$$

On the other hand, according to the Riesz theorem, $l_\varphi(f) = (f, \varphi^*)_-$ for some vector $\varphi^* \in \mathcal{H}_-$. Thus, $\|\varphi\|_+ = \|\varphi^*\|_-$. So, we get an isometric mapping

$$D_{-,+} : \mathcal{H}_+ \ni \varphi \longmapsto \varphi^* \in \mathcal{H}_-.$$

In turn, if $\mathcal{H}_+$ is given, then one can recover $\mathcal{H}_-$, since this space coincides with the closure of $\mathcal{H}_0$ with respect to the so-called negative norm

$$\|f\|_- := \sup_{\|\varphi\|_+=1} |(\varphi, f)_0|, \quad \varphi \in \mathcal{H}_+.$$

Due to (7.128), the inner product $(\cdot,\cdot)_0$ in $\mathcal{H}_0$ admits an extension to the dual pairing between $\mathcal{H}_+$ and $\mathcal{H}_-$, which is denoted by

$$\langle \omega, \varphi\rangle_{-,+} = \overline{\langle \varphi, \omega\rangle_{+,-}}, \ \omega \in \mathcal{H}_-, \ \varphi \in \mathcal{H}_+.$$

The operators

$$D_{-,+} : \mathcal{H}_+ \longrightarrow \mathcal{H}_-, \quad I_{+,-} = D_{-,+}^{-1} : \mathcal{H}_- \longrightarrow \mathcal{H}_+$$

are called the Berezansky canonical isomorphisms between $\mathcal{H}_-$ and $\mathcal{H}_+$. They satisfies the relations

$$(f,\varphi)_0 = \langle f,\varphi\rangle_{-,+} = (f, D_{-,+}\varphi)_- = (I_{+,-}f,\varphi)_+, \quad f \in \mathcal{H}_0, \ \varphi \in \mathcal{H}_+.$$

There exists a well-known connection between the triple of spaces (7.127) and the starting self-adjoint operator A on $\mathcal{H}_0$. This connection is uniquely specified by the condition $\mathfrak{D}(A) = \mathcal{H}_+$ and the mapping $D_{-,+}$. Indeed, let us consider the operator

$$L_A := D_{-,+}|\mathcal{H}_{++},$$

where

$$\mathcal{H}_{++} := \mathfrak{D}(L_A) = \{\varphi \in \mathcal{H}_+ \mid D_{-,+}\varphi \in \mathcal{H}_0\}.$$

It is easily seen that L_A is symmetric in $\mathcal{H}_0$, since for all $\varphi,\psi \in \mathfrak{D}(L_A)$

$$\begin{aligned}(L_A\varphi,\psi)_0 &= (D_{-,+}\varphi,\psi)_0 = \langle \varphi^*,\psi\rangle_{-,+} = (\varphi,\psi)_+ \\ &= \langle \varphi,\psi^*\rangle_{+,-} = (\varphi, D_{-,+}\psi)_0 = (\varphi, L_A\psi)_0,\end{aligned}$$

with φ^* defined above. In fact, L_A is self-adjoint on $\mathcal{H}_0$ because its range coincides with the whole space $\mathcal{H}_0$. Now we define $A := L_A^{1/2}$. It is clear that $\mathfrak{D}(A) = \mathcal{H}_+$, since

$$(L_A\varphi,\psi)_0 = (L_A^{1/2}\varphi, L_A^{1/2}\psi)_0 = (\varphi,\psi)_+.$$

Moreover, $A \geq 1$ because $\|\cdot\|_+ \geq \|\cdot\|_0$.

Conversely, let $A = A^* \geq 1$ be a self-adjoint unbounded operator on $\mathcal{H}_0$ with the domain $\mathfrak{D}(A)$. Starting with A one can easily introduce the rigged Hilbert space $\mathcal{H}_- \sqsupset \mathcal{H}_0 \sqsupset \mathcal{H}_+$. Let us recall this construction. We identify the space $\mathcal{H}_+$ with $\mathfrak{D}(A)$ after introducing the positive inner product

$$(\varphi,\psi)_+ := (A\varphi, A\psi)_0, \quad \varphi,\psi \in \mathfrak{D}(A).$$

Further, taking the pre-rigged chain $\mathcal{H}_0 \sqsupset \mathcal{H}_+$, we extend it to the rigged space (7.127) in the standard way described above. In this connection, we recall Theorem

4.2.1, which asserts that each rigged Hilbert space of the form (7.127) is uniquely associated with a self-adjoint operator $A = A^* \geq 1$ on $\mathcal{H}_0$ such that $\mathfrak{D}(A) = \mathcal{H}_+$ with the positive inner product $(\varphi, \psi)_+ = (A\varphi, A\psi)_0$, $\varphi, \psi \in \mathfrak{D}(A)$.

In the sequel we will need the infinite chain of Hilbert spaces

$$\{\mathcal{H}_k \equiv \mathcal{H}_k(A)\}_{k\in\mathbb{R}},$$

called the A-scale.

By the definition of the A-scale, $\mathcal{H}_k := \mathfrak{D}(A^{k/2})$, $k > 0$, with respect to the positive norm $\|\cdot\|_k$ corresponding to the inner product

$$(\varphi, \psi)_k := (A^{k/2}\varphi, A^{k/2}\psi)_0, \quad \varphi, \psi \in \mathfrak{D}(A^{k/2}).$$

The space $\mathcal{H}_{-k}$ arises as the completion of $\mathcal{H}_0$ with respect to the negative norm

$$\|f\|_{-k} := \|A^{-k/2}f\|_0, \quad f \in \mathcal{H}_0.$$

It is easily seen that every triple

$$\mathcal{H}_{-k} \sqsupset \mathcal{H}_0 \sqsupset \mathcal{H}_k, \quad k > 0 \tag{7.129}$$

yields the rigged space associated with the operator $A^{k/2}$.

Let $D_{-k,k} : \mathcal{H}_k \to \mathcal{H}_{-k}$ denote the Berezansky canonical isomorphism for the triple (7.129). It is obvious that

$$D_{-k,k} = (A^{k/2})^{\mathrm{cl}}(A^{k/2}) \equiv D_{-k,0}D_{0,k},$$

where cl stands for closure. In particular, for $k = 2$ we have: $D_{0,2} \equiv A : \mathcal{H}_2 \to \mathcal{H}_0$ and $D_{-2,0} \equiv A^{\mathrm{cl}} : \mathcal{H}_0 \to \mathcal{H}_{-2}$ (for more details see Chapter 4).

7.8.3 Again about denseness of embedded subspaces

Given a rigged Hilbert space $\mathcal{H}_- \sqsupset \mathcal{H}_0 \sqsupset \mathcal{H}_+$, assume that the positive space $\mathcal{H}_+$ is decomposed into an orthogonal sum of subspaces: $\mathcal{H}_+ = \mathcal{M}_+ \oplus \mathcal{N}_+$. The next result is the main criterion ensuring the dense embedding of the subspace $\mathcal{M}_+$ from $\mathcal{H}_+$ into $\mathcal{H}_0$. Namely, Theorem 6.1.4 states that the subspace $\mathcal{M}_+$ from the decomposition $\mathcal{H}_+ = \mathcal{M}_+ \oplus \mathcal{N}_+$ is dense in $\mathcal{H}_0$ if and only if the image of $\mathcal{N}_+$ in $\mathcal{H}_-$ under the Berezansky canonical isomorphism, i.e., the subspace $\mathcal{N}_- := D_{-,+}\mathcal{N}_+$, has a null intersection with $\mathcal{H}_0$:

$$\mathcal{H}_0 \sqsupset \mathcal{M}_+ \Longleftrightarrow \mathcal{N}_- \cap \mathcal{H}_0 = \{0\}. \tag{7.130}$$

Indeed, let $\mathcal{N}_- \cap \mathcal{H}_0 = \{0\}$. Suppose that there exists a vector $0 \neq \psi \in \mathcal{H}_0$ such that $\psi \perp \mathcal{M}_+$. Since $\mathcal{M}_+ \subset \mathcal{H}_+$ and $\psi \in \mathcal{H}_-$,

$$(\psi, \mathcal{M}_+)_0 = \langle \psi, \mathcal{M}_+ \rangle_{-,+} = (I_{+,-}\psi, \mathcal{M}_+)_+ = 0.$$

Therefore, $I_{+,-}\psi \in \mathcal{N}_+$ and $\psi \in \mathcal{N}_-$, which contradicts (7.130). Conversely, if the subspace $\mathcal{M}_+$ is dense in $\mathcal{H}_0$, then the assumption that there exists a vector $0 \neq \omega \in \mathcal{N}_- \cap \mathcal{H}_0$ also leads to a contradiction. Indeed, since $\mathcal{N}_- = D_{-,+}\mathcal{N}_+$, we have,

$$\langle \omega, \mathcal{M}_+ \rangle_{-,+} = (\omega, \mathcal{M}_+)_0 = (I_{+,-}\omega, \mathcal{M}_+)_+ = 0,$$

which contradicts the relation $\mathcal{M}_+ \sqsubset \mathcal{H}_0$.

It is easy to see that the condition (7.130) can be recast in the equivalent form:

$$\mathcal{H}_0 \sqsupset \mathcal{M}_+ \iff \mathcal{N}_0 \cap \mathcal{H}_+ = \{0\}, \text{ where } \mathcal{N}_0 := D_{0,+}\mathcal{N}_+. \tag{7.131}$$

We introduce now the extended rigged space

$$\mathcal{H}_{--} \sqsupset \mathcal{H}_- \sqsupset \mathcal{H}_0 \sqsupset \mathcal{H}_+ \sqsupset \mathcal{H}_{++},$$

where $\mathcal{H}_{--} = \mathcal{H}_{-4}(A)$ and $\mathcal{H}_{++} = \mathcal{H}_4(A) = \mathfrak{D}(A^2)$. Given a decomposition $\mathcal{H}_+ = \mathcal{N}_+ \oplus \mathcal{M}_+$, we assume that $\mathcal{H}_0 \sqsupset \mathcal{M}_+$ and define the subspace $\tilde{\mathcal{M}}_+ := \mathcal{M}_+ \cap \mathcal{H}_{++}$. It is closed in $\mathcal{H}_{++}$. Indeed, if a sequence $\varphi_n \in \tilde{\mathcal{M}}_+$ converges in $\mathcal{H}_{++}$, $\varphi_n \to \varphi \in \mathcal{H}_{++}$, it also converges in $\mathcal{H}_+$ because $\|\cdot\|_+ \leq \|\cdot\|_{++}$. Thus, $\varphi \in \mathcal{M}_+$ since $\mathcal{M}_+$ is a closed subspace in $\mathcal{H}_+$. This proves that $\tilde{\mathcal{M}}_+$ is closed in $\mathcal{H}_{++}$.

Now assume that

$$(\mathcal{N}_-)^{\mathrm{cl},--} \cap \mathcal{H}_0 = \{0\}, \tag{7.132}$$

where $\mathcal{N}_- := D_{-,+}\mathcal{N}_+$ and cl, $--$ denotes closure in $\mathcal{H}_{--}$.

Theorem 7.8.1. *If the subspace $\mathcal{M}_+$ is dense in $\mathcal{H}_0$, $\mathcal{H}_0 \sqsupset \mathcal{M}_+$, and, in addition, the condition* (7.132) *is satisfied, then the subset $\tilde{\mathcal{M}}_+$ defined as $\tilde{\mathcal{M}}_+ := \mathcal{M}_+ \cap \mathcal{H}_{++}$ is also dense in $\mathcal{H}_0$:*

$$\mathcal{H}_0 \sqsupset \tilde{\mathcal{M}}_+. \tag{7.133}$$

In particular, if $\dim \mathcal{N}_+ < \infty$, *then the condition* (7.132) *is automatically fulfilled and the subspace $\tilde{\mathcal{M}}_+$ is dense in $\mathcal{H}_0$.*

Proof. In accordance with the definition of $\tilde{\mathcal{M}}_+$ we can write

$$\tilde{\mathcal{M}}_+ = \{\varphi \in \mathcal{H}_{++} \mid (\varphi, \psi)_+ = 0, \ \psi \in \mathcal{N}_+\}.$$

From this, by the properties of the A-scale, we have

$$(\varphi, \psi)_+ = \langle \varphi, \omega \rangle_{+,-} = \langle \varphi, \omega \rangle_{++,--},$$

where $\omega = D_{-,+}\psi$, $\psi \in \mathcal{N}_+$. This implies that

$$(\mathcal{N}_-)^{\mathrm{cl},--} = \tilde{\mathcal{N}}_- := \{\omega \in \mathcal{H}_{--} \mid \langle \varphi, \omega \rangle_{++,--} = 0, \ \varphi \in \tilde{\mathcal{M}}_+\}. \tag{7.134}$$

Further, since $\mathcal{M}_+$ is dense in $\mathcal{H}_0$, from (7.132) and (7.130) it follows that $\tilde{\mathcal{N}}_- \cap \mathcal{H}_0 = \{0\}$. Therefore, $\mathcal{H}_0 \sqsupset \tilde{\mathcal{M}}_+$.

Finally, it is a trivial fact that if $\dim \mathcal{N}_0 = \dim \mathcal{N}_+ < \infty$, then the condition (7.132) is automatically fulfilled because $\mathcal{H}_0 \sqsupset \mathcal{M}_+$. □

It is clear that Theorem 7.8.1 remains valid if the condition (7.132) is replaced by

$$(\mathcal{N}_-)^{\mathrm{cl},--} \cap \mathcal{H}_- = \mathcal{N}_-. \tag{7.135}$$

However, the following question is open. What properties of $\mathcal{N}_+$ and $\mathcal{N}_-$ lead to $(\mathcal{N}_-)^{\mathrm{cl},--} \cap \mathcal{H}_0 \neq \{0\}$?

We note that similar questions were discussed in [67].

7.8.4 The operator $\breve{A}$

Here we construct and analyze the properties of the operator $\breve{A}$ described in Subsection 7.8.1. Let $\mathcal{H}_- \sqsupset \mathcal{H}_0 \sqsupset \mathcal{H}_+$ denote the usual rigged Hilbert space associated with the operator $A \geq 1$ in the sense that $\mathcal{H}_+ = \mathfrak{D}(A)$ with respect to the norm $\|\cdot\|_+ = \|A\cdot\|_0$. Then the operator A^2 coincides with the restriction of the map $D_{-,+} : \mathcal{H}_+ \to \mathcal{H}_-$ to the space $\mathcal{H}_{++} \equiv \mathcal{H}_4(A)$, i.e., $A^2 = D_{-,+} \restriction \mathcal{H}_{++}$. Let us assume that the positive space $\mathcal{H}_+$ is decomposed into the orthogonal sum $\mathcal{H}_+ = \mathcal{M}_+ \oplus \mathcal{N}_+$ such that the subspace $\mathcal{M}_+$ is dense in $\mathcal{H}_0$, i.e., $\mathcal{H}_0 \sqsupset \mathcal{M}_+$. We consider now the new rigged space

$$\breve{\mathcal{H}}_- \sqsupset \mathcal{H}_0 \sqsupset \breve{\mathcal{H}}_+, \tag{7.136}$$

where $\breve{\mathcal{H}}_+ \equiv \mathcal{M}_+$. We want to construct the self-adjoint in $\mathcal{H}_0$ operator $\breve{A}$ which is associated with the chain (7.136) so that the domain $\mathfrak{D}(\breve{A})$ coincides with $\breve{\mathcal{H}}_+$.

We recall that the negative space $\breve{\mathcal{H}}_-$ is defined as the completion of $\mathcal{H}_0$ with respect to the new norm

$$\|f\|\breve{}_- := \sup_{\|\varphi\|_+=1} |(f,\varphi)_0|, \quad f \in \mathcal{H}_0, \quad \varphi \in \mathcal{M}_+. \tag{7.137}$$

Since

$$\|f\|_- := \sup_{\|\varphi\|_+=1} |(f,\varphi)_0|,$$

where a supremum is taken over all $\varphi \in \mathcal{H}_+$, (7.137) implies the inequality

$$\|f\|\breve{}_- \leq \|f\|_-, \quad f \in \mathcal{H}_0. \tag{7.138}$$

It is clear that the space $\mathcal{H}_0$ is densely and continuously embedded in both $\mathcal{H}_-$ and $\breve{\mathcal{H}}_-$. However, it is not true that (7.138) implies that $\mathcal{H}_-$ is contained in $\breve{\mathcal{H}}_-$ as a proper subset.

Proposition 7.8.2. *The closure of the identity mapping*

$$O : \mathcal{H}_- \ni f \longmapsto f \in \breve{\mathcal{H}}_-, \quad f \in \mathcal{H}_0$$

is continuous and bounded, but has a non-trivial null-subspace:

$$\operatorname{Ker} O^{\mathrm{cl}} = \mathcal{N}_- \neq \{0\}, \quad \mathcal{N}_- = I_{-,+}\mathcal{N}_+,$$

where cl *stands for closure.*

Proof. The continuity of the mapping O follows directly from inequality (7.138). Let us show that each $\eta_- \in \mathcal{N}_- = D_{-,+}\mathcal{N}_+$ is a null-vector for O^{cl}. Suppose the sequence $f_n \in \mathcal{H}_0$ converges in $\mathcal{H}_-$ to $\eta_- \in \mathcal{N}_-$. Then due to (7.138) this sequence converges also in $\breve{\mathcal{H}}_-$. However, in the space $\breve{\mathcal{H}}_-$ this sequence necessarily converges to zero, because $\mathcal{M}_+$ is dense in $\mathcal{H}_0$. Indeed, since $\mathcal{N}_- \perp \mathcal{M}_+$ in the sense of the duality pairing, we have

$$(f_n, \varphi)_0 = \langle f_n, \varphi\rangle_{-,+} \to \langle \eta_-, \varphi\rangle_{-,+} = 0, \quad \varphi \in \mathcal{M}_+.$$

Thus, $\eta_- \in \operatorname{Ker} O^{\mathrm{cl}}$. □

We emphasize that no vector $0 \neq f \in \mathcal{H}_0$ belongs to $\operatorname{Ker} O^{\mathrm{cl}}$. Therefore, the whole space $\mathcal{H}_0$ is embedded in $\breve{\mathcal{H}}_-$ without any defect.

Proposition 7.8.3. *For each $0 \neq f \in \mathcal{H}_0$,*

$$\|f\|_-^{\breve{}} = \|P_{\mathcal{M}_-} f\|_- \neq 0, \tag{7.139}$$

where $P_{\mathcal{M}_-}$ denotes the orthogonal projection onto $\mathcal{M}_-$ in $\mathcal{H}_-$.

Proof. The equality in (7.139) follows from the definition of the norm in $\breve{\mathcal{H}}_-$ (see (7.137)) and the relation

$$(f, \varphi)_0 = \langle f, \varphi\rangle_{-,+} = \langle P_{\mathcal{M}_-} f, \varphi\rangle_{-,+}, \quad \varphi \in \mathcal{M}_+.$$

Here, we again use the orthogonality $\mathcal{M}_- \perp \mathcal{N}_+$ in the sense of the duality pairing. In addition, we remark that for all vectors $0 \neq f \in \mathcal{H}_0$,

$$P_{\mathcal{M}_-} f \neq 0, \tag{7.140}$$

since from $P_{\mathcal{M}_-} f = 0$ it follows that $f \in \mathcal{N}_-$. However, $\mathcal{N}_- \cap \mathcal{H}_0 = \{0\}$ because $\mathcal{H}_0 \sqsupset \mathcal{M}_+$. □

By Proposition 7.8.3, the restriction of the mapping O^{cl} to $\mathcal{M}_- := D_{-,+}\mathcal{M}_+$ is an isometric operator. Furthermore, O^{cl} acts as a unitary operator from $\mathcal{M}_-$ to $\breve{\mathcal{H}}_-$.

Thus, in spite of the fact that the norms in the spaces $\breve{\mathcal{H}}_-$ and $\mathcal{H}_-$ satisfy the inequality (7.138) and $\breve{\mathcal{H}}_+ \equiv \mathcal{M}_+$ is a proper subspace of $\mathcal{H}_+$, the space $\mathcal{H}_-$ is not contained in $\breve{\mathcal{H}}_-$ as a proper subset: $\breve{\mathcal{H}}_- \not\supset \mathcal{H}_-$. The reason is that all nonzero elements from $\mathcal{H}_-$, which vanish on $\mathcal{M}_+$ as linear continuous functionals on $\mathcal{H}_+$, belong to the equivalence class of zero in $\breve{\mathcal{H}}_-$.

Let $\breve{D}_{-,+} : \breve{\mathcal{H}}_+ \to \breve{\mathcal{H}}_-$ denote the Berezansky canonical isomorphism in the rigged Hilbert space (7.136). Consider the operator

$$L := \breve{D}_{-,+} \restriction \mathfrak{D}(L), \quad \mathfrak{D}(L) := \{\varphi \in \breve{\mathcal{H}}_+ \mid \breve{D}_{-,+}\varphi \in \mathcal{H}_0\}. \tag{7.141}$$

It is not hard to check (see below the proof of Theorem 7.8.4) that the operator L is symmetric and its range coincides with the whole space $\mathcal{H}_0$. Therefore, it is self-adjoint in $\mathcal{H}_0$ with the domain $\mathfrak{D}(L) \subset \mathcal{M}_+ = \breve{\mathcal{H}}_+$. The following is the main result of this subsection.

Theorem 7.8.4. *Suppose the domain $\mathfrak{D}(A)$ of the self-adjoint operator $A \geq 1$ on $\mathcal{H}_0$ is decomposed into an orthogonal sum, $\mathfrak{D}(A) = \mathcal{H}_+ = \mathcal{M}_+ \oplus \mathcal{N}_+$. Suppose also that the subspace $\mathcal{M}_+$ is dense in $\mathcal{H}_0$ and the subspace $\mathcal{N}_- := D_{-,+}\mathcal{N}_+$ satisfies the condition* (7.132). *Then the operator L defined in* (7.141) *admits an explicit description in terms of the operator A:*

$$LP_{\mathcal{M}_+}\varphi = A^2\varphi, \varphi \in \mathcal{H}_{++} \equiv \mathfrak{D}(A^2), \quad \mathfrak{D}(L) = P_{\mathcal{M}_+}\mathcal{H}_{++}, \tag{7.142}$$

where $P_{\mathcal{M}_+}$ denotes the orthogonal projection onto $\mathcal{M}_+$ in $\mathcal{H}_+$. Moreover, L coincides with the Friedrichs extension of the symmetric operator

$$\dot{L} := A^2 \restriction \tilde{\mathcal{M}}_+, \quad \tilde{\mathcal{M}}_+ := \mathcal{M}_+ \cap \mathcal{H}_{++}. \tag{7.143}$$

Furthermore, the domain of the operator $\breve{A} := L^{1/2}$ coincides with the subspace $\mathcal{M}_+$:

$$\mathfrak{D}(\breve{A}) = \mathcal{M}_+ = \breve{\mathcal{H}}_+. \tag{7.144}$$

Proof. First we show that the mapping

$$L : P_{\mathcal{M}_+}\varphi \longmapsto A^2\varphi, \quad \varphi \in \mathcal{H}_{++}$$

defines a symmetric operator in $\mathcal{H}_0$. Indeed, for all $\varphi, \psi \in \mathcal{H}_{++}$, we have:

$$\begin{aligned}
(LP_{\mathcal{M}_+}\varphi, P_{\mathcal{M}_+}\psi)_0 &= (A^2\varphi, P_{\mathcal{M}_+}\psi)_0 = \langle D_{-,+}\varphi, P_{\mathcal{M}_+}\psi\rangle_{-,+} \\
&= \langle P_{\mathcal{M}_-}D_{-,+}\varphi, P_{\mathcal{M}_+}\psi\rangle_{-,+} = \langle D_{-,+}P_{\mathcal{M}_+}\varphi, P_{\mathcal{M}_+}\psi\rangle_{-,+} \\
&= \langle P_{\mathcal{M}_+}\varphi, D_{-,+}P_{\mathcal{M}_+}\psi\rangle_{+,-} = \langle P_{\mathcal{M}_+}\varphi, P_{\mathcal{M}_-}D_{-,+}\psi\rangle_{+,-} \\
&= \langle P_{\mathcal{M}_+}\varphi, D_{-,+}\psi\rangle_{+,-} = \langle P_{\mathcal{M}_+}\varphi, A^2\psi\rangle_{+,-} \\
&= (P_{\mathcal{M}_+}\varphi, LP_{\mathcal{M}_+}\psi)_0.
\end{aligned}$$

Note that the range of L coincides with the whole Hilbert space, $\operatorname{Ran} L = \operatorname{Ran} A^2 = \mathcal{H}_0$. Thus, by the Hellinger–Toeplitz theorem, the operator L is self-adjoint.

Let us prove that the operator L defined in (7.141) is the same as L in (7.142). To this end we first show that these operators are equal on the set $\tilde{\mathcal{M}}_+$. Then we check that they coincide with the Friedrichs extension of the symmetric operator $\dot{L}$ (see (7.143)). So, we have to show that the Berezansky canonical isomorphisms $\breve{D}_{-,+}$ and $D_{-,+}$ act identically on the subspace $\tilde{\mathcal{M}}_+ = \mathcal{M}_+ \cap \mathcal{H}_{++}$ and that their ranges belong to $\mathcal{H}_0$:

$$\breve{D}_{-,+}\varphi = D_{-,+}\varphi \in \mathcal{H}_0, \quad \varphi \in \tilde{\mathcal{M}}_+. \tag{7.145}$$

It is obvious that $P_{\mathcal{M}_+}\tilde{\mathcal{M}}_+ = \tilde{\mathcal{M}}_+$, since $\tilde{\mathcal{M}}_+ := \mathcal{M}_+ \cap \mathcal{H}_{++}$. It follows that $\tilde{\mathcal{M}}_+ \subset \mathfrak{D}(\dot{L})$ and $\dot{L}\tilde{\mathcal{M}}_+ = A^2\tilde{\mathcal{M}}_+$. To prove (7.145) we recall that $\mathcal{H}_{++} \equiv \mathcal{H}_4(A) = \mathfrak{D}(A^2)$ and $\tilde{\mathcal{M}}_+ = \mathcal{M}_+ \cap \mathcal{H}_{++}$. Thus, for each $\varphi \in \mathcal{H}_{++}$ the vector $f := D_{-,+}\varphi = A^2\varphi \in \mathcal{H}_0$. Further, let us consider for some fixed $\varphi \in \tilde{\mathcal{M}}_+$ the functionals:

$$\begin{aligned}
l_\varphi(\psi) &:= \langle D_{-,+}\varphi, \psi\rangle_{-,+}, \quad \psi \in \mathcal{H}_+, \\
\breve{l}_\varphi(\psi) &:= \langle \breve{D}_{-,+}\varphi, \psi\breve{\rangle}_{-,+}, \quad \psi \in \mathcal{M}_+.
\end{aligned}$$

Clearly, the first functional $l_\varphi(\psi)$ is continuous on $\mathcal{H}_0$ since $D_{-,+}\varphi = A^2\varphi$. Moreover,

$$l_\varphi(\psi) = (f,\psi)_0 = (A^2\varphi,\psi) = (A\varphi, A\psi) = (\varphi,\psi)_+$$

because $\psi \in \mathcal{M}_+$. The second functional $\breve{l}_\varphi(\psi)$ is also continuous on $\mathcal{H}_0$ since $\mathcal{M}_+ = \breve{\mathcal{H}}_+$ and

$$\breve{l}_\varphi(\psi) = (\varphi,\psi)_+ = \langle \varphi,\psi\rangle_{\mathcal{H}_{++},\mathcal{H}_0}, \quad |\breve{l}_\varphi(\psi)| \le c\|\psi\|_0,$$

where a constant $c = \|\varphi\|_{++}$. Thus, $\breve{l}_\varphi(\psi) = (\breve{f},\psi)_0$ with some $\breve{f} \in \mathcal{H}_0$. We claim that $f = \breve{f}$. Indeed, by our constructions, $(f,\psi)_0 = (\varphi,\psi)_+ = (\breve{f},\psi)_0$ for all $\psi \in \mathcal{M}_+$. Hence, the vectors f and $\breve{f}$ coincide, since the subspace $\mathcal{M}_+$ is dense in $\mathcal{H}_0$. Therefore, (7.145) is proved.

Now we will show that the operator L in (7.141) coincides with the Friedrichs extension of the symmetric operator $\dot{L}$. Let us recall that the domain $\mathfrak{D}(\dot{L}) = \tilde{\mathcal{M}}_+$ is dense in $\mathcal{H}_0$. From (7.132) it follows that the subspace $\tilde{\mathcal{M}}_+$ is dense in $\mathcal{M}_+$. Indeed, if $\phi \in \mathcal{M}_+$ and $\phi \perp \tilde{\mathcal{M}}_+$, in $\mathcal{H}_+$ then $D_{-,+}\phi \perp \mathcal{N}_-$ and $D_{-,+}\phi \in \tilde{\mathcal{N}}_-$. Thus, $\phi \equiv 0$, since due to (7.132) $\tilde{\mathcal{N}}_- = \mathcal{N}_-$. In more detail, assume $\mathcal{M}_+ = \tilde{\mathcal{M}}_+ \oplus \tilde{\mathcal{M}}_+^\perp$ and take $\phi \in \tilde{\mathcal{M}}_+^\perp$. Then $\omega := D_{-,+}\phi \in \tilde{\mathcal{M}}_-^\perp$, where $\tilde{\mathcal{M}}_-^\perp = \mathcal{M}_- \ominus \tilde{\mathcal{M}}_-$. So, we obtain

$$\langle \omega, \tilde{\mathcal{M}}_+\rangle_{-,+} = 0 = \langle \omega, \tilde{\mathcal{M}}_+\rangle_{--,++} \Longrightarrow \omega \in \tilde{\mathcal{N}}_- = \mathcal{N}_-.$$

However, this is only possible if $\phi = 0$ because $\phi \in \mathcal{M}_+$ and $D_{-,+}\phi \perp \mathcal{N}_-$. Therefore, $\mathcal{M}_+ \sqsupset \tilde{\mathcal{M}}_+$.

Further, it is obvious that $\mathfrak{D}(\dot{L}) = \tilde{\mathcal{M}}_+$, the domain of $\dot{L}$, is closed in $\mathcal{H}_0$ since $\tilde{\mathcal{M}}_+$ is closed in $\mathcal{H}_{++}$. We claim that the range of $\dot{L}$ is also dense in $\breve{\mathcal{H}}_-$. To show this we remark that due to (7.145) the range of operator $\dot{L}$ coincides with the subspace

$$\tilde{\mathcal{M}}_- = A^2\tilde{\mathcal{M}}_+ = A^2(\mathcal{M}_+ \cap \mathcal{H}_{++}) = \mathcal{M}_- \cap \mathcal{H}_0,$$

which is dense in $\breve{\mathcal{H}}_-$ because the mapping $\breve{D}_{-,+} : \breve{\mathcal{H}}_+ \to \breve{\mathcal{H}}_-$ is isometric.

Since $\tilde{\mathcal{M}}_+ \sqsubset \mathcal{M}_+$, the space $\breve{\mathcal{H}}_+$ coincides with the closure of $\tilde{\mathcal{M}}_+$ with respect to the inner product

$$(\varphi,\psi)_{\breve{\mathcal{H}}_+} := (\dot{L}\varphi,\psi)_0 = (A\varphi, A\psi)_0 = (\varphi,\psi)_+, \quad \varphi,\psi \in \tilde{\mathcal{M}}_+.$$

Hence, we can conclude that the operator L is a self-adjoint extension of $\dot{L}$. By our construction, L is the Friedrichs extension of the operator $\dot{L}$ since we already proved that the inclusion $\tilde{\mathcal{M}}_+ \sqsubset \mathcal{M}_+$ is dense and continuous.

Finally, the equality (7.144) holds because the completion of the set $\mathfrak{D}((\breve{A})^2) = \mathfrak{D}(L)$ with respect to the norm $\|\cdot\|_+ := \|L^{1/2}\cdot\|_0$ coincides with $\mathcal{M}_+$.

Indeed, since $\tilde{\mathcal{M}}_+$ is dense in $\mathcal{M}_+$, it is sufficient to recall that $(L\varphi,\psi)_0 = (\varphi,\psi)_+$, $\varphi,\psi \in \tilde{\mathcal{M}}_+$. Thus, by the definition of L we have

$$(L\varphi,\psi)_0 = (L^{1/2}\varphi, L^{1/2}\psi)_0 = (A^2\varphi,\psi)_0 = (\varphi,\psi)_+ = ((\breve{A})^2\varphi,\psi)_0,$$

for all vectors $\varphi,\psi \in \tilde{\mathcal{M}}_+$. Thus, $\mathcal{M}_+ = \mathcal{H}_1(L)$ and moreover, $\mathcal{M}_+ = \mathcal{H}_2(\breve{A}) = \mathfrak{D}(\breve{A}) = \breve{\mathcal{H}}_+$. This completes the proof of Theorem 7.8.4. □

7.8.5 Construction of the operator $\breve{D}$

In this section we generalize the previous construction of the operator $\breve{A}$ to the case when a set $\mathcal{M}_+$ dense in $\mathcal{H}_0$ taken as a non-trivial part from $\mathcal{H}_k$, $k > 1$. In fact, we repeat the construction from Subsection 7.7.4 in an abstract form.

So, let the space $\mathcal{H}_+ = \mathcal{H}_k$, $k > 0$, be decomposed into an orthogonal sum $\mathcal{H}_+ = \mathcal{M}_+ \oplus \mathcal{N}_+$. We assume that $\mathcal{M}_+ \sqsubset \mathcal{H}_0$. Consider also

$$\mathcal{H}_- \equiv \mathcal{H}_{-k} \sqsupset \mathcal{H}_0 \sqsupset \mathcal{H}_k \equiv \mathcal{H}_+,$$

the rigged space

$$\breve{\mathcal{H}}_- \equiv (\mathcal{M}_+)_- \sqsupset \mathcal{H}_0 \sqsupset \mathcal{M}_+ \equiv \breve{\mathcal{H}}_+,$$

and the operator $\breve{D} : \breve{\mathcal{H}}_+ = \mathfrak{D}(\breve{D}) \to \mathcal{H}_0 = \mathcal{R}(\breve{D})$ associated with this space. Clearly, $\breve{D}$ is self-adjoint in $\mathcal{H}_0$ as an analog of $\breve{A}$ from Subsection 7.7.4. We want to determine how $\breve{D}$ is connected with $A^{k/2}$. Recall that $A^k : \mathcal{H}_k \to \mathcal{H}_0$ and $\mathcal{H}_k$ is the domain for A^k.

Lemma 7.8.5. *For each dense in $\mathcal{H}_0$ subspace $\mathcal{M}_+$ from $\mathcal{H}_k$ the mapping*

$$L_k : P_{\mathcal{M}_+}\varphi \longmapsto A^k\varphi, \quad \varphi \in \mathcal{H}_{2k}$$

defines a self-adjoint operator in $\mathcal{H}_0$, where $P_{\mathcal{M}_+}$ denotes the orthogonal projection of $\mathcal{H}_+$ onto $\mathcal{M}_+$.

Proof. The mapping L_k is a well-defined operator. Indeed, if $P_{\mathcal{M}_+}\varphi = 0$, then $\varphi \in \mathcal{N}_+ = \mathcal{H}_+ \ominus \mathcal{M}_+$. Now, using the embedding $\mathcal{M}_+ \sqsubset \mathcal{H}_0$, we conclude that $\mathcal{N}_+ \cap \mathcal{H}_{2k} = \{0\}$ since $\mathcal{N}_- \cap \mathcal{H}_0 = \{0\}$. Thus, $\varphi = 0$.

Let us show that L_k with domain $\mathfrak{D}(L) = P_{\mathcal{M}_+}\mathcal{H}_{2k}$ is a symmetric operator in $\mathcal{H}_0$. Really,

$$\begin{aligned}
(LP_{\mathcal{M}_+}\varphi, P_{\mathcal{M}_+}\psi)_0 &= (A^k\varphi, P_{\mathcal{M}_+}\psi)_0 = \langle (A^k)^{\mathrm{cl}}\varphi, P_{\mathcal{M}_+}\psi\rangle_{-2k,2k} \\
&= \langle P_{\mathcal{M}_-}(A^k)^{\mathrm{cl}}\varphi, P_{\mathcal{M}_+}\psi\rangle_{-2k,2k} \\
&= \langle (A^k)^{\mathrm{cl}}P_{\mathcal{M}_+}\varphi, P_{\mathcal{M}_+}\psi\rangle_{-2k,2k} \\
&= \langle P_{\mathcal{M}_+}\varphi, (A^k)^{\mathrm{cl}}P_{\mathcal{M}_+}\psi\rangle_{-2k,2k} \\
&= \langle P_{\mathcal{M}_+}\varphi, P_{\mathcal{M}_-}(A^k)^{\mathrm{cl}}\psi\rangle_{2k,-2k} \\
&= \langle P_{\mathcal{M}_+}\varphi, (A^k)^{\mathrm{cl}}\psi\rangle_{2k,-2k} \\
&= (P_{\mathcal{M}_+}\varphi, A^k\psi)_0 = (P_{\mathcal{M}_+}\varphi, L_kP_{\mathcal{M}_+}\psi)_0,
\end{aligned}$$

where $P_{\mathcal{M}_-}$ denotes the orthogonal projection of $\mathcal{H}_{-2k}$ onto subspace $\mathcal{M}_- := D_{-,+}\mathcal{M}_+$ and $(A^k)^{\rm cl}$ is the closure of $A^k : \mathcal{H}_0 \to \mathcal{H}_{-2k}$. We used also the standard relation from the theory of the rigged spaces: $(A^k)^{\rm cl} P_{\mathcal{M}_+} = P_{\mathcal{M}_-}(A^k)^{\rm cl}$. In fact, L is self-adjoint since the range $\mathcal{R}(L) = \mathcal{R}(A^k) = \mathcal{H}_0$. □

Parallel with L_k, let us consider in $\mathcal{H}_0$ the operator $\breve{L}_k$ generated by the Berezansky canonical isomorphism $\breve{D}_{-,+}$ acting from $\breve{\mathcal{H}}_+$ to $\breve{\mathcal{H}}_-$:

$$\breve{L}_k := \breve{D}_{-,+} \restriction \mathfrak{D}(\breve{L}_k), \quad \mathfrak{D}(\breve{L}_k) := \{\breve{\varphi} \in \breve{\mathcal{H}}_+ \mid \breve{D}_{-,+}\breve{\varphi} \in \mathcal{H}_0\}.$$

Lemma 7.8.6. *If the subspace $\mathcal{M}_+$ from $\mathcal{H}_k$ is dense in $\mathcal{H}_0$, $\mathcal{M}_+ \sqsubset \mathcal{H}_0$, then the operators L_k and $\breve{L}_k$ coincide.*

Proof. Let $\breve{\varphi} \in \mathfrak{D}(\breve{L}_k) \subset \mathcal{M}_+ \equiv \breve{\mathcal{H}}_+$. Then for $\mathcal{H}_0 \ni \breve{f} = \breve{D}_{-,+}\breve{\varphi}$, we have:

$$\langle \breve{D}_{-,+}\breve{\varphi}, \psi \breve{\rangle}_{-,+} = (\breve{\varphi}, \psi \breve{)}_+ = (\breve{f}, \psi)_0, \quad \psi \in \mathcal{M}_+.$$

Moreover, since $\mathcal{M}_+$ is a subspace in $\mathcal{H}_+$,

$$(\breve{\varphi}, \psi \breve{)}_+ = (\breve{\varphi}, \psi)_+ = (A^{k/2}\breve{\varphi}, A^{k/2}\psi)_0.$$

In general, a vector $\breve{\varphi} \in \mathcal{M}_+$ does not belong to the domain of A^k. However, if $\breve{\varphi} \in \mathfrak{D}(\breve{L}_k)$, then thanks of to the equality $(\breve{\varphi}, \psi \breve{)}_+ = (\breve{f}, \psi)_0$ the fact that $\mathcal{M}_+$ is dense in $\mathcal{H}_0$, there exists a vector $\varphi \in \mathcal{H}_+$ such that $\breve{f} = A^k\varphi$. For such φ we have,

$$\begin{aligned}(\breve{\varphi}, \psi)_+ &= (\breve{f}, \psi)_0 = (A^k\varphi, \psi)_0 = \langle D_{-,+}\varphi, \psi\rangle_{-,+} \\ &= (\varphi, \psi)_{-,+} = (P_{\mathcal{M}_+}\varphi, \psi)_{-,+}, \ \psi \in \mathcal{M}_+.\end{aligned}$$

This means that $\breve{\varphi} = P_{\mathcal{M}_+}\varphi$, because $\mathcal{M}_+$ is dense in $\mathcal{H}_0$. Furthermore, $\breve{L}\breve{\varphi} = \breve{D}_{-,+}\breve{\varphi} = \breve{f} = A^k\varphi = LP_{\mathcal{M}_+}\varphi$. This completes the proof of Lemma 7.8.6. □

We introduce into consideration the subspace $\tilde{\mathcal{M}}_+ := \mathcal{M}_+ \cap \mathcal{H}_{2k}$.

Proposition 7.8.7. *The operators L_k and A^k coincide after restriction to $\tilde{\mathcal{M}}_+$, i.e.,*

$$L_k \restriction \tilde{\mathcal{M}}_+ = A^k \restriction \tilde{\mathcal{M}}_+.$$

Proof. Since $P_{\mathcal{M}_+}\tilde{\mathcal{M}}_+ = \tilde{\mathcal{M}}_+$, it follows from Lemma 7.8.6 that

$$L_k \restriction \tilde{\mathcal{M}}_+ = \breve{L}_k \restriction \tilde{\mathcal{M}}_+ = A^k \restriction \tilde{\mathcal{M}}_+.$$

Moreover, since $\tilde{\mathcal{M}}_+ \sqsubset \mathcal{H}_0$, the operator L_k is a self-adjoint extension of the symmetric operator $\dot{L}_k := A^k \restriction \tilde{\mathcal{M}}_+$. That $\tilde{\mathcal{M}}_+$ is dense in $\mathcal{H}_0$ follows from a condition of type (7.132). □

Proposition 7.8.8. *In the general case, where a proper subspace $\mathcal{M}_+$ from $\mathcal{H}_k$, $k > 1$ is dense in $\mathcal{H}_0$ and $\tilde{\mathcal{M}}_+ = \mathcal{M}_+ \cap \mathcal{H}_{2k}$ is dense in $\mathcal{M}_+$, $\tilde{\mathcal{M}}_+ \sqsubset \mathcal{M}_+$, the operator L_k coincides with the Friedrichs extension of the symmetric operator $\dot{L}$.*

Proof. According to Proposition 7.8.7, the quadratic form $\gamma(\varphi,\psi) := (\dot{L}\varphi,\psi)_0$ coincides with the form $(A^k\varphi,\psi)_0 = (\varphi,\psi)_{\mathcal{M}_+}$ on vectors $\varphi,\psi \in \tilde{\mathcal{M}}_+$. This implies that the closure of γ coincides with the inner product in $\mathcal{M}_+$ thanks to $\tilde{\mathcal{M}}_+$ being dense in $\mathcal{M}_+$. Thus, $\breve{L}_k$ is the Friedrichs extension of $\dot{L}_k$. It remains to use the already proved equality $L_k = \breve{L}_k$. □

Define $\breve{D} := \breve{L}_k^{1/2} = L_k^{1/2}$. Directly from Lemmas 7.8.5, 7.8.6 and Propositions 7.8.7, 7.8.8 we get the following result.

Theorem 7.8.9. *If $\tilde{\mathcal{M}}_+ \sqsubset \mathcal{M}_+$, then the operator $\breve{D}$, defined on $\mathcal{M}_+$ which is self-adjoint in $\mathcal{H}_0$, coincides with the square root of the Friedrichs extension of the symmetric operator $\dot{L}_k := A^k \restriction \tilde{\mathcal{M}}_+$.*

Proof. It is suffices to note that the equality $\mathfrak{D}(\breve{D}) = \mathcal{M}_+$ is an immediate consequence of the denseness of $\tilde{\mathcal{M}}_+$ in $\mathcal{M}_+$. □

Finally, we make the following observations. By the above constructions, the forms

$$\breve{\gamma}(\varphi,\psi) := (\breve{L}\varphi,\psi)_0 = (\breve{D}\varphi,\breve{D}\psi)_0 = (\varphi,\psi)\breve{}_+,$$
$$\gamma(\varphi,\psi) := (A^k\varphi,\psi)_0 = (A^{k/2}\varphi,A^{k/2}\psi)_0 = (\varphi,\psi)_+$$

coincide on $\tilde{\mathcal{M}}_+$; however, their closures in $\mathcal{H}_0$ are different. Therefore, the self-adjoint operators $\breve{D}$ and $A^{k/2}$ are also different, $\breve{D} \neq A^{k/2}$. Moreover, these operators cannot coincide on any set dense in $\mathcal{H}_0$. Nevertheless, their squares, i.e., $\breve{L}$ and A^k, coincide on some set dense in $\mathcal{H}_0$. This means that $\breve{L}$ is a super-singular perturbation of the operator A^k (see Chapter 8).

Chapter 8

Super-singular Perturbations

In this chapter we deal with perturbations whose influence is concentrated outside of the domain of essential self-adjointness of the free Hamiltonian. We show that the extended rigged spaces method is applicable to such perturbations. A new step is the introduction of the scale of Hilbert spaces and involving in the consideration the unperturbed operator in some power. This allows one to treat perturbations with any order of singularity.

8.1 Idea of the method

Let $A = A^* \geq 1$ denote a self-adjoint operator on a Hilbert space $\mathcal{H}$ with the norm $\|\cdot\|$ and the inner product $(\cdot,\cdot)$. For example, one can think that

$$A = H_0 + 1, \; H_0 = -\Delta, \quad \mathcal{H} = L_2(\mathbb{R}^d, dx), \; d \geq 1.$$

Each A is associated with the A-scale of Hilbert spaces

$$\mathcal{H}_{-k} \sqsupset \mathcal{H}_0 \equiv \mathcal{H} \sqsupset \mathcal{H}_k \equiv \mathcal{H}_k(A), \quad k > 0, \tag{8.1}$$

(see Chapter 4). Here $\mathcal{H}_k = \operatorname{Dom} A^{k/2} \equiv \mathfrak{D}\left(A^{k/2}\right)$ with the norm

$$\|\varphi\|_k := \|A^{k/2}\varphi\|, \quad \varphi \in \operatorname{Dom} A^{k/2},$$

is called the positive space with respect to $\mathcal{H}_0$, and $\mathcal{H}_{-k}$ is called the negative space with respect to $\mathcal{H}_0$. It is conjugate to $\mathcal{H}_+$. This space coincides with the completion of $\mathcal{H}$ with respect to the norm

$$\|h\|_{-k} := \|A^{-k/2}h\|, \quad h \in \mathcal{H}$$

(for more details see [1, 42–45]).

In the sequel we will often use the following important fact. Each self-adjoint operator $A \geq 1$ on $\mathcal{H}$ can be uniquely recovered from a couple of spaces $\mathcal{H}_0 \sqsupset \mathcal{H}_k$

with fixed $k > 0$ (or from a dual pair $\mathcal{H}_{-k} \sqsupset \mathcal{H}_0$). So, in the case $k = 2$, the operator A can be recovered from the Berezansky canonical isomorphism D_{20} : $D_{0,2} : \mathcal{H}_2 \to \mathcal{H}_0$ (see Chapter 4 and [42]). The operator A can also be recovered from the restriction of the canonical isomorphism $D_{-1,1} : \mathcal{H}_1 \to \mathcal{H}_{-1}$:

$$A = D_{-1,1} \restriction \{\varphi \in \mathcal{H}_1 \mid D_{-1,1}\varphi \in \mathcal{H}\}.$$

Similarly, in the case $k > 2$, the operator A^k coincides with the Berezansky canonical isomorphism $D_{0,2k} : \mathcal{H}_{2k} \to \mathcal{H}_0$ treated as a self-adjoint operator in $\mathcal{H}_0$. Therefore, the operator A can be uniquely recovered by the formula, $A = \sqrt[k]{D_{0,2k}}$, where the kth root is defined by the spectral theorem. Finally, the operator A can be recovered by using of the mapping $D_{k,-k} : \mathcal{H}_k \to \mathcal{H}_{-k}$ after its restriction:

$$D_k = D_{-k,k} \restriction \mathfrak{D}_k, \ \mathfrak{D}_k \equiv \mathcal{H}_{2k} = \{\varphi \in \mathcal{H}_k | D_{-k,k}\varphi \in \mathcal{H}_0\}.$$

Namely, $A = \sqrt[k]{D_k}$, where D_k is treated as a positive self-adjoint operator in $\mathcal{H}$ (for more detail see [1]).

It is clear that the same connections exist between the perturbed operators $\tilde{A}$ and the associated $\tilde{A}$-scales of Hilbert spaces. Indeed, let $\tilde{A}$ be a self-adjoint operator in $\mathcal{H}$ which is singularly perturbed with respect to A. We assume $\tilde{A} \geq 1$. Then from $\tilde{A}$ it is easy to construct the associated scale of Hilbert spaces

$$\tilde{\mathcal{H}}_{-k} \sqsupset \mathcal{H}_0 \equiv \mathcal{H} \sqsupset \tilde{\mathcal{H}}_k = \mathcal{H}_k(\tilde{A}), \quad \tilde{\mathcal{H}}_k = \operatorname{Dom} \tilde{A}^{k/2}, \ k > 0. \tag{8.2}$$

Using this scale in the same way as above one can reconstruct $\tilde{A}$ from any pair of spaces $\mathcal{H}_0 \sqsupset \tilde{\mathcal{H}}_k$, $k \geq 2$, or from the rigged space $\tilde{\mathcal{H}}_{-2k} \sqsupset \mathcal{H}_0 \sqsupset \tilde{\mathcal{H}}_{2k}$, $k \geq 1$. Namely, one can define $\tilde{A} = \sqrt[k]{\tilde{D}_{0k}}$ or, equivalently, $\tilde{A} = \sqrt[k]{\tilde{D}_k}$ (one should always take a positive root), where

$$\tilde{D}_k = \tilde{D}_{-k,k} \restriction \tilde{\mathfrak{D}}_k,$$
$$\tilde{\mathfrak{D}}_k \equiv \tilde{\mathcal{H}}_{2k} = \{\varphi \in \tilde{\mathcal{H}}_k \mid \tilde{D}_{-k,k}\varphi \in \mathcal{H}_0\}.$$

The similar idea of using the method of rigged Hilbert spaces for constructing super-singular perturbations can be explained as follows.

Let S denote an arbitrary singular perturbation of an operator A. Now, any order of singularity for S is admissible. In particular, one can deal with the so-called super-singular perturbations (see below a precise definition). In general, it is rather difficult and even impossible to approach such singular perturbations directly. For example, the sum $A + S$ can be only defined on a zero vector. Therefore, any attempt to chose $A + S$ in the role of the singularly perturbed operator does not make sense. The idea of the method for developed below for treating singular perturbations of the form S consists in the following. First we use S to change the norm in one of the spaces $\mathcal{H}_k, k \neq 0$ from the A-scale. Then we construct a new scale of Hilbert spaces. And finally, we define the perturbed operator $\tilde{A}$ as the operator associated with this scale. In particular, on A and S we can construct one of the spaces $\tilde{\mathcal{H}}_k$, $\tilde{\mathcal{H}}_{-k}$, or $\tilde{\mathcal{H}}_{2k}$, $\tilde{\mathcal{H}}_{-2k}$. The choice of k depends on the order of singularity of the operator S. In general, it can be arbitrary. Then we define $\tilde{A}^k$ as the operator associated with the rigged space $\tilde{\mathcal{H}}_{-k} \sqsupset \mathcal{H}_0 \sqsupset \tilde{\mathcal{H}}_k$, or with $\tilde{\mathcal{H}}_{-2k} \sqsupset \mathcal{H}_0 \sqsupset \tilde{\mathcal{H}}_{2k}$. To this end we introduce the perturbed variant of the Berezansky canonical isomorphism $\tilde{D}_{0,k}$, or the above-described restriction of the mapping $\tilde{D}_{-k,k}$, which we denote by $\tilde{D}_k$. Finally, we define the perturbed operator $\tilde{A}$ as a positive kth root of the self-adjoint in $\mathcal{H}_0$ operator $\tilde{D}_{0,k}$. Instead of $\tilde{D}_{0,k}$ one can take the operator $\tilde{D}_k$.

In fact, the method of rigged spaces described above generalizes the well-known form-sum method to the case of perturbations with an arbitrary order of singularity. A peculiarity is that in the form-sum method the perturbed operator $\tilde{A}$ is defined as the one associated with the triple $\tilde{\mathcal{H}}_{-1} \sqsupset \mathcal{H} \sqsupset \tilde{\mathcal{H}}_1$, without resorting to use the spectral theorem, since now $k = 1$. Thus, for the construction and the subsequent investigation of the singularly perturbed operator $\tilde{A}$ of higher order of singularity (corresponding to perturbations from the $\mathcal{H}_{-k}$-class with $k > 2$), we have to construct at least one of the perturbed spaces $\tilde{\mathcal{H}}_k$, $\tilde{\mathcal{H}}_{-k}$, $\tilde{\mathcal{H}}_{2k}$, or $\tilde{\mathcal{H}}_{-2k}$, introduce a new rigged Hilbert space, and consider the canonical Berezansky isomorphism. Its restriction to the principal Hilbert space $\mathcal{H}_0$ defines the sought-for perturbed operator. We find this operator after taking the kth positive root of the above restriction. In addition, we remark that in the sequel we will make essential use of the theory of self-adjoint extensions of symmetric operators.

8.1.1 Details of the method

Let us describe in more detail the method of rigged spaces in the theory of super-singular perturbations. We consider an extended fixed rigged space as a part of the A-scale: (8.1)

$$\mathcal{H}_{-2} \sqsupset \mathcal{H}_{-1} \sqsupset \mathcal{H}_0 \equiv \mathcal{H} \sqsupset \mathcal{H}_1 \sqsupset \mathcal{H}_2 \equiv \mathcal{H}_2(A), \tag{8.3}$$

where $\mathcal{H}_2 = \operatorname{Dom} A$ is the Hilbert space with the norm $\|\varphi\|_2 := \|A\varphi\|$ and $\mathcal{H}_1 = \operatorname{Dom} A^{1/2}$ with the norm $\|\varphi\|_1 := \|A^{1/2}\varphi\|$. The spaces $\mathcal{H}_{-2}$ and $\mathcal{H}_{-1}$ are dual to $\mathcal{H}_2$ and $\mathcal{H}_1$, respectively. Let us recall that there exists a unique correspondence between operators $A = A^* \geq 1$ on $\mathcal{H}$ and rigged Hilbert spaces of the form (8.3) (see Theorem 4.2.1 and also [1]).

Let us repeat in detail how the spaces $\mathcal{H}_{-1}$ and $\mathcal{H}_1$ are connected to one another. We consider the linear functional $l_\omega(\varphi) := \langle \varphi, \omega \rangle_{1,-1}$ defined by a fixed $\omega \in \mathcal{H}_{-1}$. We recall that here and everywhere in the text, $\langle \cdot, \cdot \rangle_{k,-k}$, $k > 0$ denotes the duality pairing between the spaces $\mathcal{H}_k$ and $\mathcal{H}_{-k}$. Clearly, the functional l_ω is continuous on $\mathcal{H}_1$. Hence, by the Riesz theorem, it admits the representation: $l_\omega(\varphi) = (\varphi, \psi)_1$, where the vector $\psi \in \mathcal{H}_1$ is defined by ω. According to the definition of norms in $\mathcal{H}_k$ and $\mathcal{H}_{-k}$, it is easy to show that $\|\psi\|_1 = \|\omega\|_{-1}$. Let $D_{-1,1} : \mathcal{H}_1 \ni \psi \mapsto \omega \in \mathcal{H}_{-1}$ denote the Berezansky canonical isomorphism (see [1, 42, 44]).

The following restriction of this isomorphism defines an operator in $\mathcal{H}$:

$$D_1 := D_{-1,1} \restriction \mathfrak{D}_1, \quad \mathfrak{D}_1 := \{\varphi \in \mathcal{H}_1 \mid D_{-1,1}\varphi \in \mathcal{H}\}.$$

It is not hard to show that $\mathfrak{D}_1 = \mathcal{H}_2$ with respect to the norm $\|\varphi\|_2 = \|A\varphi\|$, where $\varphi \in \mathfrak{D}_1$. Hence, $A = D_1$. Repeating the same constructions with $k = 2$, we get the equality $A^2 = D_2$. Therefore, $A = \sqrt{D_2}$. In general, for arbitrary $k \geq 1$ we have $A = (D_k)^{1/k}$.

We recall again that each rigging $\mathcal{H}_{-k} \sqsupset \mathcal{H} \sqsupset \mathcal{H}_k$, as well as the whole A-scale can be reconstructed from any couple $\mathcal{H}_{-k} \sqsupset \mathcal{H}$ or $\mathcal{H} \sqsupset \mathcal{H}_k$ which generates a pre-rigging (see [42, 44, 51]).

Now, replacing the inner product in one of the spaces $\mathcal{H}_k$ or $\mathcal{H}_{-k}$, $k \geq 1$, i.e., changing $(\cdot,\cdot)_k$ to $(\cdot,\cdot)_{\tilde{k}}$ or $(\cdot,\cdot)_{-k}$ on $(\cdot,\cdot)_{-\tilde{k}}$, we obtain one of the pre-rigging couples $\mathcal{H} \sqsupset \tilde{\mathcal{H}}_k$ or $\tilde{\mathcal{H}}_{-k} \sqsupset \mathcal{H}$. In the second step we extend this couple to a new chain of spaces:

$$\tilde{\mathcal{H}}_{-2k} \sqsupset \tilde{\mathcal{H}}_{-k} \sqsupset \mathcal{H} \sqsupset \tilde{\mathcal{H}}_k \sqsupset \tilde{\mathcal{H}}_{2k}, \tag{8.4}$$

and then to a whole infinite scale of the form (8.2).

We say that the operator $\tilde{A}$ associated with (8.4) is *singularly perturbed* with respect to A *in the wide sense* if there exists $k \geq 1$ such that the operators $\tilde{A}^k$ and A^k coincide on some dense in $\mathcal{H}_{k-1}$ set $\mathcal{M}_k \subset \tilde{\mathcal{H}}_k \cap \mathcal{H}_k$. It is obvious that in this case the inner products in $\mathcal{H}_k$ and $\tilde{\mathcal{H}}_k$ also partly coincide:

$$(\varphi,\psi)_k = (\varphi,\psi)_{\tilde{k}}, \quad \varphi,\psi \in \mathcal{M}_k.$$

Note that for all $k \geq 3$ the set $\mathcal{M}_k$ with $k \geq 3$ is dense in $\mathcal{H}_2 = \operatorname{Dom} A$. Therefore, the symmetric operator $\mathbf{A} := A \restriction \mathcal{M}_k$ is essentially self-adjoint. In this case it is impossible to use directly the method of self-adjoint extensions. It is precisely this kind of situation that arises when one considers super-singular perturbations. Then resorting to the method of rigged Hilbert spaces is inevitable from our point of view. In all cases with $k > 2$ the perturbed operator $\tilde{A}$ can be defined by the formula $\tilde{A} = \sqrt[k]{\tilde{D}_k}$, where $\tilde{D}_k$ denotes the restriction of the Berezansky canonical isomorphism $\tilde{D}_{-k,k} : \tilde{\mathcal{H}}_k \to \tilde{\mathcal{H}}_{-k}$ to the set

$$\tilde{\mathfrak{D}}_k = \tilde{\mathcal{H}}_{2k} = \{\varphi \in \tilde{\mathcal{H}}_k \mid \tilde{D}_{-k,k}\varphi \in \mathcal{H}\}.$$

It is worth explaining that in the case when for the bounded from below operator $A \geq \tilde{m}$ one has that $\tilde{m} < 1$, one has to make some additional changes in the construction of spaces (8.1), (8.2). Namely, the norms $\|\cdot\|_{\pm k}$ should be defined as $\|\cdot\|_{\pm k,c} := \|(A + c)\cdot\|_{\pm k}$ with $c = 1 - \tilde{m}$.

It should be also noted that the above-described general construction of the singularly perturbed operator $\tilde{A}$ has a certain deficiency. Namely, according to the formula $\tilde{A} = \sqrt[k]{\tilde{D}_k}$ the perturbed operator arises as a positive root of the self-adjoint operator. Hence, it is impossible to get $\tilde{A}$ explicitly, but only by the use the spectral theorem and the functional operator calculus. Nevertheless, the method of rigged spaces is useful at least because it gives a way for the precise definition of the super-singularly perturbed operator in the case an arbitrary order of singularity. In addition, in this way, one can also investigate the spectral properties of the perturbed operators similarly to the method of self-adjoint extensions and to the method of abstract boundary triples [73, 74].

In particular, in specific problems using the Kreĭn formula for the resolvent and by the explicit form of the integral kernel $(1-\Delta)^{-k/2}$ in terms of the Bessel functions (see [29, 157]), one can derive conditions which ensure that new eigenvalues arise for the perturbed operator $\tilde{A}$. Moreover, in this way we can produce explicit integral kernels for perturbed resolvents. They describe the generalized resolvent of the operator $\tilde{A}^{-k/2}$ in the case when the perturbations are given by δ-like potentials concentrated on sets of dimension smaller than d in $\mathbf{R}^d$. The same results hold true for perturbations given by singular measures supported on sets of zero C_k-capacity, with $k>2$.

8.2 New riggings by means of singular quadratic forms

According to the idea described above, for the construction of the perturbed operator $\tilde{A}$ it is necessary to have any couple of spaces from the $\tilde{A}$-scale, say, $\tilde{\mathcal{H}}_{-k} \sqsupset \mathcal{H}_0$ or $\mathcal{H}_0 \sqsupset \tilde{\mathcal{H}}_k$. Then one can extend this couple to the whole $\tilde{A}$-scale of Hilbert spaces and define $\tilde{A}$ as the operator associated with a new rigged space from the chain,

$$\tilde{\mathcal{H}}_{-2} \sqsupset \tilde{\mathcal{H}}_{-1} \sqsupset \mathcal{H}_0 \equiv \mathcal{H} \sqsupset \tilde{\mathcal{H}}_1 \sqsupset \tilde{\mathcal{H}}_2. \tag{8.5}$$

In this subsection we analyze in detail the construction of the rigged Hilbert spaces of the type (8.5) and their connections with quadratic forms γ interpreted as singular perturbations of the $\mathcal{H}_{-2}(A)$-class. These forms have the representation

$$\gamma(\varphi,\psi) = \langle S\varphi,\psi\rangle_{-2,2}, \quad \gamma=\gamma_S, \quad S\in\mathcal{T}_{-2}(A).$$

In particular, we will show that the whole family of positive singularly perturbed operators of the $\mathcal{H}_{-2}(A)$-class can be constructed by the method of generalized form-sum.

We begin with the rigging (8.1) associated with an unperturbed operator $A=A^*\geq 1$ on $\mathcal{H}_0$. We consider five spaces

$$\mathcal{H}_- \equiv \mathcal{H}_{-2} \sqsupset \mathcal{H}_{-1} \sqsupset \mathcal{H}_0 \sqsupset \mathcal{H}_1 \sqsupset \mathcal{H}_2 \equiv \mathcal{H}_+ (= \operatorname{Dom} A). \tag{8.6}$$

Let the bounded quadratic form $\gamma\geq 0$ given on $\mathcal{H}_2$, $Q(\gamma)=\mathcal{H}_2$ belong to the $\mathcal{H}_{-2}(A)$-class. We shall treat γ as a form which generates a perturbation of the inner product in the space $\mathcal{H}_{-1}$. Using γ we introduce a new perturbed inner product by the formula:

$$(h_1,h_2)_{-1}^{\sim} := (A^{-1}h_1,h_2)_0 + \gamma(A^{-1}h_1,A^{-1}h_2), \; h_1,h_2\in\mathcal{H}_0. \tag{8.7}$$

Note that the inner product in (8.7) is well defined since the operator A^{-1} maps the whole space $\mathcal{H}_0$ into $\mathcal{H}_2 \equiv \mathcal{H}_+$, and $A^{-1}h \in Q(\gamma)$ for all $h \in \mathcal{H}_+$. It is obvious that this inner product is not equivalent to the initial inner product in $\mathcal{H}_{-1}$. If the form γ satisfies the condition

$$\|f\|_1^2 \leq \gamma[f] \leq \|f\|_2^2, \quad f \in \mathcal{H}_2 = \operatorname{Dom} A, \tag{8.8}$$

then by Theorem 8.3.3 (see below), the operator $\tilde{A} \in \mathcal{P}_{\mathrm{ss}}(A)$ is associated with the rigged space (8.5). Therefore, the inequality (8.8) gives a necessary condition for the construction of singularly perturbed operators by the generalizes form-sum method. We should explain that by the generalized form-sum method we mean the way in which a singularly perturbed operator arises after adding of a quadratic form γ to any other inner product in (8.6), not necessarily to the inner product in $\mathcal{H}_{-1}$. The details of this approach presented below allow us to cover a wide class of singular perturbations.

Let $\tilde{\mathcal{H}}_{-1}$ denote the completion of $\mathcal{H}_0$ with respect to the norm

$$\|h\|_{-1}^{\sim} = (\|A^{-1/2}h\|_0^2 + \gamma[A^{-1}h])^{1/2}, \quad h \in \mathcal{H}_0. \tag{8.9}$$

It is not hard to verify that the space $\tilde{\mathcal{H}}_{-1}$ admits the orthogonal sum decomposition:

$$\tilde{\mathcal{H}}_{-1} = \mathcal{H}_{-1} \oplus \tilde{\mathcal{N}}_{-1}, \tag{8.10}$$

where $\tilde{\mathcal{N}}_{-1}$ is the completion of $\mathcal{H}_0$ with respect to the norm corresponding to the quasi-inner product $\gamma(A^{-1}\cdot, A^{-1}\cdot)$. The orthogonal decomposition (8.10) results from the fact that the quadratic forms $(\cdot,\cdot)_{-1}$ and $\gamma(A^{-1}\cdot, A^{-1}\cdot)$ are mutually orthogonal (see Chapter 5, Theorem 5.3.7). Let us recall that the singularity of the quadratic form $\gamma[A^{-1}\cdot]$ in the space $\mathcal{H}_{-1}$ follows from the fact that the set $\mathcal{M}_0 = A\mathcal{M}_+$ is dense in $\mathcal{H}_{-1}$, where $\mathcal{M}_+ = \operatorname{Ker}\gamma$. In turn, the subspace $\mathcal{M}_+$ ($\mathcal{H}_+ = \mathcal{M}_+ \oplus \mathcal{N}_+$) is dense in $\mathcal{H}_1$ since the form γ belongs to the $\mathcal{H}_{-2}$-class. Therefore, $\mathcal{M}_0 = \operatorname{Ker}(\gamma[A^{-1}\cdot])$ and thus, $\mathcal{M}_0 \sqsubset \mathcal{H}_{-1}$.

In order to construct a new (perturbed) rigged space of the form (8.5) starting with a couple $\mathcal{H}_0$, $\tilde{\mathcal{H}}_{-1}$, it is necessary to ensure that the condition

$$\|\cdot\|_{-1}^{\sim} \leq \|\cdot\|_0 \tag{8.11}$$

is satisfied Note that the null subspace of the embedding operator $J : \mathcal{H}_0 \ni h \mapsto h \in \tilde{\mathcal{H}}_{-1}$ reduces to zero, $\operatorname{Ker} J = \{0\}$, since $\|h\|_{-1} > 0$ for any $0 \neq h \in \mathcal{H}_0$. Then formally we have $\mathcal{H}_0 \subset \tilde{\mathcal{H}}_{-1}$. However, we know that $\mathcal{H}_0 \sqsubset \mathcal{H}_{-1}$ and $\tilde{\mathcal{H}}_{-1} = \mathcal{H}_{-1} \oplus \tilde{\mathcal{N}}_{-1}$. Therefore, the dense and continuous embedding

$$\tilde{\mathcal{H}}_{-1} \sqsupset \mathcal{H}_0 \tag{8.12}$$

is not evident. Moreover, (8.7) and (8.9) imply the opposite to (8.11) inequality: $\|\cdot\|_{-1}^{\sim} \geq \|\cdot\|_{-1}$. So, the continuity of the embedding of $\mathcal{H}$ into $\tilde{\mathcal{H}}_{-1}$ needs to be checked or ensured.

Proposition 8.2.1. *Given $\gamma \in \mathcal{H}_{-2}$-class, let $\tilde{\mathcal{H}}_{-1}$ be constructed from $\mathcal{H}_0$ with the inner product* (8.7). *Then the dense and continuous embedding* (8.12) *holds truc if and only if the quadratic form γ satisfies the condition*

$$0 \le \gamma[\varphi] \le \|\varphi\|_2^2 - \|\varphi\|_1^2, \quad \varphi \in \mathcal{H}_2. \tag{8.13}$$

Proof. In fact, it is not hard to understand that under the condition $\operatorname{Ker} J = \{0\}$, (8.12) is equivalent to (8.11). So, we have to prove the equivalence of (8.13) and (8.11). Since each vector $\varphi \in \mathcal{H}_2$ can be written as $\varphi = A^{-1}h$, $h \in \mathcal{H}_0$, we can recast the condition (8.13) in terms of vectors from $\mathcal{H}_0$:

$$0 \le \gamma[A^{-1}h] \le \|h\|_0^2 - (A^{-1}h, h)_0, h \in \mathcal{H}_0.$$

Or, equivalently,

$$0 \le \|h\|_{-1}^2 + \gamma[A^{-1}h] \le \|h\|_0^2.$$

Now note that the last inequality, in fact, coincides with (8.11). □

Now, having (8.12), to construct the perturbed operator we have to perform the standard operations. First we need to extend the pre-rigging (8.12) to the rigged space

$$\tilde{\mathcal{H}}_{-1} \sqsupset \mathcal{H}_0 \sqsupset \tilde{\mathcal{H}}_1, \tag{8.14}$$

and then to define the operator associated with (8.14) by

$$\tilde{A} := \tilde{D}_{-1,1} \restriction \{f \in \tilde{\mathcal{H}}_1 \mid D_{-1,1}f \in \mathcal{H}_0\}, \tag{8.15}$$

using the Berezansky canonical isomorphism $\tilde{D}_{-1,1} : \tilde{\mathcal{H}}_1 \to \tilde{\mathcal{H}}_{-1}$. It is clear that $\tilde{A} = \tilde{A}^* \ge 1$, since from (8.11) it follows automatically that

$$(\tilde{A}f, f) = (\|f\|_1^{\sim})^2 \ge \|f\|_0^2.$$

Of course, if the triplet (8.14) is extended to a five-space rigging of the form (8.5), then the space $\tilde{\mathcal{H}}_2$ coincides with $\operatorname{Dom} \tilde{A}$ in the norm $\|\cdot\|_2^{\sim} = \|\tilde{A}\cdot\|_0$.

Now it is time to ask whether the constructed operator $\tilde{A}$ belongs to the family $\mathcal{P}_{\mathrm{ss}}(A)$.

Proposition 8.2.2. *Suppose the quadratic form $\gamma \in \mathcal{H}_{-2}(A)$-class satisfies the condition* (8.13). *Then the operator $\tilde{A}$ defined in* (8.15) *which is associated with* (8.14) *in accordance with the method of rigged spaces, belongs to the family $\mathcal{P}_{\mathrm{ss}}(A)$, i.e., it is strongly singularly perturbed with respect to A.*

Proof. Since $\tilde{A} \ge 1$, the inverse operator $\tilde{A}^{-1}$ exists, and is bounded and defined on the whole space $\mathcal{H}_0$. For vectors $g \in \mathcal{M}_0 = A\mathcal{M}_+$ we have

$$(\tilde{A}^{-1}g, h)_0 = (g, h)_{-1}^{\sim} = (A^{-1}g, h)_0, \quad h \in \mathcal{H}_0,$$

because $\gamma(A^{-1}g, h) = 0$ and $A^{-1}g \in \mathcal{M}_+ = \operatorname{Ker}\gamma$. Thus, $\tilde{A}\varphi = A\varphi$, $\varphi \in \mathcal{M}_+$. This means that $\tilde{A}$ is a self-adjoint extension of the symmetric operator $\mathbf{A} = A \restriction \mathcal{M}_+$. In addition, $A = A_\infty$, since $\mathcal{M}_+ \sqsubset \mathcal{H}_1$ due to $\gamma \in \mathcal{H}_{-2}(A)$-class. Thus, $\tilde{A} \in \mathcal{P}_{\mathrm{ss}}(A)$. □

We emphasize that operators $\tilde{A}$ from the family $\mathcal{P}_{\rm ss}(A)$ are constructed usually by the method of self-adjoint extensions (see, e.g., [7]). Proposition 8.2.2 shows in fact that strongly singularly perturbed operators can be constructed also by the method of generalized form-sum. However, in this approach one does not perturb the inner product in $\mathcal{H}_1(A)$, as in the usual form-sum method (see, for example, the KLMN-theorem), but the inner product in the negative space $\mathcal{H}_{-1}(A)$ in a specific way (see (8.7)). That is why we call this approach the method of rigged Hilbert spaces.

The natural question arises there. Whether can be received the whole family of operators $\mathcal{P}_{\rm ss}(A)$ by the method of rigged Hilbert spaces? We obtain the positive answer due to the use of the following proposition.

Proposition 8.2.3. *Let*

$$\tilde{\chi}_2[\varphi] := \chi_2[P_{\mathcal{N}_+}\varphi] = \|P_{\mathcal{N}_+}\varphi\|_+^2, \quad \varphi \in \mathcal{H}_+. \tag{8.16}$$

Assume that for a positive quadratic form $\gamma \in \mathcal{H}_{-2}(A)$-class its regular component $\gamma_{\rm r}$ coincides with $\tilde{\chi}_2$, i.e,

$$\gamma = \tilde{\chi}_2 + \gamma_{\rm s}, \tag{8.17}$$

where $\gamma_{\rm s}$ is the singular component of γ in $\mathcal{H}$. Then (8.17) *is equivalent to* (8.8).

Proof. From (8.16) and (8.17), since $\gamma \geq 0$, it follows that

$$\gamma_{\rm s}[\varphi] \leq \|\varphi\|_2^2, \quad \varphi \in \mathcal{N}_2.$$

Therefore, the right-hand inequality in (8.8) is fulfilled. The left-hand inequality in (8.8) follows from (8.13) if we rewrite the latter condition as

$$\|\varphi\|_2^2 - \gamma[\varphi] \leq \|\varphi\|_2^2 - \|\varphi\|_1^2, \quad \varphi \in \mathcal{N}_2.$$

The implication (8.8) $\Longrightarrow$ (8.17) is proved by a similar argument. □

Corollary 8.2.4. *If a form $\gamma \in \mathcal{H}_{-2}$-class satisfies the condition* (8.13), *then the corresponding perturbed operator $\tilde{A} \in \mathcal{P}_{\rm ss}(A)$, $\tilde{A} \geq 1$ can be constructed by the method of rigged Hilbert spaces by taking $\tilde{\mathcal{H}}_{-1}$ as the completion of $\mathcal{H}_0$ with respect to the norm*

$$\|\cdot\|_{-1}^{\sim} = (\chi_{-1}[\cdot] + \gamma[A^{-1}\cdot])^{1/2}.$$

Now we will show that each $\tilde{A} \in \mathcal{P}_{\rm ss}(A)$ can also be constructed by the perturbation of the inner product in $\mathcal{H}_+ = \mathcal{H}_2(A)$. In this way the method of rigged spaces is also necessary.

So, let $\tilde{A} \in \mathcal{P}_{\rm ss}(A)$, $\tilde{A} \geq 1$. Then $\operatorname{Dom}\tilde{A} = \tilde{\mathcal{H}}_+$ with the inner product $(f,g)_+ = (\tilde{A}f, \tilde{A}g)_0$. In accordance with the definition of the family $\mathcal{P}_{\rm ss}(A)$, there exists a linear set $\mathfrak{D}$ dense in $\mathcal{H}_0$ such that

$$(f,g)_+ = (f,g)_+^{\sim}, \quad f,g \in \mathfrak{D}. \tag{8.18}$$

Since $A \neq \tilde{A}$, the set $\mathfrak{D}$ is a proper subspace in both Hilbert spaces $\mathcal{H}_+$ and $\tilde{\mathcal{H}}_+$:

$$\mathcal{H}_+ = \mathcal{M}_+ \oplus \mathcal{N}_+, \quad \tilde{\mathcal{H}}_+ = \tilde{\mathcal{M}}_+ \oplus \tilde{\mathcal{N}}_+, \tag{8.19}$$

where

$$\mathcal{M}_+ = \tilde{\mathcal{M}}_+ = \mathfrak{D} \sqsubset \mathcal{H}_0. \tag{8.20}$$

In turn, from (8.19), (8.20) it follows that $\mathcal{H} = \mathcal{M}_0 \oplus \mathcal{N}_0$, where

$$\mathcal{M}_0 = A\mathcal{M}_+ = \tilde{A}\tilde{\mathcal{M}}_+, \quad \mathcal{N}_0 = A\mathcal{N}_+ = \tilde{A}\tilde{\mathcal{N}}_+. \tag{8.21}$$

The next proposition provides more specific features of the decompositions

$$\mathcal{H}_- = \mathcal{M}_- \oplus \mathcal{N}_+, \quad \tilde{\mathcal{H}}_- = \tilde{\mathcal{M}}_- \oplus \tilde{\mathcal{N}}_-. \tag{8.22}$$

Proposition 8.2.5. *Let $\tilde{A} \in \mathcal{P}_{\mathrm{ss}}(A)$, $\tilde{A} \geq 1$. Then the negative spaces $\mathcal{H}_- = \mathcal{H}_{-2}(A)$ and $\tilde{\mathcal{H}}_- = \mathcal{H}_{-2}(\tilde{A})$ admit the orthogonal decompositions* (8.22) *such that*

$$\mathcal{M}_- = \tilde{\mathcal{M}}_- \tag{8.23}$$

and

$$\mathcal{N}_- \cap \mathcal{H}_0 = \{0\} = \tilde{\mathcal{N}}_- \cap \mathcal{H}_0. \tag{8.24}$$

Proof. Let $D_{-,+}$ and $\tilde{D}_{-,+}$ be the Berezansky canonical isomorphisms in the rigged spaces (8.5) and (8.6), respectively. Applying $D_{-,+}$ and $\tilde{D}_{-,+}$ to (8.19) we get (8.22). To prove (8.23) we argue as follows. Each couple of vectors ω, φ and $\tilde{\omega}$, φ, which is connected by the mappings $D_{-,+}$ and $\tilde{D}_{-,+}$, i.e., such that

$$\omega = D_{-,+}\varphi, \quad \tilde{\omega} = \tilde{D}_{-,+}\varphi, \quad \varphi \in \mathfrak{D},$$

have equal norms:

$$\|\omega\|_- = \|D_{-,+}\varphi\|_- = \|\varphi\|_+ = \|\tilde{\omega}\|_-^{\sim} = \|\tilde{D}_{-,+}\varphi\|_-^{\sim}$$

and the same values of duality pairing with appropriate vectors from the spaces $\mathcal{H}_+$ and $\tilde{\mathcal{H}}_+$:

$$\langle \omega, \psi \rangle_{-,+} = (\varphi, \psi)_+ = \langle \tilde{\omega}, \psi \rangle_{-,+}^{\sim}, \quad \psi \in \mathcal{M}_+ \tag{8.25}$$

$$\langle \omega, \eta \rangle_{-,+} = 0 = \langle \tilde{\omega}, \tilde{\eta} \rangle_{-,+}^{\sim}, \quad \eta \in \mathcal{N}_+,\ \tilde{\eta} \in \tilde{\mathcal{N}}_+. \tag{8.26}$$

Therefore, one can identify the vectors ω and $\tilde{\omega}$. This proves (8.23). Finally, (8.24) immediately follows from the denseness of $\mathfrak{D} = \mathcal{M}_+$ in $\mathcal{H}_0$ (see the main denseness criterion in Theorem 6.1.4). □

Let us recall that each operator $\tilde{A} \in \mathcal{P}_{\mathrm{ss}}(A)$, $\tilde{A} \geq 1$, defined by the method of self-adjoint extensions, can be written as $\tilde{A}^{-1} = A^{-1} + \tilde{B}$, where $\tilde{B} = B^{-1}P_{\mathcal{N}_0}$, with a bounded self-adjoint operator $B = B^* \geq 0$, on the subspace $\mathcal{N}_0 = \mathcal{M}_0^{\perp}$, $\mathcal{M}_0 = A\mathcal{M}_+ = A\mathfrak{D}$. Furthermore, the domain of $\tilde{A}$ has the following description:

$$\operatorname{Dom}\tilde{A} = \{g \in \mathcal{H}_0 \mid g = f + BP_{\mathcal{N}_0}Af,\ f \in \operatorname{Dom}A = \mathcal{H}_+\}. \tag{8.27}$$

Proposition 8.2.6. *For each operator* $\tilde{A} \in \mathcal{P}_{\rm ss}(A)$, $\tilde{A} \geq 1$, *its domain* $\operatorname{Dom} \tilde{A} = \tilde{\mathcal{H}}_+$ *admits the orthogonal decomposition*

$$\tilde{\mathcal{H}}_+ \equiv \tilde{\mathcal{H}}_2 = \tilde{\mathcal{M}}_+ \oplus \tilde{\mathcal{N}}_+, \tag{8.28}$$

where $\tilde{\mathcal{M}}_+ = \mathcal{M}_+ = \mathfrak{D}$ *(see* (8.18) *with* $\mathcal{H}_0 \sqsupset \mathfrak{D}$*), and* $\tilde{\mathcal{N}}_+$ *is connected with* $\mathcal{N}_+$ *via*

$$\tilde{\mathcal{N}}_+ = \{\theta_+ \in \mathcal{H}_0 \mid \theta_+ = \eta_+ + BA\eta_+,\ \eta_+ \in \mathcal{N}_+\}, \tag{8.29}$$

$$\|\theta_+\|_+^{\sim} = \|\eta_+\|_+, \quad (\theta_{+,1}, \theta_{+,2})_+^{\sim} = (\eta_{+,1}, \eta_{+,2})_+, \tag{8.30}$$

$$\theta_{+,i} \in \tilde{\mathcal{N}}_+,\ \eta_{+,i} \in \mathcal{N}_+, \quad i = 1, 2.$$

Proof. All statements are immediate consequences of the properties of the spaces $\mathcal{H}_+$ and $\tilde{\mathcal{H}}_+$ (see (8.19), (8.20)). □

After comparing the orthogonal sum decompositions of the spaces $\mathcal{H}_+$ and $\tilde{\mathcal{H}}_+$, we conclude that in order to construct $\tilde{A}$ by the method of rigged Hilbert spaces it is necessary to replace each vector $\eta_+ \in \mathcal{N}_+$ by $\theta_+ = \eta_+ + BA\eta_+$, where $BA\eta_+ \in \mathcal{N}_0$ (recall that $\mathcal{N}_0 \cap \mathcal{H}_+ = \{0\}$). That is,

$$(\theta_+, \theta'_+)_+^{\sim} = (\eta_+, \eta'_+)_+$$

for arbitrary $\eta_+, \eta'_+ \in \mathcal{N}_+$, and corresponding $\theta_+, \theta'_+ \in \tilde{\mathcal{N}}_+$. The action of the operator $\tilde{A}$ on vectors from the subspace $\tilde{\mathcal{N}}_+$ is defined by the equality

$$\tilde{\mathcal{N}}_+ \theta_+ = A\eta_+.$$

However the action of the perturbed operator on vectors from $\mathcal{M}_+ = \mathfrak{D}$ is the same as for the initial operator.

Summarizing the previous analysis we arrive to the important result.

Theorem 8.2.7. *Each operator* $\tilde{A} \in \mathcal{P}_{\rm ss}(A)$ *with the property* $\operatorname{Ker} \tilde{A} = \{0\}$ *is uniquely determined by a subspace* $\mathcal{N}_0 \subset \mathcal{H}_0$ *satisfying the condition*

$$\mathcal{N}_0 \cap \mathfrak{D}(A) = \{0\},$$

and an operator $B = B^* \geq 0$ *on* $\mathcal{N}_0$ *with* $\operatorname{Ker} B = \{0\}$:

$$\tilde{A} g = Af, \quad g \in \mathfrak{D}(\tilde{A}), \quad g = f + BP_{\mathcal{N}_0} Af, \quad f \in \mathfrak{D}(A).$$

In addition, the domain $\mathfrak{D}(\tilde{A})$, *as the positive Hilbert space* $\tilde{\mathcal{H}}_+$ *with the inner product* $(g_1, g_2)_+^{\sim} = (\tilde{A}g_1, \tilde{A}g_2) = (f_1, f_2)_+$, *admits a decomposition into the orthogonal sum* (8.28) *(see Proposition* 8.2.6*). The inner product in the negative space* $\tilde{\mathcal{H}}_{-2}$ *can be represented in the form:*

$$(\cdot,\cdot)_{-2}^{\sim} = (\cdot,\cdot)_{-2} + \tau(\cdot,\cdot), \tag{8.31}$$

where the singular in $\mathcal{H}_{-2}$ *quadratic form* τ *is defined by* B *as*

$$\begin{aligned}\tau(\cdot,\cdot) := &(B^{-1}P_{\mathcal{N}_0}\cdot, B^{-1}P_{\mathcal{N}_0}\cdot)_0 \\ &+ (A^{-1}\cdot, B^{-1}P_{\mathcal{N}_0}A\cdot)_0 + (B^{-1}P_{\mathcal{N}_0}\cdot, A^{-1}\cdot)_0.\end{aligned} \tag{8.32}$$

Proof. All statements of this theorem are already established. We remark only that the representation (8.31) of the inner product in $\tilde{\mathcal{H}}_{-2}$ can be obtained by using the inverse perturbed operator:

$$(h_1, h_2)^{\sim}_{-2} = (\tilde{A}^{-1}h_1, \tilde{A}^{-1}h_2)_0 = (A^{-1}h_1, A^{-1}h_2)_0 + \tau(h_1, h_2).$$

The form τ defined in (8.32) is Hermitian. Since the null-subspace $\operatorname{Ker}\tau = \mathcal{M}_+ = \mathfrak{D}$ and is dense in $\mathcal{H}_{-1}$, $\tilde{A}$ belongs to the family $\mathcal{P}_{\rm ss}(A)$. □

Given a positive quadratic form γ defined on $\operatorname{Dom} A$, we assume that $\operatorname{Ker}\gamma \sqsubset \mathcal{H}_1$ and

$$0 < \gamma[\varphi] \leq \|\varphi\|_2^2 - \|\varphi\|_1^2, \quad \varphi \in \mathcal{H}_2. \tag{8.33}$$

Using γ we define a new inner product on $\mathcal{H}_2$ by

$$\tilde{\chi}_1(\cdot,\cdot) = (\cdot,\cdot)_1 + \gamma(\cdot,\cdot).$$

Let $\tilde{\mathcal{H}}_1$ denote the corresponding Hilbert space. Since $\operatorname{Ker}\gamma \sqsubset \mathcal{H}_1$ and $\gamma(\cdot,\cdot) \geq 0$, the space $\tilde{\mathcal{H}}_1$ admits an orthogonal sum decomposition $\tilde{\mathcal{H}}_1 = \mathcal{H}_1 \oplus \mathcal{H}_\gamma$, where the Hilbert spaces $\tilde{\mathcal{H}}_1$ and $\mathcal{H}_\gamma$ arise as the completion of $\operatorname{Dom} A$ with respect to the inner products $\tilde{\chi}_1(\cdot,\cdot)$ and $\gamma(\cdot,\cdot)$, respectively. Thus,

$$\tilde{A} \in \mathcal{P}^1_{\rm ss}(A) \Longleftrightarrow \tilde{\mathcal{H}}_1 = \mathcal{H}_1 \oplus \tilde{\mathcal{N}}_1.$$

A similar result is valid for any $k > 2$. Let A be associated with the scale

$$\mathcal{H}_{-k} \sqsupset \mathcal{H}_{-k/2} \sqsupset \mathcal{H}_0 \sqsupset \mathcal{H}_{k/2} \sqsupset \mathcal{H}_k = \operatorname{Dom} A^{k/2}. \tag{8.34}$$

Then each positive quadratic form $\gamma \in \mathcal{H}_{-k}$-class defines, similarly to (8.31), a new inner product on $\mathcal{H}_0$:

$$(h_1, h_2)^{\sim}_{-k/2} := (A^{-k/2}h_1, h_2)_0 + \gamma(A^{-k/2}h_1, A^{-k/2}h_2), \quad h_1, h_2 \in \mathcal{H}_0. \tag{8.35}$$

In addition, if γ satisfies the condition

$$-\|f\|^2_{k/2} \leq \gamma[f] \leq \|f\|^2_k - \|f\|^2_{k/2}, \quad f \in \mathcal{H}_k, \tag{8.36}$$

then one can construct a new scale of the Hilbert spaces,

$$\tilde{\mathcal{H}}_{-k} \sqsupset \tilde{\mathcal{H}}_{-k/2} \sqsupset \mathcal{H} \sqsupset \tilde{\mathcal{H}}_{k/2} \sqsupset \tilde{\mathcal{H}}_k. \tag{8.37}$$

The operator $\tilde{A}$ associated with this scale is singularly perturbed with respect to A in the wide sense. In this way one can treat a wide class of singular perturbations of higher orders. In particular, the following theorem is true.

Theorem 8.2.8. *For each strongly singularly perturbed in the wide sense operator $\tilde{A} \geq 1$, the inner product $(\cdot,\cdot)^{\sim}_{-k}$, $k \geq 2$, in the space $\tilde{\mathcal{H}}_{-k}$ can be treated as a singular perturbation of the inner product in $\mathcal{H}_{-k}$:*

$$(\cdot,\cdot)^{\sim}_{-k} = (\cdot,\cdot)_{-k} + \tau_k(\cdot,\cdot), \tag{8.38}$$

where the Hermitian, singular in $\mathcal{H}_0$ quadratic form τ_k has the representation

$$\tau_k(\cdot,\cdot) := (A^{-k}\cdot, \tilde{B}\cdot)_0 + (\tilde{B}\cdot, A^{-k}\cdot)_0 + (\tilde{B}\cdot, \tilde{B}\cdot)_0.$$

Proof. In fact, the proof is the same as in the case $k = 2$. The only difference is that we have to use the representation by Kreĭn's formula for a negative power of the operator $\tilde{A}$, i.e., for $(\tilde{A})^{-k}$ with $k \neq 1$. □

If the quadratic form γ is defined on the whole space $\mathcal{H}_k = \operatorname{Dom} A^{k/2}$, its null subspace satisfies the condition $\operatorname{Ker}\gamma \sqsubset \mathcal{H}_{k/2}$, and

$$0 < \gamma[\varphi] \leq \|\varphi\|^2_k - \|\varphi\|^2_{k/2}, \quad \varphi \in \mathcal{H}_k, \tag{8.39}$$

then the quadratic form

$$\tilde{\chi}_{k/2}(\cdot,\cdot) = (\cdot,\cdot)_{k/2} + \gamma(\cdot,\cdot)$$

defines a new inner product on $\mathcal{H}_k$. Therefore, in much the same way as for $k = 2$, one can introduce the space $\tilde{\mathcal{H}}_{k/2}$ and the scale of spaces of type (8.37). The operator $\tilde{A}$ associated with this scale is singularly perturbed in the wide sense with respect to A. However, we note that the construction of the embedding $\mathcal{H}_0 \sqsupset \tilde{\mathcal{H}}_{k/2}$ should be started with the domain $\mathfrak{D}(\tilde{A}^{k/2})$, whose vectors are viewed as $g = f + B^{-1}P_{\mathcal{N}_0}Af$, $f \in \mathfrak{D}(A^{k/2})$.

8.3 Parametrization of super-singular perturbations

In this section we establish a parametrization of super-singular perturbations of self-adjoint operators by using the method of rigged spaces and scales of Hilbert spaces. In Chapter 9 we shall use this approach to study spectral properties of singularly perturbed operators.

We provide the parametrization of operators $\tilde{A}$ in terms of self-adjoint operators $S : \mathcal{H}_k \to \mathcal{H}_{-k}$ acting in the A-scale of Hilbert spaces

$$\mathcal{H}_{-k} \sqsupset \mathcal{H} \sqsupset \mathcal{H}_k = \operatorname{Dom} A^{k/2}, \quad k > 0.$$

The operators S possess the typical property that their null-subspaces $\operatorname{Ker} S$ are dense in the space $\mathcal{H}_{k/2}$.

Let us fixe an operator $A = A^*$ on $\mathcal{H}$. We assume that the lower bound of A is equal to one, i.e.,

$$\inf_{\|f\|=1}(Af, f) = 1.$$

We recall that an operator $\tilde{A} \neq A$ on $\mathcal{H}$ is said to be singularly perturbed with respect to A if the linear set

$$\mathfrak{D} := \{f \in \mathfrak{D}(A) \cap \mathfrak{D}(\tilde{A}) \mid Af = \tilde{A}f\} \tag{8.40}$$

is dense in $\mathcal{H}$. Therefore, the operators A and $\tilde{A}$ have some common Hermitian part $\mathbf{A}$ which is the symmetric operator with the domain $\mathfrak{D}(\mathbf{A}) := \mathfrak{D}$, i.e., $\mathbf{A} = A \restriction \mathfrak{D} = \tilde{A} \restriction \mathfrak{D}$. Since $\tilde{A}$ and A are distinct, the symmetric operator $\mathbf{A}$ has non-trivial deficiency indices. In turn, the operators A and $\tilde{A}$ are different self-adjoint extensions of this symmetric operator $\mathbf{A}$.

If the condition (8.40) is fulfilled for some fixed power $k > 2$ of operators A and $\tilde{A}$, i.e., if the set

$$\mathfrak{D}_k := \{f \in \mathfrak{D}(A^{k/2}) \cap \mathfrak{D}(\tilde{A})^{k/2} \mid A^{k/2}f = \tilde{A}^{k/2}f\}, \quad k > 2, \tag{8.41}$$

is dense in $\mathcal{H}$, then we say that the operator $\tilde{A}$ is *super-singularly perturbed* with respect to A in the wide sense. Note, that $\tilde{A}$ may have another lower bound, different from one.

Definition 8.3.1. A bounded from below self-adjoint on $\mathcal{H}$ operator $\tilde{A} \geq \tilde{m} > -\infty$,

$$\tilde{m} := \inf_{\|f\|=1} (\tilde{A}f, f),$$

is said to be super-singularly perturbed with respect to A if for some $k > 2$ the set

$$\mathcal{M}_k := \{\varphi \in \mathcal{H}_k \mid A_c^{k/2}\varphi = \tilde{A}_c^{k/2}\varphi\} \tag{8.42}$$

is dense in $\mathcal{H}_{k/2}$,

$$\mathcal{H}_{k/2} \sqsupset \mathcal{M}_k, \tag{8.43}$$

where $A_c := A + c$, $\tilde{A}_c := \tilde{A} + c$ with $c = 0$ if $\tilde{m} \geq 1$ and $c = 1 - \tilde{m}$, if $\tilde{m} < 1$. We denote by $\mathcal{P}_{\mathrm{s}}(A^{k/2})$ the class of so-defined operators.

From this definition it follows directly that for $\tilde{A} \in \mathcal{P}_{\mathrm{s}}(A^{k/2})$ the domain of the operator $A^{k/2}$ admits the orthogonal decomposition

$$\mathcal{H}_k = \mathcal{M}_k \oplus \mathcal{N}_k, \quad \mathcal{N}_k \neq \{0\},$$

and that the operators $A_c^{k/2}$ and $\tilde{A}_c^{k/2}$ are different self-adjoint extensions of the same symmetric operator

$$\mathbf{A}_c^{k/2} := A_c^{k/2} \restriction \mathcal{M}_k = \tilde{A}_c^{k/2} \restriction \mathcal{M}_k.$$

In what follows we consider the super-singularly perturbed operators $\tilde{A} \in \mathcal{P}_{\mathrm{s}}(A^{k/2})$ with $k \geq 4$ (see, also, [23, 24]). It should be noted that in (8.42) and (8.43), the power k and the constant c are assumed to be minimal.

The above definitions together with constructions presented in Subsection 7.4 allow one to establish a classification of all singularly and super-singularly perturbed operators (also in the wide sense) in the terms of auxiliary singular operators acting in the A-scale of Hilbert spaces.

Here, it is convenient to change some notations. Namely, we will denote by B the operators B^{-1} in the Kreĭn formula.

Let us consider an operator S which acts in the A-scale, $S : \mathcal{H}_k \to \mathcal{H}_{-k}$, $k \geq 2$. We assume that S self-adjoint as an operator in a pair of spaces, i.e.,

$$\langle S\varphi, \psi\rangle_{-k,k} = \langle \varphi, S\psi\rangle_{k,-k}, \quad \varphi, \psi \in \mathfrak{D}(S) = \mathfrak{D}(S^*) \subseteq \mathcal{H}_k.$$

We will interpret both operators of this kind (having also some additional properties) and the associated with quadratic forms

$$\gamma_S(\varphi, \psi) := \langle S\varphi, \psi\rangle_{-k,k}, \quad \varphi, \psi \in \operatorname{Dom}\gamma_S \subseteq \mathcal{H}_k$$

as a singular perturbation of the operator A.

Definition 8.3.2. We say that a self-adjoint in a pair spaces operator $S : \mathcal{H}_k \to \mathcal{H}_{-k}$ (similarly, the associated quadratic form γ_S) belongs to the $\mathcal{H}_{-k}(A)$-class, $k \geq 2$, of singular perturbations of an operator A, if the set

$$\operatorname{Ker} S := \{\varphi \in \mathfrak{D}(S) \subseteq \mathcal{H}_k \mid S\varphi = 0\} = \operatorname{Ker}\gamma_S$$

is dense in $\mathcal{H}_{k/2}$, i.e.,

$$\mathcal{H}_{k/2} \sqsupset \operatorname{Ker} S. \tag{8.44}$$

Surely, the set $\operatorname{Ker} S$ will also be dense in $\mathcal{H}$. Thus, we extended the well-known family of singular perturbations of $\mathcal{H}_{-2}$-class (see [7]). Similar perturbations with $k \geq 4$ were already discussed in [23, 24].

The main additional difficulty arising in the discussion of super-singular perturbations with $k \geq 4$ is that the set $\operatorname{Ker} S$ is dense in $\mathcal{H}_2 = \operatorname{Dom} A$. For this reason, the symmetric operator $\mathbf{A} = A \restriction \operatorname{Ker} S$ is essentially self-adjoint. Until now we have not any procedure for getting of the uniquely defined perturbed operator $\tilde{A}$ under super-singular perturbation of the mentioned type.

The next theorem extends the main result of Subsection 7.4. It establishes a one-to-one correspondence between a fixed set of singular perturbations of the $\mathcal{H}_{-2}$-class and a certain family of singularly perturbed operators. In fact, it is the simplest version of the general result (see below Theorem 8.3.5).

Theorem 8.3.3 (The case $c = 0$, $k = 2$). *Given $A \geq 1$ on $\mathcal{H}$ there exists a one-to one correspondence between the family of the singularly perturbed operators*

$$\tilde{A} \in \mathcal{P}_{\mathrm{ss}}(A), \quad \tilde{A} \geq 1, \quad \inf_{\|f\|=1} (\tilde{A}f, f) = 1$$

and the set of the positive singular perturbations $S \in \mathcal{H}_{-2}(A)$-class satisfying the condition:

$$\|\varphi\|_1^2 \leq \langle S\varphi, \varphi\rangle_{-2,2} \equiv \gamma_S[\varphi] \leq \|\varphi\|_2^2, \quad \varphi \in \mathcal{N}_2 := \mathcal{H}_2 \ominus \operatorname{Ker} S. \tag{8.45}$$

Proof. Let an operator $\tilde{A} \geq 1$ belong to the family $\mathcal{P}_{\rm ss}(A)$. Then, by Definition 8.3.1 (see (8.42)), there exists a subspace $\mathcal{M}_2$ in $\mathcal{H}_2$ which is dense in $\mathcal{H}_1$. That is, the actions of the operators $\tilde{A}$ and A coincide on this subspace. Thus, $\tilde{A}$ and A are different self-adjoint extensions of the symmetric operator $\mathbf{A} := \tilde{A} \restriction \mathcal{M}_2 = A \restriction \mathcal{M}_2$. Moreover, since $\mathcal{M}_2$ is dense in $\mathcal{H}_1$, the Friedrichs extension A_∞ of the operator $\mathbf{A}$ coincides with A. Thus, A is the maximal positive extension of the symmetric operator $\mathbf{A}$. Therefore, any other positive extension of $\mathbf{A}$ satisfies the inequality $\tilde{A} \leq A$. Thus, the difference of the inverse operators defines a bounded non-negative operator on $\mathcal{H}_0$:

$$\tilde{B} := \tilde{A}^{-1} - A^{-1} \geq 0. \tag{8.46}$$

Clearly, this operator is equal to zero on the subspace $\mathcal{M}_0 := A\mathcal{M}_2$ and its restriction $B := \tilde{B} \restriction \mathcal{N}_0$ to the subspace $\mathcal{N}_0 := \mathcal{H} \ominus \mathcal{M}_0$ satisfies the inequalities

$$0 < B < 1, \quad 0 < B \leq 1 - A^{-1}. \tag{8.47}$$

So, on $\mathcal{N}_0$ we have

$$0 < A^{-1} \leq s_0 < 1, \quad s_0 := 1 - B. \tag{8.48}$$

Now we are able to define the operator $S : \mathcal{H}_2 \to \mathcal{H}_{-2}$. Put S to be zero on $\mathcal{M}_2$ and $S = A^{\rm cl} s_0 A$ on $\mathcal{N}_2$, where $A^{\rm cl}$ denotes a closure of the mapping $A : \mathcal{H} \to \mathcal{H}_{-2}$. It is obvious that S is bounded and self-adjoint. Since $\mathcal{M}_2$ is dense in $\mathcal{H}_1$, the operator S belongs to the $\mathcal{H}_{-2}(A)$-class. Further, (8.47) and (8.48) imply that the relations

$$0 < (A^{-1}h, h) \leq (s_0 h, h) < \|h\|^2, \quad h \in \mathcal{N}_2,\ h \neq 0,$$

$$(A\varphi, \varphi) = \|\varphi\|_1^2 \leq (s_0 A\varphi, A\varphi) = \langle S\varphi, \varphi\rangle_{-2,2} < \|\varphi\|_2^2, \quad \varphi = A^{-1}h \in \mathcal{N}_2,$$

hold, which prove (8.45).

Conversely, suppose we are given a bounded non-negative self-adjoint operator $S : \mathcal{H}_2 \to \mathcal{H}_{-2}$ from the $\mathcal{H}_{-2}(A)$-class. We assume that it satisfies the condition (8.45). Using S, define the operator

$$s_0 := {A^{\rm cl}}^{-1}(S \restriction \mathcal{N}_0)A^{-1}, \quad \mathcal{N}_0 := (A\mathrm{Ker}\, S)^{\perp}.$$

Thanks to (8.45), this operator has the properties (8.48). Therefore, we can define the operator $B := 1 - s_0$. We denote its extension by zero on $\mathcal{M}_0$ as $\tilde{B}$. Now it is easy to see that the operator $\tilde{A}$ given by the Kreĭn formula $\tilde{A}^{-1} := A^{-1} + \tilde{B}$ belongs to the family $\mathcal{P}_{\rm ss}(A)$. Indeed, by (8.45) the set $\mathcal{M}_2 := \mathrm{Ker}\, S$ is dense in $\mathcal{H}_1$, and due to (8.47) we have $\tilde{A} \geq 1$. □

We consider the case when the lower bound

$$\tilde{m} = \inf_{\|f\|=1} (\tilde{A}f, f)$$

of the operator $\tilde{A} \in \mathcal{P}_{\rm ss}(A)$ is strictly less than one, $\tilde{m} < 1$. In this case, the formulation of the previous theorem is more complicated. First, we assume $c = 1 - \tilde{m} > 0$ and define the space $\mathcal{H}_{k,c}$ as $\operatorname{Dom} A^{k/2}$ in the norm

$$\|f\|_{k,c} := \|(A + c)^{k/2} f\|, \quad f \in \operatorname{Dom} A^{k/2}.$$

It is clear that this norm is equivalent to the norm $\|f\|_k$.

Theorem 8.3.4 (The case $c > 0$, $k = 2$). *The Kreĭn formula establishes a bijective correspondence between the family of operators*

$$\tilde{A} \in \mathcal{P}_{\rm ss}(A), \quad \tilde{A} \geq \tilde{m}, \quad 0 < \tilde{m} < 1$$

and the set of bounded non-negative perturbations $S \in \mathcal{H}_{-2}(A)$-class satisfying the conditions

$$\|\varphi\|^2_{1,c} \leq \langle S\varphi, \varphi\rangle_{-2,2} \equiv \gamma_S[\varphi] \leq \|\varphi\|^2_{2,c}, \quad \varphi \in \mathcal{N}_{2,c}, \tag{8.49}$$

$$\inf_{\varphi \in \mathcal{N}_{2,c},\ \|\varphi\|_{1,c}=1} \gamma_S[\varphi] = 1, \tag{8.50}$$

where $\mathcal{N}_{2,c} := \mathcal{H}_{2,c} \ominus \mathcal{M}_{2,c}$, $\mathcal{M}_{2,c} \equiv \mathcal{M}_2$, $c = 1 - \tilde{m}$,

Proof. Due to (8.42), for each operator $\tilde{A} \geq \tilde{m}$ belonging to the family $\mathcal{P}_{\rm ss}(A)$ there exists a subspace $\mathcal{M}_2 \subset \mathcal{H}_2$ which is dense in $\mathcal{H}_1$. That is, the operators $\tilde{A}$ and A coincide on $\mathcal{M}_2$. Obviously, the same is true for the operators $\tilde{A}_c = \tilde{A} + c$ and $A_c = A + c$ with $c = 1 - \tilde{m}$. Notice that the lower bound of $\tilde{A}_c$ is exactly equal to zero. Now we change the norm $\|\cdot\|_k$ in the space $\mathcal{H}_k$, $k = 1, 2$ to the equivalent norms $\|f\|_{k,c} := \|(A + c)^{k/2} f\|$, $f \in \operatorname{Dom} A^{k/2}$. In this way we arrive at the space $\mathcal{H}_{k,c}$ which, as a set, coincides with $\operatorname{Dom} A^{k/2}$. This ensures that the embedding $A_c^{k/2} : \mathcal{H}_{k,c} \to \mathcal{H}$ is isometric. Further, we have to change the notations. Instead $\mathcal{M}_2$, for the set $\operatorname{Ker} S$ we will use the notation $\mathcal{M}_{2,c}$. Thus, taking into account that now $k = 2$, we can assert that $\tilde{A}_c$ and A_c are different self-adjoint extensions of a positive symmetric operator, i.e.,

$$\mathbf{A}_c := \tilde{A}_c \restriction \mathcal{M}_{2,c} = A_c \restriction \mathcal{M}_{2,c}.$$

Since the subspace $\mathcal{M}_{2,c}$ is dense in $\mathcal{H}_{1,c}$, the Friedrichs extension of $\mathbf{A}_c$ coincides with A_c. So, A_c is the maximal positive extension of $\mathbf{A}$. Therefore, $\tilde{A}_c \leq A_c$. Now one can define a bounded non-negative operator by

$$\tilde{B}_c := \tilde{A}_c^{-1} - A_c^{-1}. \tag{8.51}$$

It is obvious that $\tilde{B}_c = 0$ on $\mathcal{M}_0 := A_c \mathcal{M}_{2,c}$, and in $\mathcal{N}_0 := \mathcal{H} \ominus \mathcal{M}_0$ the operator $B_c := \tilde{B}_c \restriction \mathcal{N}_0$ satisfies the inequalities

$$0 < B_c < 1, \quad 0 < B_c \leq 1 - A_c^{-1}. \tag{8.52}$$

In other terms,

$$0 < A_c^{-1} \leq s_0 < 1, \quad s_0 := 1 - B_c. \tag{8.53}$$

Define now the operator $S : \mathcal{H}_2 \to \mathcal{H}_{-2}$. It is equal to zero on $\mathcal{M}_{2,c}$ and $S = A_c^{\rm cl} s_0 A_c$ on $\mathcal{N}_{2,c} = \mathcal{H}_{2,c} \ominus \mathcal{M}_{2,c}$, where $A_c^{\rm cl}$ denotes the closure of the mapping $A_c : \mathcal{H} \to \mathcal{H}_{-2}$. The operator S is singular in $\mathcal{H}$ since the subspace $\mathcal{M}_{2,c} := \operatorname{Ker} S$ is dense in $\mathcal{H}$.

Directly from (8.52) and (8.53) we get

$$\begin{aligned}&(A_c^{-1}h, h) \le (s_0 h, h) < \|h\|^2, \quad h \in \mathcal{N}_0,\\ &(A_c\varphi, \varphi) \equiv \|\varphi\|_{1,c}^2 \le (s_0 A_c\varphi, A_c\varphi) = \langle S\varphi, \varphi\rangle_{-2,2} < \|\varphi\|_{2,c}^2,\\ &\qquad \varphi = A_c^{-1} h \in \mathcal{N}_{2,c},\end{aligned}$$

which proves (8.49).

Further, since the lower bound of $\tilde{A}_c$ is 1, $\sup_{\|h\|=1}(\tilde{A}_c^{-1}h, h) = 1$ also. That is,

$$1 = \sup_{\|h\|=1} ((A_c^{-1} + \tilde{B}_c)h, h) = \sup_{h\in\mathcal{N}_0, \|h\|=1} ((A_c^{-1} + B_c)h, h).$$

Replacing B_c by $1 - s_0$ we obtain

$$1 = \sup_{h\in\mathcal{N}_0,\|h\|=1} (\|h\|_{-1,c}^2 + \|h\|^2 - (s_0 h, h).$$

This means that

$$\sup_{h\in\mathcal{N}_0,\|h\|=1} (\|h\|_{-1,c}^2 - (s_0 h, h)) = 0.$$

Since the operator $A_c : \mathcal{H}_{2,c} \to \mathcal{H}$ is isometric, then taking $h = A_c\varphi, \varphi \in \mathcal{N}_{2,c}$ we obtain

$$\sup_{\varphi\in\mathcal{N}_{2,c},\|\varphi\|_{2,c}=1} (\|\varphi\|_{1,c}^2 - \gamma_S[\varphi]) = 0. \tag{8.54}$$

This equality is equivalent to (8.50). Indeed, (8.54) means that there exists a sequence $\varphi_n \in \mathcal{N}_{2,c}$, $\|\varphi_n\|_{2,c} = 1$, such that

$$\lim_{n\to\infty} \|\varphi_n\|_{1,c}^2 = \lim_{n\to\infty} \gamma_S[\varphi_n].$$

One can replace φ_n by $\varphi_n' = c_n\varphi_n$ so that $\|\varphi_n'\|_{1,c} = 1$. Then

$$1 = \|\varphi_n'\|_{1,c} \le \gamma_S[\varphi_n'].$$

Now from (8.54) it follows that the sequence φ_n' minimizes the quadratic form γ_S. This proves (8.50).

Conversely, let $S : \mathcal{H}_2 \to \mathcal{H}_{-2}$ be a positive operator from the $\mathcal{H}_{-2}(A)$-class. Let us assume that the set $\operatorname{Ker} S$ is dense in $\mathcal{H}_1$. Then the quadratic form γ_S associated with S, is singular in $\mathcal{H}$. We assume the conditions (8.49) and (8.50) are fulfilled. Then one can construct the operator $s_0 := (A_c{}^{\rm cl})^{-1} S A_c^{-1} \restriction \mathcal{N}_0$, where the subspace $\mathcal{N}_0 := (A_c \operatorname{Ker} S)^{\perp}$. Due to (8.49), the operator s_0 satisfies the inequalities (8.53). Therefore, we can define the positive operator $B_c := 1 - s_0$. We denote its extension by zero on $\mathcal{M}_0$ by $\tilde{B}$. Now it is clear that $\tilde{A}_c$, defined by the Kreĭn formula $\tilde{A}_c^{-1} := A_c^{-1} + \tilde{B}$, belongs to the family $\mathcal{P}_{\rm ss}(A)$, since the operators $\tilde{A}_c$ and A_c coincide on the dense subset $\mathcal{M}_{2,c}$. In addition, it is obvious that the exact below bound of $\tilde{A}_c$ is 1. Therefore, the exact lower bound of the operator $\tilde{A} = \tilde{A}_c - c$ equals $\tilde{m} = 1 - c < 1$. □

The next theorem gives the most general version of the main statement of this section.

Theorem 8.3.5 (The case $c \geq 0$, $k \geq 2$). *Let $\tilde{A} \geq \tilde{m} > -\infty$ be a bounded from below singularly perturbed operators belonging to the family $\mathcal{P}_{\rm ss}(A^{k/2})$, $k \geq 2$. Let $S : \mathcal{H}_k \to \mathcal{H}_{-k}$ be a bounded positive operator from the $\mathcal{H}_{-k}(A)$-class. Assume that the quadratic form γ_S associated with S satisfies the inequalities*

$$\begin{gathered}\|\varphi\|_{k/2,c}^2 \leq \gamma_S[\varphi] \equiv \langle S\varphi, \varphi\rangle_{-k,k} < \|\varphi\|_{k,c}^2,\\ \varphi \in \mathcal{N}_{k,c} := \mathcal{H}_{k,c} \ominus \operatorname{Ker} S.\end{gathered} \tag{8.55}$$

and has the property

$$\inf_{\|\varphi\|_{k/2,c}=1} \gamma_S[\varphi] = 1, \quad \varphi \in \mathcal{N}_{k,c}, \tag{8.56}$$

where $c = 0$ if $\tilde{m} \geq 1$, and $c = 1 - \tilde{m} > 0$ if $\tilde{m} < 1$. Then the Kreĭn formula

$$\begin{gathered}((\tilde{A} + c)^{k/2})^{-1} = ((A + c)^{k/2})^{-1} + \tilde{B}_{c,k},\\ \tilde{B}_{c,k} = 1 - (A_c^{-k/2})^{\rm cl} S A_c^{-k/2},\end{gathered} \tag{8.57}$$

establishes the bijective correspondence between the operators $\tilde{A}$ and S.

Proof. We only have to consider the case with $c > 0$ and $k > 2$. Let an operator $\tilde{A} \geq \tilde{m} > -\infty, \tilde{m} < 1$ belong to the family $\mathcal{P}_{\rm ss}(A^{k/2})$, $k > 2$. According to Definition 8.3.1, there exists a subspace $\mathcal{M}_k \sqsubset \mathcal{H}_{k/2}$ in $\mathcal{H}_k$ (see (8.42)) on which the operators $\tilde{A}_c^{k/2}$ and $A_c^{k/2}$, $c = 1 - \tilde{m}$, coincide. After changing the norm in $\mathcal{H}_k$ to $\|\cdot\|_{c,k}$, we denote the above subspace as $\mathcal{M}_{c,k}$. Thus, the operators $\tilde{A}_c^{k/2}$ and $A_c^{k/2}$ are different self-adjoint extensions of the positive symmetric operator

$$\mathbf{A}_{(k,c)} := \tilde{A}_c^{k/2} \restriction \mathcal{M}_{c,k} = A_c^{k/2} \restriction \mathcal{M}_{c,k}.$$

Since the subspace $\mathcal{M}_{c,k}$ is dense in $\mathcal{H}_{c,k/2}$, the Friedrichs extension of the operator $\mathbf{A}_{(k,c)}$ coincides with $A_c^{k/2}$. This operator is a maximal positive extension. It follows that $\tilde{A}_c^{k/2} \leq A_c^{k/2}$. Therefore, one can define the bounded positive operator

$$\tilde{B}_{c,k} := \tilde{A}_c^{-k/2} - A_c^{-k/2}. \tag{8.58}$$

It is clear that $\tilde{B}_{c,k} = 0$ on $\mathcal{M}_0 := A_c^{k/2}\mathcal{M}_{c,k}$, and on the subspace $\mathcal{N}_0 := \mathcal{H} \ominus \mathcal{M}_0$ the operator $B_{c,k} := \tilde{B}_{c,k} \restriction \mathcal{N}_0$ satisfies the inequalities

$$0 < B_{c,k} < 1, 0 < B_{c,k} \leq 1 - A_c^{-k/2}. \tag{8.59}$$

By this for $s_{0c,k} := 1 - B_{c,k}$ in $\mathcal{N}_0$, we have:

$$0 < A_c^{-k/2} \le s_{0c,k} < 1. \tag{8.60}$$

Now, one can define the operator S. It is zero on $\mathcal{M}_{c,k}$ and represented by the equality

$$S = (A_c^{-k/2})^{\rm cl} s_{0c,k} A_c^{k/2}$$

on $\mathcal{N}_{c,k}$.

From (8.59) and (8.60) we conclude that

$$(A_c^{-k/2}h, h) \le (s_{0c,k}h, h) < \|h\|^2, \quad h \in \mathcal{N}_0,$$

or

$$\begin{aligned}\|\varphi\|^2_{k/2,c} &= (A_c^{k/2}\varphi, \varphi) \le (s_{0c,k}A_c^{k/2}\varphi, A_c^{k/2}\varphi) = \langle S\varphi, \varphi\rangle_{-k,k} \\ &< \|\varphi\|^2_{k,c} = (A_c^{k/2}\varphi, A_c^{k/2}\varphi), \quad \varphi = A_c^{-k/2}h \in \mathcal{N}_{c,k},\end{aligned}$$

which proves (8.55).

Further, since the lower bound of $\tilde{A}_c$ is 1,

$$\sup_{\|h\|=1} (\tilde{A}_c^{-k/2}h, h) = 1.$$

That is

$$1 = \sup_{\|h\|=1} ((A_c^{-k/2} + \tilde{B}_{c,k})h, h) = \sup_{h\in\mathcal{N}_0, \|h\|=1} ((A_c^{-k/2} + B_{c,k})h, h).$$

Replacing $B_{c,k}$ by $1 - s_{0c,k}$ we obtain

$$1 = \sup_{h\in\mathcal{N}_0, \|h\|=1} (\|h\|^2_{-k/2,c} + \|h\|^2 - (s_{0c,k}h, h).$$

This means that

$$\sup_{h\in\mathcal{N}_0, \|h\|=1} (\|h\|^2_{-k/2,c} - (s_{0c,k}h, h)) = 0.$$

From this, for $h = A_c^{k/2}\varphi$, $\varphi \in \mathcal{N}_{k,c}$ we get

$$\sup_{\varphi\in\mathcal{N}_{k,c}, \|\varphi\|_{k,c}=1} (\|\varphi\|^2_{k/2,c} - \gamma_S[\varphi]) = 0, \tag{8.61}$$

because the operator $A_c^{k/2} : \mathcal{H}_{k,c} \to \mathcal{H}$ is isometric. In fact, the condition (8.61) is equivalent to (8.56), which can be proved in the same way as the equivalence of (8.54) and (8.50).

Conversely, beginning with an operator $S : \mathcal{H}_k \to \mathcal{H}_{-k}$ which belongs to the $\mathcal{H}_{-k}(A)$-class and satisfies the conditions (8.55), (8.56)), we construct the operator $s_{0c,k} := (A_c^{-k/2})^{\rm cl} S A_c^{-k/2} \restriction \mathcal{N}_0$ in the space $\mathcal{N}_0 := (A_c^{k/2}\mathrm{Ker}\, S)^\perp$. It is obvious that this operator satisfies the condition (8.60). Now we put $B_{c,k} := 1 - s_{0c,k}$. We denote its extension by zero on $\mathcal{M}_{0ck} = \mathcal{M}_0 = \mathcal{H}_0 \ominus \mathcal{N}_0$ by $\tilde{B}_{c,k}$. Now it is easily seen that the operator $\tilde{A} = \tilde{A}_c - c$, where $\tilde{A}_c$ is defined by the equality$\tilde{A}_c^{-k/2} := A_c^{-k/2} + \tilde{B}_{c,k}$, belongs to the family $\mathcal{P}_{\rm ss}(A)$. The lower bound of the operator $\tilde{A}$ obeys the inequalities $1 > \tilde{m} = 1 - c > -\infty$. □

Thus, the formula (8.57) establishes a bijective correspondence between operators $\tilde{A} \in \mathcal{P}_{\rm ss}(A^{k/2})$, $k \geq 2$, $\tilde{A} \geq \tilde{m} > -\infty$, and super-singular perturbations $S \in \mathcal{H}_{-k}(A)$-class satisfying additional requirements.

Example 8.3.6 (The model $d^4/dx^4 + \delta - \delta''$). In this example we construct a rank-one singular perturbation of the operator d^4/dx^4 on $L_2(\mathbb{R})$. The perturbation is given by the vector $\omega = \delta - \delta'' \in W_2^{-3}(\mathbb{R})$. Formally, the perturbed operator is written as

$$d^4/dx^4 + \beta(\delta - \delta'') = d^4/dx^4 + \beta\gamma_\omega,$$

where $\gamma_\omega(\cdot,\cdot) = \langle\cdot,\omega\rangle\langle\omega,\cdot\rangle$ with $\beta \in \mathbb{R}$. Specifically, we build the operator $\tilde{A} = d^4/dx^4 + \beta(\delta - \delta'')$ by means of the method of rigged spaces. To this end we introduce the perturbed scale of Sobolev spaces

$$\tilde{W}_2^{-4}(\mathbb{R}) \sqsupset \tilde{W}_2^{-2}(\mathbb{R}) \sqsupset L_2(\mathbb{R}) \sqsupset \tilde{W}_2^{2}(\mathbb{R}) \sqsupset \tilde{W}_2^{4}(\mathbb{R}), \tag{8.62}$$

and then we define $\tilde{A}$ as the operator associated with this scale.

In accordance with our general approach, the operator $\tilde{A}$ can be defined as the restriction of the Berezansky canonical isomorphism:

$$\tilde{D}_{-2,2} \restriction \{\varphi \in \tilde{W}_2^2 \mid \varphi^{(4)}(x) + \beta(\varphi(0)\delta(x) - \varphi''(0)\delta''(x)) \in L_2\},$$

where $\tilde{D}_{-2,2} : \tilde{W}_2^2 \to \tilde{W}_2^{-2}$ acts as the differential operator with derivatives in the generalized sense. The chain (8.62) can be obtained by starting with the pre-rigged pair $\tilde{W}_2^{-2}(\mathbb{R}) \sqsupset L_2(\mathbb{R})$, where $\tilde{W}_2^{-2}(\mathbb{R})$ is defined as the completion of $L_2(\mathbb{R})$ with respect to the inner product

$$(f,g)_{-2}^{\sim} := (f,g)_{W_2^{-2}} + \beta\gamma_\omega((1 - d^2/dx^2)^{-2}f, (1 - d^2/dx^2)^{-2}g).$$

Note that the element $\omega = \delta - \delta''$, regarded as a vector from $W_2^{-4}(\mathbb{R})$ admits the representation

$$\omega = (1 - d^2/dx^2)\delta = (1 - d^2/dx^2)^2\eta,$$

where

$$\eta(x) = \frac{1}{2}e^{-|x|} \in L_2(\mathbb{R}).$$

Further, the operator $\tilde{A}$ can be also written as the formal expression

$$d^4/dx^4 + \beta(\delta - \delta'') = (1 - d^2/dx^2)^2 + 2d^2/dx^2 - 1 + \beta(\delta - \delta'').$$

A precise definition uses the Kreĭn formula:

$$[(1 - d^2/dx^2)^2 + \beta(\delta - \delta'')]^{-1} := (1 - d^2/dx^2)^{-2} + \frac{\beta}{2(2+\beta)}(\cdot, \eta)_0\eta, \tag{8.63}$$

where the constant β satisfies the inequality

$$0 < \beta \le 1 - ((1 - d^2/dx^2)^{-1}\eta, \eta)_{L_2}.$$

The domain of the operator $d^4/dx^4 + \beta(\delta - \delta'')$ has the following description:

$$\begin{aligned} &\mathrm{Dom}(d^4/dx^4 + \beta(\delta - \delta'')) \\ &\quad = \left\{ g \in L_2 \mid g(x) = \varphi(x) + \frac{\beta}{2}(\varphi(0) - \varphi''(0))e^{-|x|},\ \varphi \in W_2^4 \right\}. \end{aligned}$$

Example 8.3.7 (Perturbation of the Bessel potential by δ-functions). In this example we give an unambiguous meaning to the formal expression $-\Delta + \beta\delta(x)$ with $x \in \mathbb{R}^d$, $d > 3$. We will use for the operator $A = (1 - \Delta)^{-k/2}$, $k > 0$, the representation in terms of a Bessel integral kernel.

So, let us consider in $L_2(\mathbb{R}^d, dx)$ the operator $A = (1 - \Delta)^{k/2}$, $k > 0$ defined by the spectral theorem, where Δ denotes the usual Laplace operator. It is well known (see [29, 182]) that the inverse of A is an integral operator with a Bessel kernel

$$G_k = \mathfrak{F}^{-1}((1 + |\xi|^2)^{-k/2}),$$

where $\mathfrak{F}^{-1}$ denotes the inverse Fourier transformation. We recall that the explicit representation of G_k has a form

$$\begin{aligned} G_k(x) &= \frac{1}{(2\pi)^d} \int_{\mathbb{R}^d} \frac{e^{ix\xi}}{(1 + |\xi|^2)^{k/2}} d\xi \\ &= (2\pi)^{-d/2} |x|^{-(d-2)/2} \int_0^\infty \frac{t^{d/2}}{(1 + t^2)^{k/2}} \mathfrak{J}_{(d-2)/2}(|x|t) dt, \end{aligned}$$

where $\mathfrak{J}_\nu$ denotes the Bessel function of order ν. For other properties of G_k, see, e.g., [29, 182].

The function

$$\varphi = G_k * h = (1 - \Delta)^{-k/2} h$$

is known as the Bessel potential. If h runs over the space $L_2(\mathbb{R}^d, dx)$, then the set of Bessel potentials $\varphi \in \mathrm{Dom}(1 - \Delta)^{k/2}$ form the Sobolev space

$$W_2^k(\mathbb{R}^d) = \mathcal{H}_k := \{\varphi = G_k * h,\ h \in L_2\}.$$

We take $\frac{d}{2} < k < d$ for fixed $d > 3$. Then by the Sobolev embedding theorem, the functions φ are continuous. Hence $W_2^k \subset C(\mathbb{R}^d)$ and the Dirac delta-function

$\delta(x) \in W_2^{-k}$. Therefore, the δ-perturbation of the Bessel potential can be written as follows. Formally, the perturbed integral kernel is of the form

$$\tilde{G}_k = G_k + \beta(\cdot, \eta_\delta)\eta_\delta, \quad \beta > 0,$$

where the function $\eta_\delta = G_k * \delta \in L_2$. The set of the δ-perturbed Bessel potentials

$$\tilde{\varphi} = \tilde{G}_k * h, \quad h \in L_2(\mathbb{R}^d, dx)$$

forms a new Hilbert space $\tilde{W}_2^k$, which consists of the functions of the form

$$\tilde{\varphi} = \varphi + \beta\varphi(0)\eta_\delta, \quad \varphi \in W_2^k.$$

By the above assumptions,

$$0 < \beta \leq 1 - (G_k * \eta_\delta, \eta_\delta)_{L_2}.$$

So, the space $\tilde{W}_2^k \sqsubset L_2$ is a positive space with respect to L_2. Moreover, the subspace

$$\mathcal{M}_k = \tilde{W}_k^2 \cap W_k^2 = \{\varphi \mid \varphi(0) = 0\}$$

is dense in L_2. Consequently, the mapping

$$\tilde{A}_\beta : \tilde{W}_k^2 \ni \tilde{\varphi} \longmapsto h = (1 - \Delta)^{k/2}\varphi \in L_2$$

is a self-adjoint extension in L_2 of the symmetric operator

$$\mathbf{A}_k := (1 - \Delta)^{k/2} \restriction \mathcal{M}_k.$$

This operator $\tilde{A}_\beta$ gives a precise meaning to the formal expression

$$(1 - \Delta)^{\frac{k}{2}} + \beta\delta.$$

Finally, by the spectral theorem we define the operator $(\tilde{A}_\beta)^{-2/k} - 1$ and treat it as a unique version of the formal expression $-\Delta + \beta\delta(x)$ with $x \in \mathbb{R}^d$, $d > 3$.

Example 8.3.8 (Perturbations of the Bessel potential by measures with fractal support). Let us consider a measure μ on $\mathbb{R}^d$ with compact support $\Gamma \subset \mathbb{R}^d$ of fractal structure (for the precise definitions see [20, 139, 180]). We assume that Γ has a zero Lebesgue measure, $\lambda(\Gamma) = 0$, but for some $k > 0$ its C_k-capacity is strongly positive, $C_k(\Gamma) > 0$. Also, $C_{k/2}(\Gamma) = 0$. We recall that the notion of C_α-capacity, $\alpha > 0$ has discussed in Subsection 6.3 (see also [29, 180]). In accordance with [29],

$$C_\alpha(\Gamma) = \inf_\nu \int_{\mathbb{R}^d}\int_{\mathbb{R}^d} G_\alpha(x - y)d\nu(x)d\nu(y),$$

where the infimum is taken over all Borel measures ν satisfying

$$\Gamma \subset \operatorname{supp}\nu, \quad \nu(\Gamma) = 1.$$

Here G_α denotes the Bessel integral kernel.

Using μ we define the quadratic form

$$\gamma_\mu[\varphi] := \int |\varphi|^2 d\mu, \quad \varphi \in W_2^k.$$

We assume that γ_μ is bounded on W_2^k. Then the associated operator $S_\mu : W_2^k \to W_2^{-k}$ is bounded. It acts as multiplication by μ. Since $C_{k/2}(\Gamma) = 0$, we can consider the operator S_μ as a strongly singular perturbation of the operator $A = (1-\Delta)^{k/2}$. By the same procedure as in the previous example, we introduce the perturbed version of the Bessel kernel:

$$\tilde{G}_{k,\mu} = G_k + B_\mu,$$

where B_μ denotes the integral operator with the kernel

$$B_\mu(x,y) = \int_{\mathbb{R}^d} \int_{\mathbb{R}^d} G_k(x-z)G_k(y-z')\mu(dz)\mu(dz').$$

Using $\tilde{G}_{k,\mu}$, we define the set of Bessel perturbed potentials:

$$\tilde{\varphi} = \tilde{G}_{k,\mu} * h, \quad h \in L_2(\mathbb{R}^d, dx).$$

This set forms the positive space $\tilde{\mathcal{H}}_{k,\mu} \sqsubset L_2$. Further, we define the singularly perturbed operator

$$\tilde{A}_\mu : \tilde{\mathcal{H}}_{k,\mu} \ni \tilde{\varphi} \longmapsto h = (1-\Delta)^{k/2}\varphi \in L_2, \quad \varphi \in W_2^k.$$

It is clear that $\tilde{A}_\mu$ acts as $(-\Delta+1)^{k/2}$ on functions φ from W_2^k which are zero on the support of the measure μ (i.e., $\varphi(x) = 0, x \in \Gamma$). The set of these functions is dense in L_2. Finally, we define $\tilde{A}_\mu^{-2/k} - 1$ as a version of the operator corresponding to the expression $-\Delta + \mu$.

Chapter 9

Some Aspects of Spectral Theory

The spectral theory of operators is one of the central and attractive branches of the mathematical physics. This is evidenced, in particular, by the large number of publications devoted to the spectral theory of singularly perturbed operators. List a few sources: [3–6, 8–10, 12–14, 17, 66, 67, 76, 132, 136, 139–141, 144, 158, 163, 176]

In this chapter we discuss the eigenvalue problem for the purely singular finite-rank perturbations of a self-adjoint operator A on a Hilbert space $\mathcal{H}$. The perturbed operators $\tilde{A}$ will be defined in two ways: by the Kreĭn's formula for resolvents

$$(\tilde{A}-z)^{-1} = (A-z)^{-1} + B_z, \quad \mathrm{Im}(z) \neq 0,$$

where B_z is a finite-rank operator such that $\mathrm{Dom}\, B_z \cap \mathrm{Dom}\, A = \{0\}$, or by the form-sum method, $\tilde{A} = A \dot{+} T$.

We construct the singularly perturbed operator $\tilde{A}$ solving the eigenvalue problem

$$\tilde{A}\psi_i = \lambda_i \psi_i, \quad \lambda_i \in \mathbb{R}^1, \quad i = 1, \ldots, n.$$

for a system of orthonormal vectors $\{\psi_i\}_{i=1}^{n<\infty}$ which satisfies the condition

$$\mathrm{span}\{\psi_i\} \cap \mathrm{Dom}\, A = \{0\}.$$

We prove the uniqueness of $\tilde{A}$ under the condition $\mathrm{rank}\, B_z = n$,

9.1 The point spectrum of singularly perturbed operators

Let us consider the problem described above in more detail. Let an unbounded self-adjoint operator $A = A^*$ with the domain $\mathfrak{D}(A) \equiv \mathrm{Dom}\, A$ be given on a complex separable Hilbert space $\mathcal{H}$ with the inner product $(\cdot,\cdot)$ and the norm $\|\cdot\|$.

Recall that a self-adjoint operator $\tilde{A} \neq A$ on $\mathcal{H}$ [135] is called the purely singularly perturbed with respect to A (compare with [24, 89]), if the set

$$\mathfrak{D} := \{f \in \mathfrak{D}(A) \cap \mathfrak{D}(\tilde{A}) \mid Af = \tilde{A}f\} \tag{9.1}$$

is dense in $\mathcal{H}$ (we write $\tilde{A} \in \mathcal{P}_{\mathrm{s}}(A)$). It is clear that for each $\tilde{A} \in \mathcal{P}_{\mathrm{s}}(A)$ there exists a densely defined symmetric operator

$$\mathbf{A} := A \upharpoonright \mathfrak{D} = \tilde{A} \upharpoonright \mathfrak{D}, \quad \mathfrak{D}(\mathbf{A}) = \mathfrak{D} \tag{9.2}$$

with deficiency indices $\mathbf{n}^{\pm}(\mathbf{A}) = \dim \operatorname{Ker}(\mathbf{A} \pm i)^* \neq 0$. In what follows we consider the subclass of operators $\tilde{A} \in \mathcal{P}_{\mathrm{s}}^n(A)$ for which $n = \mathbf{n}^+(\mathbf{A}) = \mathbf{n}^-(\mathbf{A}) < \infty$.

We are interested in construction of the singularly perturbed operator $\tilde{A} \in \mathcal{P}_{\mathrm{s}}^n(A)$ which solves the eigenvalue problem

$$\tilde{A}\psi_i = \lambda_i \psi_i, \quad i = 1, \ldots, n \tag{9.3}$$

for given real numbers λ_i and a system of orthonormal vectors $\{\psi_i\}_{i=1}^n$ satisfying the condition $\operatorname{span}\{\psi_i\} \cap \operatorname{Dom} A = \{0\}$.

An investigation of the spectrum, particularly, the point spectrum of self-adjoint extensions of symmetric operators with finite deficiency indices was already carried out by Kreĭn in [148]. Kreĭn proved the existence of at least one extension with given eigenvalues from the field of regularity of a symmetric operator (see also [3–17, 55, 56]. We also mention the works [63, 70–74, 91, 93], where in terms of boundary values spaces and the Weyl function the existence of an arbitrary type of the spectrum in gaps of symmetric operators was proved (see in addition [76, 77, 81, 142, 144, 149, 156, 158]).

In sequel we consider the eigenvalue problem for self-adjoint extensions of the symmetric operator $\mathbf{A}$ by using the singular perturbation theory. The points λ_i in (9.3) are assumed to be arbitrary, in particular, they can lie in the spectrum of the operator A. Note that results presented in [140] concern the case when the operator A is positive and points $\lambda_i \leq 0$.

9.1.1 On the point spectrum arising under singular finite-rank perturbations

The main result of this subsection is the following theorem.

Theorem 9.1.1. *For an unbounded self-adjoint operator A on a Hilbert space $\mathcal{H}$ there exists a unique purely singularly perturbed operator $\tilde{A} \in \mathcal{P}_{\mathrm{s}}^n(A)$ which solves the eigenvalue problem (9.3) with arbitrary given real numbers $\lambda_i \in \mathbb{R}^1$, $i = 1, \ldots, n < \infty$ and an arbitrary set of orthonormal vectors $\{\psi_i\}_{i=1}^n$ satisfying*

$$\operatorname{span}\{\psi_i\}_{i=1}^n \cap \mathfrak{D}(A) = \{0\}. \tag{9.4}$$

Note that this theorem admits a constructive proof. The resolvent of the operator $\tilde{A}$ will be constructed sequentially, by using a singular rank-one perturbation at each step.

Let $\dot{A} \subset \dot{A}^*$ be a closed symmetric operator with a domain $\mathfrak{D}(\dot{A})$ that is dense in $\mathcal{H}$, and with deficiency indices $\mathbf{n}^{\pm}(\dot{A}) = 1$. Then we have a decomposition $\mathcal{H} = \mathfrak{M}_z \oplus \mathfrak{N}_z$, $\mathrm{Im}(z) \neq 0$, where $\mathfrak{M}_z = (\dot{A} - z)\mathfrak{D}(\dot{A})$ is the domain of the operator $\dot{A} - z$, and $\mathfrak{N}_z := \mathfrak{M}_{\bar{z}}^{\perp} = \mathrm{Ker}(\dot{A}^* - \bar{z})$ is the deficiency subspace, $\dim \mathfrak{N}_z = 1$.

Let $\mathcal{A}(\dot{A})$ denote the set of all self-adjoint extensions of the operator $\dot{A}$. Let us fix some self-adjoint extension $A \in \mathcal{A}(\dot{A})$. It is clear that each operator $\tilde{A} \neq A$ in the set $\mathcal{A}(\dot{A})$ belongs also to the family $\mathcal{P}_{\mathrm{s}}^1(A)$. Therefore, the set $\mathfrak{D}$ in (9.1) coincides with $\mathfrak{D}(\dot{A})$. It is known [28, 148] that $\mathfrak{N}_z \cap \mathfrak{D}(A) = \{0\}$.

Theorem 9.1.2 ([32, 148]). *The resolvent of any self-adjoint operator $\tilde{A} \in \mathcal{A}(\dot{A})$, $\tilde{A} \neq A$, is defined by via the Kreĭn formula in the form:*

$$(\tilde{A} - z)^{-1} = (A - z)^{-1} + b_z^{-1}(\cdot, \eta_{\bar{z}})\eta_z, \tag{9.5}$$

where the vector-function η_z with values in $\mathfrak{N}_z$ satisfies the equality

$$\eta_z = (A - \xi)(A - z)^{-1}\eta_\xi, \quad \mathrm{Im}(z),\ \mathrm{Im}(\xi) \neq 0, \tag{9.6}$$

and values of the scalar function b_z are connected by the relation

$$b_z = b_\xi + (\xi - z)(\eta_\xi, \eta_{\bar{z}}), \quad \mathrm{Im}(z),\ \mathrm{Im}(\xi) \neq 0, \tag{9.7}$$

$$\bar{b}_z = b_{\bar{z}} \tag{9.8}$$

Theorem 9.1.2 yields a description of all operators $\tilde{A} \in \mathcal{P}_{\mathrm{s}}^1(A)$.

Theorem 9.1.3. *The self-adjoint on $\mathcal{H}$ operator $\tilde{A} \neq A$ belongs to the set $\mathcal{P}_{\mathrm{s}}^1(A)$ if and only if for an arbitrary $z_0 \in \mathbb{C}$, $\mathrm{Im}(z_0) \neq 0$ (and also for all such z) there exists a subspace*

$$\mathfrak{N}_{z_0} \subset \mathcal{H}, \quad \dim \mathfrak{N}_{z_0} = 1, \quad \mathfrak{N}_{z_0} \cap \mathfrak{D}(A) = \{0\} \tag{9.9}$$

and a number

$$b_{z_0} \in \mathbb{C}, \quad \mathrm{Im}(b_{z_0}) = -\mathrm{Im}(z_0), \tag{9.10}$$

such that

$$(\tilde{A} - z_0)^{-1} = (A - z_0)^{-1} + b_{z_0}^{-1}(\cdot, \eta_{\bar{z}_0})\eta_{z_0}, \tag{9.11}$$

where $\eta_{z_0} \in \mathfrak{N}_{z_0}$, $\|\eta_{z_0}\| = 1$, and $\eta_{\bar{z}_0} = (A - z_0)(A - \bar{z}_0)^{-1}\eta_{z_0}$.

For an arbitrary point $z \in \mathbb{C}$, $\mathrm{Im}(z) \neq 0$, the resolvent of the operator $\tilde{A}$ is defined by the formula (9.5), where the functions

$$\eta_z = (A - z_0)(A - z)^{-1}\eta_{z_0}, \tag{9.12}$$

$$b_z = b_{z_0} + (z_0 - z)(\eta_{z_0}, \eta_{\bar{z}}) \tag{9.13}$$

satisfy the relations (9.6)–(9.8).

Proof. Necessity. It is clear that if $\tilde{A} \in \mathcal{P}^1_{\mathrm{s}}(A)$ then $\tilde{A} \in \mathcal{A}(\dot{A})$ with $\dot{A} := A \restriction \mathfrak{D}$, where $\mathfrak{D}$ is defined by (9.1). The conditions (9.9)–(9.11) and the relations (9.12), (9.13) are valid due to (9.5)–(9.8). In particular, (9.10) follows from (9.7) and (9.8). Indeed, due to (9.7) we have

$$b_{\bar{z}_0} = b_{z_0} + (z_0 - \bar{z}_0)(\eta_{z_0}, \eta_{z_0}).$$

Hence, by (9.8),

$$\bar{b}_{z_0} - b_{z_0} = -2i \operatorname{Im}(b_{z_0}) = 2i \operatorname{Im}(z_0),$$

i.e., $\operatorname{Im}(b_{z_0}) = -\operatorname{Im}(z_0)$, where, without loss of generality, the vector η_{z_0} is normalized to 1.

Sufficiency. We have to show that the right-hand side of (9.11) defines a self-adjoint operator $\tilde{A} \in \mathcal{P}^1_{\mathrm{s}}(A)$. To this end, we consider the operator-valued function

$$\tilde{R}(z) = (A - z)^{-1} + b_z^{-1}(\cdot, \eta_{\bar{z}})\eta_z, \tag{9.14}$$

where the vector-valued function η_z and the scalar function b_z are given by (9.12) and (9.13). Then we show that $\tilde{R}(z)$ is the resolvent of the self-adjoint operator $\tilde{A} \in \mathcal{P}^1_{\mathrm{s}}(A)$, i.e., $(\tilde{A} - z)^{-1} = \tilde{R}(z)$.

First, we verify that $\tilde{R}(z)$ is a pseudo-resolvent (see [107]) that satisfies the Hilbert identity

$$\tilde{R}(z) - \tilde{R}(\xi) = (z - \xi)\tilde{R}(z)\tilde{R}(\xi), \quad \operatorname{Im}(z),\ \operatorname{Im}(\xi) \neq 0. \tag{9.15}$$

Using (9.14), we rewrite the identity (9.15) in the form

$$\begin{aligned} R(z) + b_z^{-1}(\cdot, \eta_{\bar{z}})\eta_z - R(\xi) - b_\xi^{-1}(\cdot, \eta_{\bar{\xi}})\eta_\xi \\ = (z - \xi)[R(z) + b_z^{-1}(\cdot, \eta_{\bar{z}})\eta_z] \cdot [R(\xi) + b_\xi^{-1}(\cdot, \eta_{\bar{\xi}})\eta_\xi], \end{aligned} \tag{9.16}$$

where $R(z) = (A - z)^{-1}$. Further, due to the Hilbert identity for the self-adjoint operator A, we obtain

$$\begin{aligned} b_z^{-1}(\cdot, \eta_{\bar{z}})\eta_z - b_\xi^{-1}(\cdot, \eta_{\bar{\xi}})\eta_\xi = (z - \xi)b_\xi^{-1}(\cdot, \eta_{\bar{\xi}})R(z)\eta_\xi \\ + (z - \xi)b_z^{-1}(\cdot, R(\bar{\xi})\eta_{\bar{z}})\eta_z \\ + (z - \xi)b_z^{-1}b_\xi^{-1}(\cdot, \eta_{\bar{\xi}})(\eta_\xi, \eta_{\bar{z}})\eta_z. \end{aligned} \tag{9.17}$$

In addition, (9.12) yields

$$\eta_z - \eta_\xi = (z - \xi)(A - z)^{-1}\eta_\xi, \quad \operatorname{Im}(z),\ \operatorname{Im}(\xi) \neq 0.$$

Hence (9.17) can be written in the simpler form

$$0 = b_\xi^{-1}(\cdot, \eta_{\bar{\xi}})\eta_z - b_z^{-1}(\cdot, \eta_{\bar{\xi}})\eta_z + (z - \xi)b_z^{-1}b_\xi^{-1}(\cdot, \eta_{\bar{\xi}})(\eta_\xi, \eta_{\bar{z}})\eta_z. \tag{9.18}$$

On the other hand, the right-hand side of (9.18) is equal zero due to the equality (9.13). Therefore, (9.15) is really valid.

Now we will use the well-known criterion (see [107], and also [166], Theorem 7.7.1) asserting that a pseudo-resolvent $\tilde{R}(z)$ is the resolvent of some densely defined closed operator, if and only if $\operatorname{Ker} \tilde{R}(z_0) = \{0\}$ at least for one point $z_0 \in \mathbb{C}$, $\operatorname{Im}(z_0) \neq 0$. In our case for all $0 \neq f \perp \eta_{\bar{z}_0}$ (9.11) implies that

$$\tilde{R}(z_0)f = (A - z_0)^{-1} f \neq 0.$$

Moreover, for the vector $\eta_{\bar{z}_0}$ it holds that

$$\tilde{R}(z_0)\eta_{\bar{z}_0} = (A - z_0)^{-1}\eta_{\bar{z}_0} + b_{z_0}^{-1}\|\eta_{\bar{z}_0}\|^2\eta_{z_0} \neq 0,$$

since $(A - z_0)^{-1}\eta_{\bar{z}_0} \in \mathfrak{D}(A)$, and $\eta_{z_0} \notin \mathfrak{D}(A)$ due to (9.9). Therefore, the expression $\tilde{R}(z) = (\tilde{A} - z)^{-1}$ is the resolvent of a closed operator $\tilde{A}$ on $\mathcal{H}$. In fact, $\tilde{A}$ is self-adjoint. To see this, we have to show that

$$(\tilde{R}(z))^* = \tilde{R}(\bar{z}) \tag{9.19}$$

(see [166], Theorem 7.7.3 and also [107]). But here (9.19) holds because the expression (9.8) follows from (9.10) and, in turn, it leads to the representation:

$$(\tilde{R}(z))^* = (A - z)^{-1} + b_{\bar{z}}^{-1}(\cdot, \eta_z)\eta_{\bar{z}} = \tilde{R}(\bar{z}).$$

Therefore, we proved that for the resolvent $\tilde{R}(z)$ of the self-adjoint operator $\tilde{A}$ the formula (9.11) is true. It remains to prove that the domain $\mathfrak{D}$, defined in accordance with (9.1), is dense in $\mathcal{H}$. Let us denote

$$\mathfrak{M}_{z_0} := (\tilde{A} - z_0)\mathfrak{D} = (A - z_0)\mathfrak{D}. \tag{9.20}$$

From (9.11) it follows that $\mathfrak{M}_{z_0}^{\perp} = \mathfrak{N}_{z_0}$. Let $\varphi \perp \mathfrak{D}$. Then, by (9.20), we have, for all $f \in \mathfrak{D}$, $f = (A - z_0)^{-1}h$, $h \in \mathfrak{M}_{z_0}$:

$$0 = (\varphi, f) = (\varphi, (A - z_0)^{-1}h) = ((A - \bar{z}_0)^{-1}\varphi, h).$$

This means that $(A - \bar{z}_0)^{-1}\varphi \in \mathfrak{N}_{z_0}$, which in view (9.9) is possible only for $\varphi = 0$. Thus we proved that $\tilde{A} \in \mathcal{P}_{\mathrm{s}}^1(A)$. □

Theorem 9.1.1 in the case $n = 1$ is formulated as follows (compare with Theorem 2 from [140]).

Theorem 9.1.4. *For an arbitrary self-adjoint unbounded operator A on a Hilbert space $\mathcal{H}$, there exists a unique purely singularly perturbed operator $\tilde{A} \in \mathcal{P}_{\mathrm{s}}^1(A)$, which solves the problem*

$$\tilde{A}\psi = \lambda\psi \tag{9.21}$$

for a given vector $\psi \in \mathcal{H} \setminus \mathfrak{D}(A)$ and a number $\lambda \in \mathbb{R}^1$.

Proof. Let us fix $z_0 \in \mathbb{C}$, $\mathrm{Im}(z_0) \neq 0$, and put

$$\eta_{z_0} := (A-\lambda)(A-z_0)^{-1}\psi, \tag{9.22}$$

$$b_{z_0} := (\lambda - z_0)(\psi, \eta_{\bar{z}_0}). \tag{9.23}$$

For an arbitrary $z \in \mathbb{C}$, $\mathrm{Im}(z) \neq 0$, the functions η_z, b_z are defined by formulas (9.12) and (9.13). The functional calculus yields

$$\eta_z = (A-\lambda)(A-z)^{-1}\psi = \psi + (z-\lambda)(A-z)^{-1}\psi, \tag{9.24}$$

$$b_z := (\lambda - z)(\psi, \eta_{\bar{z}}). \tag{9.25}$$

Let us consider the operator-valued function $\tilde{R}(z)$ given by (9.14). Let us verify that it really is the resolvent of a self-adjoint operator. To this end we will use Theorem 9.1.3. It is enough to show that the functions η_z and b_z satisfy the relations (9.6)–(9.8). The identity (9.6) follows immediately from (9.22) and (9.24). Equality (9.7) is checked directly by using the Hilbert identity for resolvents of the operator A. From formulas (9.24) and (9.25) it follows that

$$\bar{b}_z = (\lambda - \bar{z})(\eta_{\bar{z}}, \psi) = (\lambda - \bar{z})((A-\lambda)(A-\bar{z})^{-1}\psi, \psi) = b_{\bar{z}}.$$

Thus, (9.8) is also valid. Hence, $\tilde{R}(z) = (\tilde{A} - z)^{-1}$, where $\tilde{A}$ is a self-adjoint operator on $\mathcal{H}$. The fact that $\tilde{A}$ belongs to $\mathcal{P}^1_{\mathrm{s}}$, is checked in the following way. Let us put $\mathfrak{N}_{z_0} := \{c\eta_{z_0}\}_{c\in\mathbb{C}}$. The condition $\mathfrak{N}_{z_0} \cap \mathfrak{D}(A) = \{0\}$ follows from the representation (9.24) and the fact that

$$\eta_z = \psi + (z-\lambda)(A-z)^{-1}\psi \notin \mathfrak{D}(A),$$

since $\psi \notin \mathfrak{D}(A)$. The equality (9.10) follows from (9.13) and the fact that $\tilde{A}$ is self-adjoint if the vector ψ is normed, as follows $\|\eta_{z_0}\| = 1$. According to Theorem 9.1.3 $\tilde{A} \in \mathcal{P}^1_{\mathrm{s}}(A)$. Its resolvent has the form

$$(\tilde{A} - z)^{-1} = (A-z)^{-1} + \frac{1}{(\lambda - z)(\psi, \eta_{\bar{z}})}(\cdot, \eta_{\bar{z}})\eta_z, \tag{9.26}$$

where η_z is defined by the vector ψ in accordance with (9.24).

From (9.26) we deduce, due to (9.24), that

$$\begin{aligned}(\tilde{A} - z)^{-1}\psi &= (A-z)^{-1}\psi + \frac{1}{\lambda - z}\eta_z \\ &= (A-z)^{-1}\psi + \frac{1}{\lambda - z}(\psi + (z-\lambda)(A-z)^{-1}\psi) \\ &= \frac{1}{\lambda - z}\psi.\end{aligned}$$

Hence, $\tilde{A}$ solves the problem (9.21).

The operator $\tilde{A} \in \mathcal{P}_s^1$ solving the problem (9.21) is uniquely determined because the representation (9.11) and the equation

$$(\tilde{A} - z_0)^{-1}\psi = \frac{1}{\lambda - z_0}\psi$$

determine a single number b_{z_0} and a unique vector η_{z_0} (up to a phase factor $\mathrm{e}^{-\theta}, 0 \leq \theta < 2\pi$). □

Proof of Theorem 9.1.1. We will use induction, relying on Theorem 9.1.4. We introduce, in a similar way to (9.14), the operator-valued function $R_1(z)$, changing notations:

$$R_1(z) = (A - z)^{-1} + b_1^{-1}(z)(\cdot, \eta_1(\bar{z}))\eta_1(z), \quad \mathrm{Im} z \neq 0, \tag{9.27}$$

where $R_1(z) \equiv \tilde{R}(z)$ (see (9.26)) is the resolvent of the operator $A_1 = \tilde{A}$, and (see (9.24), (9.25))

$$\eta_1(z) \equiv \eta_z = (A - \lambda_1)(A - z)^{-1}\psi_1, \tag{9.28}$$
$$b_1(z) \equiv b_z = (\lambda_1 - z)(\psi_1, \eta_1(\bar{z})), \tag{9.29}$$

with $\psi = \psi_1$ and $\lambda = \lambda_1$. According to the proof of Theorem 9.1.4, the operator-valued function

$$\tilde{R}(z) := R_2(z) = R_1(z) + b_2^{-1}(z)(\cdot, \eta_2(\bar{z}))\eta_2(z),$$

is the resolvent of a unique self-adjoint operator $A_2 \in \mathcal{P}_s^1(A_1)$. There $\eta_2(z)$ and $b_2(z)$ are defined by (9.28) and (9.29), in terms of λ_2, ψ_2. The expression for A_1 consists of λ_1, ψ_1 and A. The operator $A_2 \in \mathcal{P}_s^1(A_1)$ solves the problem $A_2\psi_2 = \lambda_2\psi_2$, only if $\psi_2 \notin \mathfrak{D}(A_1)$. The last fact follows from the condition (9.4). Indeed, from (9.27) we obtain the description of the domain of the self-adjoint operator A_1 as

$$\mathfrak{D}(A_1) = \{h \in \mathcal{H} \mid h = f + c(z)\eta_1(z),\ f \in \mathfrak{D}(A)\},$$

where

$$c(z) = b_1^{-1}(z)((A - \lambda_1)f, \psi_1).$$

If we now assume that the vector $\psi_2 = f + c(z)\eta_1(z)$, then, since

$$\eta_1(z) = \psi_1 + (z - \lambda_1)(A - z)^{-1}\psi_1,$$

we get

$$\psi_2 - c(z)\psi_1 = f + (z - \lambda_1)c(z)(A - z)^{-1}\psi_1 \in \mathfrak{D}(A).$$

However, this contradicts (9.4). Hence, $\psi_2 \notin \mathfrak{D}(A_1)$ and $A_2 \in \mathcal{P}_s^1(A_1)$. We claim that $A_2\psi_1 = \lambda_1\psi_1$. Indeed, from (9.27), due to $A_1\psi_1 = \lambda_1\psi_1$, we have

$$(A_2 - z)^{-1}\psi_1 = \frac{1}{z - \lambda_1}\psi_1,$$

since the fact that $\psi_1 \perp \psi_2$ implies

$$(\psi_1, \eta_2(\bar{z})) = \big((A_1 - \lambda_2)\psi_1, (A_1 - z)^{-1}\psi_2\big) = 0.$$

It is easy to see that similar considerations are valid for any kth step, $1 < k \leq n$. By induction, the operator-valued function

$$\tilde{R}(z) \equiv R_n(z) = (A_{n-1} - z)^{-1} + b_n^{-1}(z)(\cdot, \eta_n(\bar{z}))\eta_n(z),$$

with $\eta_n(z)$ and $b_n(z)$, defined via (9.28) and (9.29), by ψ_n, λ_n, and the operator A_{n-1}, is the resolvent of an operator $A_n \in \mathcal{P}_{\mathrm{s}}^1(A_{n-1})$, that solves the problem (9.3).

It remains to prove that A_n is a uniquely determined operator from the family $\in \mathcal{P}_{\mathrm{s}}^n(A)$.

According to our construction,

$$(A_n - z)^{-1} = (A - z)^{-1} + B_n(z), \tag{9.30}$$

where $\operatorname{rank} B_n(z) = n$. Indeed,

$$B_n(z) = \sum_{k=1}^{n} b_k^{-1}(z)(\cdot, \eta_k(\bar{z}))\eta_k(z), \tag{9.31}$$

where $b_k(z)$ and $\eta_k(z)$ are defined, via (9.28), and (9.29) or (9.24) and (9.25), by ψ_k, λ_k, and the operator A_{k-1}. It is easy to verify that, thanks to the condition (9.4), all vectors $\eta_k(z)$ are linearly independent and belong to $\mathfrak{D}(A)$. Then, by Theorem A1 in [99], the set

$$\mathfrak{D} = (A - z)^{-1}\mathrm{Ker}\, B_n(z) = (A_n - z)^{-1}\mathrm{Ker}\, B_n(z)$$

is dense in $\mathcal{H}$ and the symmetric operator

$$\dot{A} = A \restriction \mathfrak{D} = A_n \restriction \mathfrak{D}$$

has deficiency indices $\mathbf{n}^+(\dot{A}) = \mathbf{n}^-(\dot{A}) = n$. Therefore, $A_n \in \mathcal{P}_{\mathrm{s}}^n(A)$. The uniqueness of A_n follows from (9.30) and (9.31), since the operator $B_n(z)$ has a fixed value on the vectors ψ_i:

$$B_n(z)\psi_i = \frac{1}{\lambda_i - z}\psi_i - (A - z)^{-1}\psi_i, \quad i = 1, \ldots, n.$$

Here $\{\psi_i\}_{i=1}^n$ is a set of linearly independent vectors such that

$$\mathrm{span}\{\psi_i\} \cap \mathrm{Ker}\, B_n(z) = \{0\}.$$

In addition, the resolvent $R_n(z)$ coincides with $R(z)$ on the subspace $\mathrm{Ker}\, B_n(z)$. $\square$

9.2 The inverse eigenvalue problem

Here we consider a problem closely related to the previous one. Let $A \geq 0$ be a self-adjoint unbounded operator on a Hilbert space $\mathcal{H}$ and let $\mathcal{H}_{-1} \sqsupset \mathcal{H} \sqsupset \mathcal{H}_1$ denote the rigging associated with the operator A, where the norm in $\mathcal{H}_{\pm 1}$ is defined as follows $\| \cdot \|_{\pm 1} = ((A+I)^{\pm 1} \cdot, \cdot)^{1/2}$, where I denotes the identity operator. Let us fix a sequence of orthonormal vectors $\psi_j \in \mathcal{H}_1 \setminus \mathfrak{D}(A)$ and a sequence of negative numbers $E_j < 0$, $j = 1, \dots, N$. One can ask:does a weakly singularly perturbed operator $\tilde{A} \in \mathcal{P}^n_{\mathrm{ws}}(A)$, which solves the eigenvalue problem $\tilde{A}\psi_j = E_j \psi_j$ exist? The answer is affirmative. We will show that for given ψ_j and E_j there always exists an operator $T : \mathcal{H}_1 \to \mathcal{H}_{-1}$ such that for the perturbed operator $\tilde{A} = A \tilde{+} T$, defined as a generalized operator sum, the numbers E_j belong to its point spectrum, and ψ_j are the corresponding eigenvectors. Moreover, we will give a description of all operators $T' : \mathcal{H}_1 \to \mathcal{H}_{-1}$ such that $\tilde{A}' = A \tilde{+} T'$ solves the same eigenvalue problem. Additionally, we will show that T is uniquely determined if $\operatorname{rank} T = N$.

9.2.1 A general construction

Let $A \geq 0$ be a positive unbounded self-adjoint operator on the complex separable Hilbert space $\mathcal{H}$ with the scalar product $(\cdot, \cdot)$ and the norm $\| \cdot \|$. We suppose that $(Af, f) > 0$, $f \neq 0$ and $\inf_{\|f\|=1}(Af, f) = 1$, hence A has a bounded everywhere defined inverse operator.

Let $\{\mathcal{H}_k(A)\}_{k \in \mathbb{R}^1}$ denote, as usually, the A-scale of Hilbert spaces (see details in Chapters 4 and 6). Actually below we mainly use only a part of the scale:

$$\mathcal{H}_{-1} \sqsupset \mathcal{H}_0 \equiv \mathcal{H} \sqsupset \mathcal{H}_1, \tag{9.32}$$

where $\mathcal{H}_1 \equiv \mathcal{H}_1(A)$ coincides with the domain $\mathfrak{D}(A^{1/2})$ with the norm $\|\varphi\|_1 := \|(A+I)^{1/2}\varphi\|$, and $\mathcal{H}_{-1} \equiv \mathcal{H}_{-1}(A)$ is the dual space of $\mathcal{H}_1$ space ($\mathcal{H}_{-1}$, is the completion of $\mathcal{H}$ with respect to the norm $\|f\|_{-1} := \|(A+I)^{-1/2} f\|$). It is obvious that A is a bounded mapping from $\mathcal{H}_1$ into $\mathcal{H}_{-1}$. Thus, the expression $\langle \varphi, A\psi \rangle$ is well defined for all $\varphi, \psi \in \mathcal{H}_1$, where $\langle \cdot, \cdot \rangle$ denotes the duality pairing between these spaces $\mathcal{H}_1$ and $\mathcal{H}_{-1}$ (see Chapter 4).

Let $T : \mathcal{H}_1 \mapsto \mathcal{H}_{-1}$ be a closed symmetric operator that acts in the A-scale. It is singular with respect to $\mathcal{H}$ if the range $\operatorname{Ran} T$ contains elements that do not belong to $\mathcal{H}$. In the sequel, we consider purely singular operators T such that $\operatorname{Ran} T \cap \mathcal{H} = \{0\}$. More precisely, we assume that T belongs to the $\mathcal{H}_{-1}(A)$-class (see Chapter 5).

Further, from the given A and T, we construct the singularly perturbed operator $\tilde{A} = A \tilde{+} T$ by using the generalized operator sum method, which generalizes the well-known form-sum procedure (see Chapter 7, and also [42, 44, 45, 98, 121, 149]). According to the definition, the *generalized operator sum* $\tilde{A} = A \tilde{+} T$ is the restriction of the operator sum $A^{\mathrm{cl}} + T : \mathcal{H}_1 \mapsto \mathcal{H}_{-1}$ onto $\mathcal{H}$, where A^{cl} denotes the closure of A as an operator from $\mathcal{H}_1$ into $\mathcal{H}_{-1}$. Thus,

$$\begin{aligned} \mathfrak{D}(\tilde{A}) &= \{\varphi \in \mathcal{H}_1 \cap \mathfrak{D}(T) : A^{\mathrm{cl}}\varphi + T\varphi \in \mathcal{H}\}, \\ \tilde{A}\varphi &= A^{\mathrm{cl}}\varphi + T\varphi. \end{aligned} \tag{9.33}$$

In this subsection we consider a variant of the inverse eigenvalue problem (cf. with [4]).

Let $\psi_j \in \mathcal{H}_1(A) \setminus \mathfrak{D}(A)$, $j = 1, \dots, N$ be an arbitrary orthonormal sequence of vectors in $\mathcal{H}$, $(\psi_j, \psi_k) = \delta_{jk}$, and let $E_j < 0$, $j = 1, \dots, N$ be a sequence of negative numbers. We suppose additionally that $\mathrm{span}\{\psi_j\}_{j=1}^N \cap \mathfrak{D}(A) = \{0\}$. We want to find a singular operator $T : \mathcal{H}_1 \mapsto \mathcal{H}_{-1}$ (more precisely, an operator of $\mathcal{H}_{-1}$-class) such that the perturbed operator $\tilde{A} = A \tilde{+} T$ is self-adjoint on $\mathcal{H}$ and solves the eigenvalue problem

$$\tilde{A}\psi_j = E_j\psi_j, \quad \psi_j \in \mathcal{H}_1(A) \setminus \mathfrak{D}(A), \ E_j < 0 \quad j = 1, \dots, N \tag{9.34}$$

where $\mathrm{span}\{\psi_j\}_{j=1}^N \cap \mathfrak{D}(A) = \{0\}$.

We will construct T sequentially, starting with a rank-one singular perturbation. Let us consider a regular rank-one perturbation of the operator A,

$$A_1 = A + \alpha_1(\cdot, \omega_1)\omega_1,$$

with

$$\omega_1 = (A - E_1)\psi_1, \quad \psi_1 \in \mathfrak{D}(A), \ \alpha_1 = -\frac{1}{(\psi_1, \omega_1)}.$$

By direct calculation it is easy to check that A_1 solves the eigenvalue problem $A_1\psi_1 = E_1\psi_1$. We emphasize that the number E_1 and the vector $\psi_1 \in \mathfrak{D}(A)$ are arbitrary and preassigned. They only satisfy the condition $(A\psi_1, \psi_1) \neq E_1\|\psi_1\|^2$ (although further it will be seen that this condition can be ignored).

One can repeat the construction described above for any real number E_2 and new vector $\psi_2 \in \mathfrak{D}(A)$, taking A_1 instead of A. Then the operator

$$A_2 = A_1 + \alpha_2(\cdot, \omega_2)\omega_2, \quad \omega_2 = (A_1 - E_2)\psi_2, \ \alpha_2 = -\frac{1}{(\psi_2, \omega_2)}$$

solves the problem $A_2\psi_2 = E_2\psi_2$. If $\psi_2 \perp \psi_1$, then in the second step the previous couple E_1, ψ_1 is preserved. This means that A_2 solves also the previous problem $A_2\psi_1 = E_1\psi_1$. Similarly, by induction, in the Nth step, we obtain the operator

$$A_N = A_{N-1} + \alpha_N(\cdot, \omega_N)\omega_N,$$

namely, a rank-N perturbation of the operator A, which solves the analogous problem (9.34) with the vector

$$\psi_j \in \mathfrak{D}(A), \quad (\psi_j, \psi_k) = \delta_{jk}, \ j, k = 1, \dots, N,$$

and an arbitrary number E_j.

We show below that the described method can be generalized to the case, when all $\psi_j \in \mathcal{H}_1(A) \setminus \mathfrak{D}(A)$, $\text{span}\{\psi_j\}_{j=1}^N \cap \mathfrak{D}(A) = \{0\}$, and $E_j < 0$. Moreover, it turns out that a similar result (see [182, 183]) holds true for $\psi_j \in \mathcal{H}$, $\text{span}\{\psi_j\} \cap \mathfrak{D}(A) = \{0\}$ and $E_j \in \mathbb{R}$.

Note that in the case of a rank-one perturbation that is given in the form $\tilde{A} = A \tilde{+} \alpha\langle\cdot, \omega\rangle\omega$, if $\tilde{A}$ solves the problem $\tilde{A}\psi = E\psi$, then there exists a one-to-one correspondence between the pairs $\{E, \psi\}$ and $\{\alpha, \omega\}$. Moreover, this fact gives the uniqueness in the theorem for the operator T in the presentation $\tilde{A} = A \tilde{+} T$ under the condition $\operatorname{rank} T = N$.

But, generally, the operator T is not unique for the problem (9.34). Later we will give a description of all operators $T' : \mathcal{H}_1 \mapsto \mathcal{H}_{-1}$ such that $\tilde{A}' = A \tilde{+} T'$ solves the same problem on the negative eigenvalues.

9.2.2 The eigenvalue problem for rank-one singular perturbations

We give an explicit construction of a singular rank-one perturbation $\tilde{A}$ of the operator A that which solves the eigenvalue problem

$$\tilde{A}\psi = E\psi, \quad \psi \in \mathcal{H}_1, \ E \in \mathbb{R}. \tag{9.35}$$

Let us start with the case of a weakly singular rank-one perturbation of the operator A, which can be considered as a semibounded from below one (see details in [15, 16, 23, 24, 89, 121, 123, 135, 167]).

Fix a vector $\omega \in \mathcal{H}_{-1} \setminus \mathcal{H}$, $\|\omega\|_{-1} = 1$. Associate with ω the bounded operator $T = T_\omega = \langle\cdot, \omega\rangle\omega$, which acts from $\mathcal{H}_1(A)$ into $\mathcal{H}_{-1}(A)$ according to the rule

$$T_\omega\varphi = \langle\varphi, \omega\rangle\omega, \quad \varphi \in \mathcal{H}_1(A).$$

It is known that every such operator T_ω defines a one-parameter family of self-adjoint singularly rank-one perturbed operators

$$A_{\alpha,\omega} = A \tilde{+} \alpha T_\omega = A \tilde{+} \alpha\langle\cdot, \omega\rangle\omega, \quad \alpha \in (\mathbb{R} \setminus \{0\}) \cup \infty, \tag{9.36}$$

where α is called the coupling constant.

If $\alpha \neq \infty$, then $A_{\alpha,\omega}$ is defined as a generalized operator sum (see (9.33) and [98]). In the case when $\alpha = \infty$, the operator $A_{\infty,\omega}$ is defined (cf. with [89]) as the Friedrichs extension of the densely defined symmetric operator

$$\mathbf{A} := A \restriction \mathfrak{D},$$

where the domain

$$\mathfrak{D} = \{\varphi \in \mathfrak{D}(A) \mid \langle\varphi, \omega\rangle = 0\}$$

is a maximal linear set dense in $\mathcal{H}$, on which each $A_{\alpha,\omega}$ coincides with A. In each case the resolvent of $A_{\alpha,\omega}$ has a representation by the Kreĭn formula:

$$\begin{aligned} \tilde{R}_z &= (A_{\alpha,\omega} - zI)^{-1} = (A - zI)^{-1} - \frac{1}{\alpha^{-1} + \langle\eta_z, \omega\rangle}(\cdot, \eta_{\bar{z}})\eta_z, \\ &= R_z - b_\alpha^{-1}(z)(\cdot, \eta_{\bar{z}})\eta_z, \end{aligned} \tag{9.37}$$

where

$$\begin{aligned} R_z &= (A - zI)^{-1}, & \eta_z &= R_z\omega, \\ R_z &= (A^{\text{cl}} - zI)^{-1}, & b_\alpha(z) &= \frac{1}{\alpha} + \langle \eta_z, \omega \rangle. \end{aligned} \tag{9.38}$$

The domain $\mathfrak{D}(A_{\alpha,\omega})$ lies in $\mathcal{H}_1(A)$ and has the following description. For all z in the resolvent set of the operator $A_{\alpha,\omega}$,

$$\begin{aligned} &\mathfrak{D}(A_{\alpha,\omega}) = \{\psi \in \mathcal{H}_1(A) \mid \psi = \varphi - b_\alpha^{-1}(z)\langle \varphi, \omega\rangle\, \eta_z, \varphi \in \mathfrak{D}(A)\}, \\ &(A_{\alpha,\omega} - zI)\psi = (A - zI)\varphi. \end{aligned}$$

For $\alpha = \infty$, $b_\infty(z) = \langle \eta_z, \omega\rangle$, we obtain the Friedrichs extension $A_{\infty,\omega}$ of the operator $\mathbf{A}$, which, in the case where A is positive has the following description:

$$\begin{aligned} \mathfrak{D}(A_{\infty,\omega}) &= \{\psi \in \mathcal{H}_1(A) \mid \psi = \varphi - \langle \varphi, \omega\rangle\eta,\ \varphi \in \mathfrak{D}(A)\}, \\ A_{\infty,\omega}\psi &= A\varphi + \langle \varphi, \omega\rangle \eta, \end{aligned}$$

where we denote $\eta = (A^{\text{cl}} + I)^{-1}\omega$ and use the equality $\langle \eta, \omega\rangle = \|\omega\|_{-1}^2 = 1$. Hence, each $A_{\alpha,\omega}$ is a self-adjoint extension of the operator $\mathbf{A}$.

Next, for the sake of simplicity, we assume that A is an unbounded positive operator. It is known, and it is easy to check directly (see details in [16, 23, 24, 123, 135, 136]) that the operator $A_{\alpha,\omega}$ solves the eigenvalue problem $A_{\alpha,\omega}\psi = E\psi$, if

$$\psi = (A^{\text{cl}} - E)^{-1}\omega, \quad \psi \in \mathfrak{D}(A_{\alpha,\omega}) \setminus \mathfrak{D}(A),$$

and the number $E < 0$ is the solution of the equation

$$b_\alpha(E) = \frac{1}{\alpha} + \langle \eta_E, \omega\rangle = 0. \tag{9.39}$$

The question is: is there a solution E for an arbitrary $\alpha \neq 0$, with $\omega \in \mathcal{H}_{-1} \setminus \mathcal{H}$ is fixed? We give a positive answer under the assumption $\alpha < 0$.

Let us consider the function

$$\begin{aligned} a_\omega(E) &:= \langle \eta_E, \omega\rangle = \langle (A^{\text{cl}} + E)^{-1}\omega, \omega\rangle \\ &= \int_0^\infty \frac{1}{\lambda - E}\, d\mu_\omega(\lambda), \quad E \le 0, \end{aligned} \tag{9.40}$$

where $d\mu_\omega(\lambda) = d\langle \mathbf{E}_\lambda \omega, \omega\rangle$ denotes the spectral measure of A associated with ω and $\mathbf{E}_\lambda$ is the resolution of the identity of A. In general, it is possible that $a_\omega(0) = \infty$. By its construction, the function $a_\omega(E)$ is continuous, non-negative and non-decreasing on $E \in (-\infty, 0]$; in addition,

$$\lim_{E \to -\infty} a_\omega(E) = 0.$$

By (9.39), the equality $A_{\alpha,\omega}\psi = E\psi$ holds for some vector $\psi \in \mathfrak{D}(A_{\alpha,\omega}) \setminus \mathfrak{D}(A)$ if and only if

$$\langle \omega, R_E\omega\rangle = -\frac{1}{\alpha}, \quad R_E \equiv (A^{\text{cl}} + E)^{-1}. \tag{9.41}$$

So, the answer to this question depends on whether the limit value

$$a_\omega(0) := \lim_{E\to 0} a_\omega(E)$$

is finite or not. Since $a_\omega(E)$ increases monotonically from 0 to $a_\omega(0)$ when E runs on the half-line $(-\infty, 0]$, then obviously there always exists the number $E \le 0$, that satisfies (9.41) for each fixed negative α and such that

$$-\frac{1}{\alpha} \le a_\omega(0).$$

Hence, if

$$a_\omega(0) = \int_0^\infty \frac{1}{\lambda} d\mu_\omega(\lambda) < \infty,$$

then the maximal value of the coupling constant α, which guarantees the existence of the solution $E \le 0$ is given by

$$\alpha = -\frac{1}{a_\omega(0)}.$$

However, if

$$\lim_{E\to 0} a_\omega(E) = a_\omega(0) = +\infty,$$

then for each fixed $\alpha < 0$, the operator $A_{\alpha,\omega}$ has the negative eigenvalue E. Of course, the solution E, if it exists, is unique, since the function $a_\omega(E)$ is monotonically increasing for all values $E \to 0$.

We can characterize the vectors ω, for which the operator $A_{\alpha,\omega}$ has the eigenvalue $E < 0$ for fixed $\alpha < 0$. To do this, before the statement of results, we need some additional preparations.

Let us introduce a "homogeneous" version of the positive space $\mathcal{H}_1(A)$. Let the space $H_1 \equiv H_1(A)$ denote the completion of the domain $\mathfrak{D}(A)$ with respect to the norm $\|f\|_{H_1} := (Af, f)^{1/2}$. Recall that we supposed that A has an inverse operator. It is clear that $\|\cdot\|_{H_1} \le \|\cdot\|_{\mathcal{H}_1}$ and also $\|\cdot\|_{H_{-1}} \ge \|\cdot\|_{\mathcal{H}_{-1}}$, where $H_{-1} \equiv H_{-1}(A)$ is the completion of $\mathcal{R}(A)$ with respect to the norm generated by the scalar product $(f,g)_{H_{-1}} := (A^{-1}f, g)$. Hence, we have

$$\mathcal{H}_{-1}(A) \supseteq H_{-1}(A), \quad H_1(A) \supseteq \mathcal{H}_1(A). \tag{9.42}$$

Let us remark (see (9.40)) that

$$a_\omega(0) = \|\omega\|^2_{H_{-1}}.$$

Hence $\omega \in H_{-1}$ if and only if $a_\omega(0) < \infty$. Therefore, if we suppose that $\omega \in H_{-1}$, then $\lim_{E\to 0} a_\omega(E) = a_\omega(0) < \infty$ and the solution E of the equation (9.41), does not exist for all values α of the coupling constant in the range

$$-\frac{1}{a_\omega(0)} < \alpha < 0.$$

This follows from the fact that for the equation

$$\alpha = -\langle \omega, (\mathbf{A}-E)^{-1}\omega\rangle^{-1}$$

to have a solution E, the coupling constant α must satisfy the inequality

$$-\frac{1}{\alpha} \leq a_\omega(0).$$

Thus, we have the following result (cf. with [10]).

Theorem 9.2.1. *For a fixed vector $\omega \in \mathcal{H}_{-1} \setminus \mathcal{H}$ such that $\|\omega\|_{-1} = 1$, the operator*

$$A_{\alpha,\omega} = A \tilde{+} \alpha\langle\cdot,\omega\rangle\omega, \quad \alpha < 0,$$

has exactly one negative point in the point spectrum $E < 0$ if and only if

$$\omega \in \mathcal{H}_{-1} \setminus H_{-1}.$$

In this case, the only existing eigenvalue E is the solution of the equation

$$a_\omega(E) + \alpha^{-1} = 0,$$

and $\psi = (A^{\mathrm{cl}} - E)^{-1}\omega$ is a corresponding eigenvector.

In the considered situation we have a one-to-one correspondence between the pairs $\{\alpha, \omega\}$ and $\{E, \psi\}$.

We show that such a correspondence exists in a more general situation.

Theorem 9.2.2. *For each bounded from below (but unbounded) self-adjoint operator A on $\mathcal{H}$, and an arbitrary vector $0 \neq \psi \in \mathcal{H}_1$ and a real E, there exists a uniquely defined singularly rank-one perturbed operator $\tilde{A}{=}A_{\alpha,\omega}$, with $\omega \in \mathcal{H}_{-1}$ and α given by the formulas*

$$\omega = (A^{\mathrm{cl}} - E)\psi, \quad \alpha = -\frac{1}{\langle\psi, (A^{\mathrm{cl}} - E)\psi\rangle}, \tag{9.43}$$

which solves the problem

$$\tilde{A}\psi = E\psi. \tag{9.44}$$

In other words, the theorem states that the operator

$$\tilde{A} = A_{\alpha,\omega} = A \tilde{+} \alpha\langle\cdot,\omega\rangle\omega$$

with vectors $\omega \in \mathcal{H}_{-1}$ and a coupling constant $\alpha \in (\mathbb{R}\setminus\{0\}) \cup \infty$, defined in (9.43) in term of the vectors $0 \neq \psi \in \mathcal{H}_1$ and a real number E, solves the problem (9.44). We put $\tilde{A} = A$ if $\omega = 0$, namely $A\psi = E\psi$.

Conversely, if $\tilde{A}$ is a singular rank-one perturbation of A and $\tilde{A}$ solves the problem (9.44), then such operator has the form shown above, where α and ω are connected uniquely by the relation (9.43) (precisely, the vector ω is defined up to a factor $e^{i\theta}$ by ψ and E). Thus, the expression (9.43) establishes the one-to-one correspondence between $\{E, \psi\}$ and $\{\alpha, \omega\}$ under the condition that $\tilde{A}$ solves the problem (9.44).

Proof. Let $\psi \in \mathcal{H}_1$ and $E \in \mathbb{R}$ be given. In the case where the vector $\psi \in \mathfrak{D}(A)$ and $A\psi = E\psi$, we put $\tilde{A} = A$ and then (9.44) is valid. Otherwise, $\omega \neq 0$ and $\alpha \neq 0$. Let us suppose that $\alpha \neq \infty$. Then an immediate verification shows that the operator

$$\tilde{A} = A \tilde{+} \left(-\frac{1}{\langle \psi, (A^{\text{cl}} - E)\psi \rangle} \right) \langle \cdot, (A^{\text{cl}} - E)\psi \rangle (A^{\text{cl}} - E)\psi,$$

solves the problem $\tilde{A}\psi = E\psi$. Moreover, for another operator

$$\tilde{A}' = A \tilde{+} \alpha' \langle \cdot, \omega' \rangle \omega', \quad \omega' \neq 0, \ \alpha' \neq \infty,$$

which also solves this problem, the vector ω' must have a form $\omega' = e^{i\theta}\omega$, where $\theta \in [0, 2\pi)$, and $\alpha' = \alpha$, where ω and α are given in (9.43). If ψ and E are such that $\langle \mathbf{A}\psi, \psi \rangle = E\|\psi\|^2$ ($\omega \neq 0$), then $\alpha = \infty$ and we define $\tilde{A}$ by the Kreĭn formula

$$\tilde{R}_z = (A_{\infty,\omega} - zI)^{-1} = (A - zI)^{-1} - \frac{1}{\langle \eta_z, \omega \rangle} (\cdot, \eta_{\bar{z}}) \eta_z, \tag{9.45}$$

with $\eta_z = R_z \omega$, where $\omega = (A^{\text{cl}} - E)\psi$. We recall that $A^{\text{cl}}\psi \neq E\psi$. By direct verification we find that $\tilde{R}_z \psi = \frac{1}{E-z}\psi$. Thus, the operator $\tilde{A} = A_{\infty,\omega}$ solves the problem (9.44). Conversely, if $\tilde{A}$ is the Friedrichs extension of the symmetric operator $\mathbf{A}$, constructed from the element $\omega \in \mathcal{H}_{-1} \setminus \mathcal{H}$, and $\tilde{A}$ solves the problem (9.44), then the resolvent $\tilde{A}$ has a form (9.45), where necessarily

$$\omega = (A^{\text{cl}} - E)\psi, \quad \langle \mathbf{A}\psi, \psi \rangle = E\|\psi\|^2,$$

which corresponds to $\alpha = \infty$. □

Thus, the set of rank-one perturbations $A_{\alpha,\omega}$ such that each $A_{\alpha,\omega}$ solves the problem $\tilde{A}\psi = E\psi$ with a real E and some $\psi \in \mathcal{H}_1$, establishes a one-to-one correspondence between the couples

$$\{E \in \mathbb{R}^1, \psi \in \mathcal{H}_1\} \quad \text{and} \quad \{\alpha \in (\mathbb{R}^1 \setminus \{0\}) \cup \infty, \ \omega \in \mathcal{H}_{-1}\}.$$

Note that a similar result holds true for (purely) strongly singular rank-one perturbations $\tilde{A} \in \mathcal{P}_{\text{ss}}(A)$ with vectors $\psi \in \mathcal{H}$ and $\omega \in \mathcal{H}_{-2}$. In such a case the operator $\tilde{A}$ can be defined for an arbitrarily given $E \in \mathbb{R}^1$ and $\psi \in \mathcal{H}$ by the Kreĭn formula

$$(\tilde{A} - z)^{-1} = (A - z)^{-1} + b_z^{-1} (\cdot, \eta_{\bar{z}}) \eta_z, \quad \text{Im}(z) \neq 0,$$

with

$$\eta_z = (A - E)(A - z)^{-1}\psi, \quad b_z = (E - z)(\psi, \eta_{\bar{z}}).$$

The procedure described above for constructing the operator $\tilde{A}$, which solves the problem (9.44), is called *the inverse eigenvalue problem method* in the theory of singularly perturbed operators. In the next subsection we use this method in the case of finite-rank perturbations.

9.2.3 The eigenvalue problem for singularly perturbed rank-one operators

Let $A = A^* \geq 0$ have an inverse operator. For a given sequence of vectors $\psi_j \in \mathcal{H}_1(A) \setminus \mathfrak{D}(A)$, $\operatorname{span}\{\psi_j\}_{j=1}^N \cap \mathfrak{D}(A) = \{0\}$, orthonormal on $\mathcal{H}$, and a sequence of numbers $E_j \leq 0$, $j = 1, \ldots, N$ we construct the sequence of operators $A_1, \ldots, A_N$ in the following way. In the first step we put

$$A_1 = A \tilde{+} \alpha_1 \langle \cdot, \omega_1 \rangle \omega_1, \quad \omega_1 \equiv \omega_1^0 = (A^{\mathrm{cl}} - E_1)\psi_1,$$

with

$$\alpha_1 = -\frac{1}{\langle \psi_1, (A^{\mathrm{cl}} - E_1)\psi_1 \rangle} = -\frac{1}{\langle \psi_1, A^{\mathrm{cl}}\psi_1 \rangle - E_1} = -\frac{1}{a_{11}^0 - E_1},$$

where

$$a_{11}^0 := \langle \psi_1, A^{\mathrm{cl}}\psi_1 \rangle.$$

Let us remark that by the starting assumption, the coupling constant is necessarily negative, $\infty \neq \alpha_1 < 0$ because $\langle \psi_1, A^{\mathrm{cl}}\psi_1 \rangle > 0$ and $-E \geq 0$. According to the constructions of the previous subsection, A_1 is a weakly singular rank-one perturbation of A that solves the problem $A_1\psi_1 = E_1\psi_1$ and in addition, it is uniquely determined by the given vector $\psi_1 \in \mathcal{H}_1(A) \setminus \mathfrak{D}(A)$ and by the eigenvalue $E_1 \leq 0$. Let us remark that A_1 can be decomposed as

$$A_1 = E_1 P_{\psi_1} \oplus A_1^{\perp},$$

where P_{ψ_1} denotes the orthogonal projection onto the vector ψ_1, and $A_1^{\perp}$ denotes the part of A that acts on the subspace that is orthogonal to ψ_1. It is clear that $A_1^{\perp} \geq 0$.

In the second step we construct a weakly singular perturbation of the operator A_1 by

$$A_2 = A_1 \tilde{+} \alpha_2 \langle \cdot, \omega_2^1 \rangle \omega_2^1, \quad \omega_2^1 = (A_1^{\mathrm{cl}} - E_2)\psi_2,$$

where

$$\omega_2^1 = (A_1^{\mathrm{cl}} - E_2)\psi_2 \in \mathcal{H}_{-1}(A_1),$$

since $\psi_2 \in \mathcal{H}_1(A)$ and the norm in $\mathcal{H}_1(A)$ is equivalent to the norm in $\mathcal{H}_1(A_1)$, and where

$$\alpha_2 = -\frac{1}{\langle \psi_2, (A_1^{\mathrm{cl}} - E_2)\psi_2 \rangle} = -\frac{1}{a_{22}^1 - E_2}.$$

We notice that

$$a_{22}^1 := \langle \psi_2, A_1^{\mathrm{cl}}\psi_2 \rangle = \langle \psi_2, A^{\mathrm{cl}}\psi_2 \rangle + \alpha_1 |\langle \psi_2, \omega_1 \rangle|^2 = a_{22}^0 - \frac{1}{a_{11}^0 - E_1}|a_{21}^0|^2,$$

$$a_{21}^0 := \langle \psi_2, A\psi_1 \rangle,$$

and hence

$$\alpha_2 = -\frac{1}{a_{22}^0 - E_2 - \dfrac{|a_{21}^0|^2}{a_{11}^0 - E_1}}.$$

According to the construction, A_2 solves the problem $A_2\psi_2 = E_2\psi_2$, and moreover, solves also the problem $A_2\psi_1 = E_1\psi_1$, since $\psi_1 \perp \psi_2$.

So, by the construction and Theorem 9.1.2, the operator A_2 is uniquely determined by the vectors $\psi_j \in \mathcal{H}_1(A)$ and the numbers $E_j \leq 0$, $j = 1, 2$, is the singularly rank-two perturbation of the operator A, and which solves the eigenvalue problem, namely $A_2\psi_j = E_j\psi_j$, $j = 1, 2$. Let us remark that A_2 can be written as

$$A_2 = E_1 P_{\psi_1} \oplus E_2 P_{\psi_2} \oplus A_2^{\perp},$$

where P_{ψ_j}, $j = 1, 2$, denote orthogonal projections onto the vectors ψ_j, and $A_2^{\perp}$ denotes the part of the operator A_2 that acts on the subspace that is orthogonal to the vectors ψ_1 and ψ_2. It is obvious that A_2, as a rank-two perturbation of A, has the representation

$$A_2 = A \tilde{+} T_2,$$

where the singular on $\mathcal{H}$ rank-two operator $T_2 : \mathcal{H}_1(A) \mapsto \mathcal{H}_{-1}(A)$, has the representation

$$T_2 = \alpha_1 \langle \cdot, \omega_1^0 \rangle \omega_1^0 + \alpha_2 \langle \cdot, \omega_2^1 \rangle \omega_2^1.$$

This operator can also be represented as

$$T_2 = \sum_{j,k=1}^{2} t_{jk} \langle \cdot, \omega_j \rangle \omega_k, \quad \omega_j := (A^{\mathrm{cl}} - E_j)\psi_j,$$

where

$$t_{11} = \alpha_1 + \alpha_2(\alpha_1)^2 |a_{21}^0|^2, \; t_{12} = \alpha_1\alpha_2 a_{21}^0, \; t_{21} = \alpha_1\alpha_2 a_{12}^0, \; t_{22} = \alpha_2.$$

Therefore,

$$A_2 = A \tilde{+} T_2 = A \tilde{+} \sum_{j,k=1}^{2} t_{jk} \langle \cdot, \omega_j \rangle \omega_k.$$

One can continue this construction for an arbitrary finite number of steps. On the nth step we construct the weakly singular rank-one perturbation of the operator A_{n-1},

$$A_n = A_{n-1} \tilde{+} \alpha_n \langle \cdot, \omega_n^{n-1} \rangle \omega_n^{n-1},$$

where

$$\omega_n^{n-1} = (A_{n-1}^{\mathrm{cl}} - E_n)\psi_n \in \mathcal{H}_{-1}(A_{n-1}),$$

(since the norms in $\mathcal{H}_1(A)$ and in $\mathcal{H}_1(A_{n-1})$ are equivalent and $\psi_n \in \mathcal{H}_1(A)$), and where

$$\begin{aligned} \alpha_n &= -\frac{1}{\langle \psi_n, \omega_n^{n-1} \rangle} = -\frac{1}{a_{nn}^{n-1} - E_n}, \\ a_{nn}^{n-1} &:= \langle \psi_n, A_{n-1}^{\mathrm{cl}} \psi_n \rangle = \langle \psi_n, A_{n-2}^{\mathrm{cl}} \psi_n \rangle + \alpha_{n-1} |\langle \psi_n, \omega_{n-1}^{n-2} \rangle|^2 \\ &\equiv a_{nn}^{n-2} + \alpha_{n-1} |a_{n,n-1}^{n-2}|^2, \end{aligned}$$

with

$$a_{n,n-1}^{n-2} = \langle \psi_n, A_{n-2}^{\mathrm{cl}} \psi_{n-1} \rangle.$$

Therefore,

$$\alpha_n = -\cfrac{1}{a_{nn}^{n-2} - E_n - \cfrac{|a_{n,n-1}^{n-2}|^2}{a_{n-1,n-1}^{n-3} - E_{n-1} - \cdots - \cfrac{|a_{21}^0|^2}{a_{11}^0 - E_1}}}.$$

We put

$$A_n = A \tilde{+} \alpha_1 \langle \cdot, \omega_1 \rangle \omega_1 \tilde{+} \cdots \tilde{+} \alpha_n \langle \cdot, \omega_n^{n-1} \rangle \omega_n^{n-1}. \tag{9.46}$$

According to Theorem 9.2.2, the operator A_n is uniquely determined by the vector $\psi_n \in \mathcal{H}_1(A)$ and the number $E_n \leq 0$ and it is a weakly singular rank-one perturbation of A_{n-1}, which solves the eigenvalue problem, namely $A_n \psi_n = E_n \psi_N$. Moreover, by induction, A_n is uniquely determined by the vectors $\psi_j \in \mathcal{H}_1(A)$ and by the numbers $E_j \leq 0$, $j = 1, \ldots, n$, and it is a weakly singular rank-n perturbation of the operator A, which solves the eigenvalue problem $A_n \psi_j = E_j \psi_j$ for all $E_j \leq 0$.

Let us remark that A_n can be written as

$$A_n = E_1 P_{\psi_1} \oplus \cdots \oplus E_n P_{\psi_n} \oplus A_n^{\perp},$$

where P_{ψ_j}, $j = 1, \ldots, n$, denote the orthogonal projections onto the vectors ψ_j, and $A_N^{\perp}$ is the part of A_n acting on the subspace that is orthogonal to all vectors ψ_j.

The obtained result is next formulated as a theorem, where one denotes $N = n$.

Theorem 9.2.3. *Let there be given a set of orthonormal in $\mathcal{H}$ vectors $\psi_j \in \mathcal{H}_1(A) \setminus \mathfrak{D}(A)$, $j = 1, \ldots, N$ such that $\operatorname{span}\{\psi_j\}_{j=1}^N \cap \mathfrak{D}(A) = \{0\}$, and a sequence of arbitrary numbers $E_j \leq 0$. Let $\tilde{A}$ be a rank-N weakly singular perturbation of the self-adjoint operator $A \geq 0$, namely,*

$$\tilde{A} \equiv A_N = A \tilde{+} T_N,$$

where $T_N : \mathcal{H}_1(A) \mapsto \mathcal{H}_{-1}(A)$ is a singular rank-N operator. Then the operator $\tilde{A}$ solves the eigenvalue problem

$$\tilde{A} \psi_j = E_j \psi_j, \quad E_j \leq 0, \quad j = 1, \ldots, N,$$

if and only if the operator T_N *has the representation*

$$T_N = \sum_{j=1}^{N} \alpha_j \langle \cdot, \omega_j^{j-1} \rangle \omega_j^{j-1}, \tag{9.47}$$

where α_j *are the coupling constants and the elements* ω_j^{j-1} *are uniquely determined by the vectors* ψ_j *and the numbers* E_j*, in accordance with the recursion formulas:*

$$\omega_j^{j-1} := (A_{j-1}^{\text{cl}} - E_j)\psi_j, \tag{9.48}$$

$$\alpha_j := -\frac{1}{\langle \psi_j, \omega_j^{j-1} \rangle}. \tag{9.49}$$

Thus, we have a one-to-one correspondence between the set of given vectors $\psi_j \in \mathcal{H}_1(A) \backslash \mathfrak{D}(A)$ such that $\text{span}\{\psi_j\}_{j=1}^N \cap \mathfrak{D}(A) = \{0\}$ and orthonormal in $\mathcal{H}$, and numbers $E_j \le 0$, $j = 1, \ldots, N$, on one side, and the set of rank-N weakly singularly perturbed operators $\tilde{A}$ which solve the eigenvalue problem. This correspondence is given by the formulas (9.47), (9.48), and (9.49).

Proof. We use inductively Theorem 9.2.2. In one direction, the statement of the theorem follows from the previously described construction of the operator A_n, $n = N$ (see (9.46)). In the opposite direction, the theorem was already proved in the case $N = 1$ (Theorem 9.2.2). In the general case follows from $N > 1$, one can use the following consideration. Let us suppose that $\tilde{A}$ is a rank-N weakly singular perturbation of the operator $A \ge 0$, namely $\tilde{A} = A \tilde{+} T$ with some $\mathcal{H}$ that is a singular rank-N symmetric operator $T : \mathcal{H}_1(A) \mapsto \mathcal{H}_{-1}(A)$. We assume that $\tilde{A}$ solves an eigenvalue problem. Then

$$(A_N - \tilde{A})\psi_j = 0, \quad j = 1, \ldots, N,$$

where $A_{N=n}$ is defined by (9.46), and thus

$$T_N \psi_j = T\psi_j = -\omega_j^{j-1}.$$

This means that $T_N = T$, since all vectors ω_j^{j-1} are linearly independent according to the construction and both operators T_N and T have rank-N and are self-adjoint in the couple of spaces. Thus, $\tilde{A} = A_N$. □

Let us remark that the required rank-N operator T can be constructed in a different way. Let us put

$$T = \sum_{i,j=1}^{N} t_{ij} \langle \cdot, \omega_i \rangle \omega_j, \quad \omega_j := (A^{\text{cl}} - E_j)\psi_j.$$

We claim that $\tilde{A} = A \tilde{+} T$ solves the eigenvalue problem, if $(t_{ij})_{i,j=1}^N$ is up to a sign the inverse matrix of

$$\mathbf{a} = (a_{kj})_{k,j=1}^N, \quad a_{kj} := \langle \psi_k, (A^{\text{cl}} - E_j)\psi_j \rangle.$$

Indeed, the equality $\tilde{A}\psi_k = E_k\psi_k$ can be rewritten in the form

$$A\psi_k + \sum_{i,j=1}^{N} t_{ij}\langle\psi_k,\omega_i\rangle\omega_j = E_k\psi_k,$$

or

$$\sum_{i,j=1}^{N} t_{ij}a_{ki}\omega_j = \sum_{j=1}^{N}\alpha_{jk}\omega_j = -(A^{\mathrm{cl}} - E_k)\psi_k = -\omega_k,$$

where $\alpha_{jk} = \sum_{i=1}^{N} t_{ij}a_{ki}$. Since the vectors ω_j, $j = 1,\dots,N$ are linearly independent, we conclude that the equality $\tilde{A}\psi_k = E_k\psi_k$, $k = 1,\dots,N$ is fulfilled if and only if $\alpha_{jk} = -\delta_{jk}$. This means that $t_{ij} = -(\mathbf{a}^{-1})_{ij}$, where $\mathbf{a}^{-1}$ denotes the inverse to matrix of $\mathbf{a}$. Thus, if we put

$$T = \sum_{i,j=1}^{N}(-1)(\mathbf{a}^{-1})_{ij}\langle\cdot,\omega_i\rangle\omega_j,$$

then $\tilde{A} = A\tilde{+}T$ solves the eigenvalue problem.

9.2.4 The inductive method in the inverse eigenvalue problem

Let us suppose, as before, that one is given the operator $A \geq 0$, the sequence of vectors $\psi_j \in \mathcal{H}_1(A)\backslash\mathfrak{D}(A)$ orthonormal in $\mathcal{H}$ such that $\mathrm{span}\{\psi_j\}_{j=1}^{N}\cap\mathfrak{D}(A) = \{0\}$, and the sequence of numbers $E_j < 0$, $j = 1,\dots,N$.

Here we proceed in a slightly different manner than in the previous subsections to construct explicitly the operator $T : \mathcal{H}_1 \mapsto \mathcal{H}_{-1}$ such that $\tilde{A} = A\tilde{+}T$ is self-adjoint and solves the eigenvalue problem

$$\tilde{A}\psi_j = E_j\psi_j, \quad E_j < 0, \quad j = 1,\dots,N. \tag{9.50}$$

Similarly to the previous subsection, in the next construction we consistently use at each step a singular rank-one perturbation of the form

$$\alpha\langle\cdot,\omega\rangle\omega : \mathcal{H}_1 \longrightarrow \mathcal{H}_{-1}, \quad 0 \neq \alpha \in \mathbb{R},$$

where

$$\omega \in \mathcal{N}_{-1}, \;\; \mathcal{N}_{-1} := \mathrm{span}\{(A^{\mathrm{cl}} - E_k)\psi_k\}_{k=1}^{N} \subset \mathcal{H}_{-1}.$$

We introduce the following notation:

$$A_0^k \equiv A, \quad k = 1,\dots,N$$

and we introduce the operator

$$A_1^k := A_0^k \tilde{+} t_{1k}\langle\cdot,\omega_1^k\rangle\omega_1^k,$$

where

$$t_{1k} = -\frac{1}{\langle\psi_1, \omega_1^k\rangle}, \quad \omega_1^k = ((A_0^k)^{\mathrm{cl}} - NE_k\delta_{1k})\psi_1 \in \mathcal{H}_{-1}.$$

Here we use the same notations as in the previous subsection for different vectors $\omega \in \mathcal{N}_{-1}$; in particular $(\cdot)^{\mathrm{cl}}$ denotes the closure of the relevant operator as a mapping from $\mathcal{H}_1$ into $\mathcal{H}_{-1}$. Analogously, we define the operator

$$A_j^k := A_{j-1}^k \tilde{+} t_{1k}\langle\cdot, \omega_j^k\rangle\omega_j^k, \quad j = 1, \ldots, N, \tag{9.51}$$

where

$$t_{1k} = -\frac{1}{\langle\psi_j, \omega_j^k\rangle}, \quad \omega_j^k = ((A_{j-1}^k)^{\mathrm{cl}} - NE_k\delta_{jk})\psi_j \in \mathcal{H}_{-1}.$$

Let us remark that the operator A_j^k can be represented as

$$A_j^k = A \tilde{+} T_j^k,$$

where T_j^k is a self-adjoint operator of rank-j which acts from $\mathcal{H}_1$ into $\mathcal{H}_{-1}$ (we can easily find its exact range). According to the definition,

$$A^{(k)} := A_{j=N}^k = A \tilde{+} T^{(k)}, \tag{9.52}$$

where

$$T^{(k)} := T_{j=N}^k = \sum_{j=1}^N t_{jk}\langle\cdot, \omega_j^k\rangle\omega_j^k.$$

We define

$$\tilde{A} := \frac{1}{N}\sum_{k=1}^N A^{(k)}, \tag{9.53}$$

and also $T : \mathcal{H}_1 \mapsto \mathcal{H}_{-1}$ as

$$T = \frac{1}{N}\sum_{k=1}^N T^{(k)} = \frac{1}{N}\sum_{k,j=1}^N t_{kj}\langle\cdot, \omega_j^k\rangle\omega_j^k, \tag{9.54}$$

where the vectors $\omega_j^k \in \mathcal{H}_{-1}$ are given above and all the coefficients

$$t_{jk} = -\frac{1}{\langle\psi_j, \omega_j^k\rangle} = -\frac{1}{\langle\psi_j, (A_{j-1}^k)^{\mathrm{cl}} - NE_k\delta_{jk})\psi_j\rangle} < \infty. \tag{9.55}$$

The purpose of this subsection is the proof of the following result.

Theorem 9.2.4. *Suppose given a sequence of orthonormal vectors $\psi_j \in \mathcal{H}_1(A) \setminus \mathfrak{D}(A)$ such that $\mathrm{span}\{\psi_j\}_{j=1}^N \cap \mathfrak{D}(A) = \{0\}$ in $\mathcal{H}$, and a sequence of negative numbers $E_j < 0$, $j = 1, \ldots, N$. Then the self-adjoint operator $\tilde{A}$, constructed in accordance with the prescription (9.51)–(9.55), solves the eigenvalue problem (9.50). In addition, if we suppose that there exists a vector ψ such that $\tilde{A}\psi = E\psi$, $E < 0$, and $E \neq E_j$, then $\psi \equiv 0$.*

We begin the proof by the use of the next statement.

Proposition 9.2.5. *The operator* $\tilde{A}$ *defined by* (9.53) *coincides with the generalized operator sum*

$$\tilde{A} = A \tilde{+} T. \tag{9.56}$$

Proof. Using previous notations and definitions (see (9.51)–(9.55)), we have

$$\begin{aligned} A^{(1)} \equiv A^1_{j=N} &= A^1_{j=N-1} \tilde{+} t_{N1} \langle \cdot, \omega^1_N \rangle \omega^1_N \\ &= A^1_{N-2} \tilde{+} t_{N-1,1} \langle \cdot, \omega^1_{N-1} \rangle \omega^1_{N-1} \tilde{+} t_{N1} \langle \cdot, \omega^1_N \rangle \omega^1_N \\ &= A \tilde{+} \sum_{j=1}^{N} t_{j1} \langle \cdot, \omega^1_j \rangle \omega^1_j. \end{aligned}$$

Analogously, for each $k \geq 1$,

$$\begin{aligned} A^{(k)} \equiv A^k_{j=N} &= A^k_{j=N-1} \tilde{+} t_{Nk} \langle \cdot, \omega^k_N \rangle \omega^k_N \\ &= A \tilde{+} \sum_{j=1}^{N} t_{jk} \langle \cdot, \omega^k_j \rangle \omega^k_j \equiv A \tilde{+} T^{(k)}. \end{aligned}$$

Thus,

$$\tilde{A} = \frac{1}{N} \sum_{k=1}^{N} A^{(k)} = A \tilde{+} \frac{1}{N} \sum_{k=1}^{N} T^{(k)} = A \tilde{+} T. \qquad \square$$

Let, as above, denote the closures of the operators A and A^k_j, as mappings from $\mathcal{H}_1$ into $\mathcal{H}_{-1}$, be

$$\begin{aligned} &(A^k_0)^{\mathrm{cl}} : \mathcal{H}_1 \longrightarrow \mathcal{H}_{-1}, \\ &(A^k_j)^{\mathrm{cl}} : \mathcal{H}_1 \longrightarrow \mathcal{H}_{-1}, \quad k, j = 1, \ldots, N. \end{aligned}$$

Proposition 9.2.6. *All operators* $(A^k_0)^{\mathrm{cl}}$, $(A^k_j)^{\mathrm{cl}}$ *are self-adjoint in the sense of the couple of spaces.*

Proof. Let us recall that

$$A^k_0 \equiv A, \quad A^k_j = A^k_{j-1} \tilde{+} t_{jk} \langle \cdot, \omega^k_j \rangle \omega^k_j.$$

According to the definition, all these operators, regarded as mappings from $\mathcal{H}_1$ into $\mathcal{H}_{-1}$, are densely defined, symmetric, and bounded. So, their closures are self-adjoint. $\square$

Proposition 9.2.7. *If* $l \leq j < k$, *then*

$$A^k_j \psi_l = 0.$$

Proof. According to the construction, for $j = 1$ and any $k > 1$, we have

$$A_1^k\psi_1 = (A_1^k)^{\text{cl}}\psi_1 = (A_1^k)^{\text{cl}}\psi_1 - \frac{1}{\langle\psi_1, A^{\text{cl}}\psi_1\rangle}\langle\psi_1, A^{\text{cl}}\psi_1\rangle A^{\text{cl}}\psi_1 = 0. \tag{9.57}$$

Analogously, for $j = 1$ and any $k > 2$,

$$A_2^k\psi_2 = (A_2^k)^{\text{cl}}\psi_2 = (A_1^k)^{\text{cl}}\psi_2 - \frac{1}{\langle\psi_2, (A_1^k)^{\text{cl}}\psi_2\rangle}\langle\psi_2, (A_1^k)^{\text{cl}}\psi_2\rangle(A_1^k)^{\text{cl}}\psi_2 = 0,$$

and also

$$A_2^k\psi_1 = (A_2^k)^{\text{cl}}\psi_1 = (A_1^k)^{\text{cl}}\psi_1 - \frac{1}{\langle\psi_2, (A_1^k\psi_2\rangle}\langle\psi_1, (A_1^k)^{\text{cl}}\psi_2\rangle(A_1^k)^{\text{cl}}\psi_2 = 0,$$

since $(A_1^k)^{\text{cl}}\psi_1 = 0$, and

$$\langle\psi_1, (A_1^k)^{\text{cl}}\psi_2\rangle = \langle(A_1^k)^{\text{cl}}\psi_1, \psi_2\rangle = 0,$$

due to (9.57), where we used the self-adjoint operator $(A_1^k)^{\text{cl}}$ in the A-scale (see the previous proposition). By induction, for an arbitrary $l \le j < k$, we have

$$A_j^k\psi_l = (A_j^k)^{\text{cl}}\psi_l = (A_{j-1}^k)^{\text{cl}}\psi_l - \frac{1}{\langle\psi_j, (A_{j-1}^k)^{\text{cl}}\psi_j\rangle}\langle\psi_l, (A_{j-1}^k)^{\text{cl}}\psi_j\rangle(A_{j-1}^k)^{\text{cl}}\psi_j = 0,$$

since $(A_{j-1}^k)^{\text{cl}}\psi_l = 0$, if $l \le j-1 < k$, and

$$\langle\psi_l, (A_{j-1}^k)^{\text{cl}}\psi_j\rangle = \langle(A_{j-1}^k)^{\text{cl}}\psi_l, \psi_j\rangle = 0$$

too, where we used that the operators $(A_{j-1}^k)^{\text{cl}}$ are self-adjoint in the A-scale. □

Proposition 9.2.8. *Let $l \le j = k$, then*

$$A_j^k\psi_l = \delta_{jk}NE_k\psi_l.$$

Proof. Let $l < j = k$. Then, similarly to the previous arguments, we have

$$\begin{aligned} A_j^k\psi_l &= (A_j^k)^{\text{cl}}\psi_l \\ &= (A_{j-1}^k)^{\text{cl}}\psi_l - \frac{1}{\langle\psi_{k-1}, (A_{k-1}^k)^{\text{cl}}\psi_{k-1}\rangle}\langle\psi_l, (A_{k-1}^k)^{\text{cl}}\psi_{k-1}\rangle(A_{k-1}^k)^{\text{cl}}\psi_{k-1} = 0, \end{aligned}$$

since

$$\langle\psi_l, (A_{k-1}^k)^{\text{cl}}\psi_{k-1}\rangle = \langle(A_{k-1}^k)^{\text{cl}}\psi_l, \psi_{k-1}\rangle$$

and $(A_{k-1}^k)^{\text{cl}}\psi_l = 0$, $l \le j = k-1$. In the case where $l = j = k$, we have

$$\begin{aligned} A_k^k\psi_k &= (A_{k-1}^k)^{\text{cl}}\psi_k \\ &= A_{k-1}^k\psi - \frac{1}{\langle\psi_k, ((A_{k-1}^k)^{\text{cl}} - NE_k)\psi_k\rangle} \\ &\quad \times \langle\psi_k, ((A_{k-1}^k)^{\text{cl}} - NE_k)\psi_k\rangle((A_{k-1}^k)^{\text{cl}}v - NE_k)\psi_k \\ &= (A_{k-1}^k)^{\text{cl}}\psi_k - ((A_{k-1}^k)^{\text{cl}} - NE_k)\psi_k = NE_k\psi_k. \end{aligned}$$

□

Proposition 9.2.9. *Let $k < j$. Then*

$$A_j^k \psi_k = N E_k \psi_k, \quad \text{if} \quad k < j, \tag{9.58}$$

$$A_j^k \psi_l = 0, \qquad \text{if} \quad l \le j, \quad l \ne k. \tag{9.59}$$

Proof. According to the definition, we have

$$\begin{aligned} A_j^k \psi_k &= (A_j^k)^{\text{cl}} \psi_k \\ &= (A_{j-1}^k)^{\text{cl}} \psi_k - \frac{1}{\langle \psi_j, ((A_{j-1}^k)^{\text{cl}} - N E_k \delta_{j-1,k}) \psi_j \rangle} \\ &\quad \times \langle \psi_k, ((A_{j-1}^k)^{\text{cl}} - N E_k \delta_{j-1,k}) \psi_j \rangle ((A_{j-1}^k)^{\text{cl}} - N E_k \delta_{j-1,k}) \psi_j. \end{aligned}$$

We suppose $k = j - 1$. Then obviously $A_{j-1}^k \psi_k = N E_k \psi_k$. In particular,

$$\langle \psi_k, (A_{j-1}^k)^{\text{cl}} \psi_j \rangle = \langle (A_{j-1}^k)^{\text{cl}} \psi_k, \psi_j \rangle = E_k \langle \psi_k, \psi_j \rangle = 0,$$

since $\psi_k \perp \psi_j, k \ne j$. Therefore, $A_j^k \psi_k = N E_k \psi_k$, if $j = k + 1$. A similar relation holds for each $j > k$ by induction, i.e., (9.58) is proved. To show (9.59) it is necessary only to consider the case $l \le j$, $l > k$, since the case $l < k$ was already proved in the previous proposition.

In the first step we put $j = k + 1 = l$. Then we have

$$\begin{aligned} A_{k+1}^k \psi_k &= (A_{k+1}^k)^{\text{cl}} \psi_k \\ &= A_k^k \psi_k - \frac{1}{\langle \psi_{k+1}, (A_k^k)^{\text{cl}} \psi_{k+1} \rangle} \langle \psi_{k+1}, (A_k^k)^{\text{cl}} \psi_{k+1} \rangle (A_k^k)^{\text{cl}} \psi_{k+1} = 0. \end{aligned}$$

In the second step we take $j = k + 2$ and consider $l = k + 1, k + 2$. By direct calculation we obtain

$$\begin{aligned} A_{k+2}^k \psi_{k+1} &= (A_{k+2}^k)^{\text{cl}} \psi_{k+1} \\ &= A_{k+1}^k \psi_{k+1} - \frac{1}{\langle \psi_{k+2}, (A_{k+1}^k)^{\text{cl}} \psi_{k+2} \rangle} \langle \psi_{k+1}, (A_{k+1}^k)^{\text{cl}} \psi_{k+2} \rangle (A_{k+1}^k)^{\text{cl}} \psi_{k+2} = 0, \end{aligned}$$

since

$$\langle \psi_{k+1}, (A_{k+1}^k)^{\text{cl}} \psi_{k+2} \rangle = \langle (A_{k+1}^k)^{\text{cl}} \psi_{k+1}, \psi_{k+2} \rangle = 0$$

and $A_{k+1}^k \psi_{k+1} = 0$ similarly to the previous arguments. In the case $l = k + 2$, we have

$$\begin{aligned} A_{k+2}^k \psi_{k+2} &= (A_{k+2}^k)^{\text{cl}} \psi_{k+2} \\ &= A_{k+1}^k \psi_{k+2} - \frac{1}{\langle \psi_{k+2}, (A_{k+1}^k)^{\text{cl}} \psi_{k+2} \rangle} \langle \psi_{k+2}, (A_{k+1}^k)^{\text{cl}} \psi_{k+2} \rangle (A_{k+1}^k)^{\text{cl}} \psi_{k+2} = 0. \end{aligned}$$

Similarly, we get the same expressions for the indices $j = k + 3$, $l = k + 1$, $k + 2$, $k + 3$. And so on. □

As a corollary we get

Proposition 9.2.10.

$$A^{(k)}\psi_l = NE_k\delta_{kl}\psi_k, \qquad \textit{if} \quad k,l = 1,\ldots,N. \tag{9.60}$$

Proof. To test this claim, it is convenient to represent the operators in question in tabular form:

$$\begin{pmatrix} A_0^1 = A \\ A_2^1 = A \\ \vdots \\ A_N^1 = A \end{pmatrix} \begin{pmatrix} A_1^1 & A_2^1 & \cdots & A_N^1 = A^{(1)} \\ A_1^2 & A_2^2 & \cdots & A_N^2 = A^{(2)} \\ \vdots & \vdots & \ddots & \vdots \\ A_1^N & A_2^N & \cdots & A_N^N = A^{(N)} \end{pmatrix}$$

and successively verify this statement for each operator by using the already proved statements. Note that

$$A_1^2 = A_1^3 = \cdots = A_1^N, \quad A_2^3 = A_2^4 = \cdots = A_2^N, \ldots,$$

and so on. □

We emphasize that each of the operators $A^{(k)}$ is independent of the order, in which the couples $\{\psi_j, \delta_{jk}NE_j\}_{j=1}^N$, are used in the construction of this operators. In other words,

$$A^{(k)} = A_{j=N}^k = A_{P(j)=N}^k,$$

where $P : \{1,\ldots,N\} \mapsto \{i_1,\ldots,i_N\}$ denotes a permutation of the indices $\{1, \ldots, N\}$ that preserves an index k. Thus, every $A^{(k)}$ can be written in the form

$$A^{(k)} = A^{(k,0)} \tilde{+}\, t_{kk}^0 \langle \cdot, \omega_2^{k,0}\rangle \omega_2^{k,0},$$

where $A^{(k,0)}$ is constructed by the inverse eigenvalue problem method using the sequence of couples $\{\psi_j, 0\}_{j=1,j\neq k}^N$, and

$$\omega_2^{k,0} = ((A^{(k,0)})^{\mathrm{cl}} - NE_k)\psi_k, \quad t_{kk}^0 = -\frac{1}{\langle \psi_k, \omega_2^{k,0}\rangle}.$$

We used here the additivity property of the generalized sum. It is clear that each $A^{(k,0)}$ is positive,

$$A^{(k,0)} \geq 0, \tag{9.61}$$

and, in addition, it solves the problem

$$A^{(k,0)}\psi_j = 0, \quad j \neq k. \tag{9.62}$$

Proof of Theorem 9.2.4. It follows directly from (9.53), (9.56) and from previous proposition we obtain that $\tilde{A}$ is a self-adjoint operator and (9.50) is valid. Thus, we have to prove that the operator $\tilde{A}$, defined in 9.53, has no negative eigenvalues.

Let us suppose the contrary, namely, that $\tilde{A}\psi = E\psi$, $E < 0$ for some vector ψ. It is clear that $\psi \notin \mathcal{N}_0$, $\mathcal{N}_0 := \operatorname{span}\{\psi_k\}_{k=1}^N$ and, moreover, the vector ψ must be orthogonal to this subspace. But each operator $A^{(k)}$ is positive on the subspace $\mathcal{H} \ominus \mathcal{N}_0$ (see (9.61), (9.62)). Then $\psi = 0$ and $\tilde{A}$ solves the problem (9.50) in the exact sense. □

Let us remark that the operator T, defined in (9.54), has the property that

$$\operatorname{rank} T \leq 2N.$$

Inded, according to the construction, all elements ω_j^k in (9.54) belong to the subspace

$$\mathcal{N}_{-1} = \operatorname{span}\{(A)^{\mathrm{cl}}\psi_k, ((A)^{\mathrm{cl}} - NE_j)\psi_j\}_{k,j=1}^N.$$

In other words, for the range of the operator T we have

$$\operatorname{Ran} T \subset \mathcal{N}_{-1} \subset \mathcal{H}_{-1}.$$

Clearly, $\dim \mathcal{N}_{-1} \leq 2N$. Then $\operatorname{rank} T \leq 2N$ too.

Another exact solution of the problem (9.35) for the operator T of rank-$2N$ can be obtained as follows. Suppose for the moment that all vectors $\psi_j \in \mathfrak{D}(A)$. Then in $\mathcal{H}$ we fix the N-dimensional subspace $\mathcal{N} = \operatorname{span}\{\psi_j\}_{j=1}^N$ and the operator

$$T' = \sum_{i=1}^N E_i(\cdot, \psi_i)\psi_i.$$

Let $P = \sum_{i=1}^N (\cdot, \psi_i)\psi_i$ be the orthogonal projection onto $\mathcal{N}$ in $\mathcal{H}$. Then the operator

$$\begin{aligned} \tilde{A} &= (I-P)A(I-P) + T' = A + T, \\ T &= PAP - PA - AP + T', \end{aligned}$$

obviously solves the problem (9.35).

In the general case, if $\psi_j \in \mathcal{H}_1(A)$, then the exact solution of the problem 9.35 can be expressed as a sum $\tilde{A} = A \tilde{+} T$, where T has the form

$$T = T' + \sum_{i=1}^N \left[\sum_{k=1}^N (\cdot, \psi_k)\langle (A)^{\mathrm{cl}}\psi_k, \psi_i\rangle\psi_i - \langle \cdot, (A)^{\mathrm{cl}}\psi_i\rangle\psi_i - \langle \cdot, \psi_i\rangle (A)^{\mathrm{cl}}\psi_i \right].$$

Finally, we note that each operator $\tilde{A}$ that solves the eigenvalue problem with negative values (9.35), in our construction (see (9.53)–(9.56)), has very specific properties. Namely, each operator $A^{(k)}$, $k = 1, \ldots, N$ solves such a problem for negative eigenvalues: $A^{(k)}\psi_j = E'_j\,\psi_j$, where the vectors ψ_j are the same as in (9.35), while $E'_k = NE_k$ and $E'_j = 0$, if $j \neq k$. Let us remark that each $A^{(k)}$ is a rank-N perturbation of the operator A (see (9.51), (9.52)), and according to our construction $A^{(k)}$ is the last of A_j^k for $j \to k$, where at each step the operator

A_j^k (the rank-j perturbation of the operator A) also solves the negative eigenvalue problem. Moreover, each operator of the form

$$A^{(k_1,\dots,k_i)} = A^{(k_1)} + \dots + A^{(k_i)}, \quad k_i \le N$$

also solves the negative eigenvalue problem, $A^{(k_1,\dots,k_i)}\psi_j = E_j'\psi_j$, where $E_j' = NE_j$, if at least one of the indices $k_1, \dots, k_i$ coincides with j, and $E_j' = 0$ in the other case.

Example 9.2.11 (The eigenvalue problem for super-singular perturbations). Let us consider a super-singular rank-one perturbation of the operator A on $\mathcal{H}$, given by the quadratic form

$$\gamma_\omega[\varphi] = \langle \varphi, \omega\rangle_{k,-k}\langle \omega, \varphi\rangle_{-k,k}, \quad k > 2,$$

where $\omega \in \mathcal{H}_{-k}\backslash\mathcal{H}_{k/2}$. The associated operator with γ_ω

$$S_\omega : \mathcal{H}_k \ni \varphi \longmapsto \langle \varphi, \omega\rangle_{k,-k}\omega \in \mathcal{H}_{-k}$$

belongs to $\mathcal{S}_{-k}$-class, since the set

$$\operatorname{Ker} S_\omega = \operatorname{Ker}\gamma_\omega = \{\varphi \in \mathcal{H}_k \mid \langle \varphi, \omega\rangle_{k,-k} = 0\}$$

is dense in $\mathcal{H}_{-k/2}$ because $\omega \notin \mathcal{H}_{-k/2}$. In particular, if $k > 4$, then $\operatorname{Ker} S_\omega$ is dense in $\mathcal{H}_2 = \operatorname{Dom} A$, and the usual method cannot be used for the construction of a perturbed operator $\tilde{A}$. We propose to consider γ_ω or S_ω as a perturbation of $A^{k/2} : \mathcal{H}_k \mapsto \mathcal{H}$. Thus, we define $\tilde{A}$ as $(A^{-k/2} + B_\omega)^{2/k}$, where B_ω is the rank-one operator on $\mathcal{H}$ given by

$$B_\omega = \beta_\omega(\cdot, \eta_0)\eta_0, \eta_0 := \mathbf{A}^{-k/2}\omega,$$

with the corresponding coupling constant β_ω. Namely, if $\beta_\omega = 1 - c_\omega$ (which corresponds to the representation $b = 1 - s$), then the constant c_ω must satisfy the equality $\|\eta_0\|_{-1}^2 \le c_\omega < 1$. Only in this case we have $\tilde{A} \ge 1$ (see Example 3.1 in [1]).

Let

$$\omega = c_\omega(\mathbf{A}^{k/2}\psi - \lambda\psi), \ \psi \in \mathcal{H} \setminus \mathcal{H}_{k/2}, \ \|\psi\| = 1, \ \lambda \in \mathbb{R}.$$

Then, the Kreĭn formula for resolvents yields

$$(\tilde{A}^{k/2} - z)^{-1} = (A^{k/2} - z)^{-1} + B_\omega(z),$$

where

$$B_\omega(z) = \beta_{\omega,z}(\cdot, \eta_{\bar{z}})\eta_z,$$

and

$$\eta_z = (A^{k/2} - \lambda)(A^{k/2} - z)^{-1}\psi, \quad \beta_{\omega,z} = \frac{1}{(\lambda - z)(\psi, \eta_{\bar{z}})},$$

we need to solve the eigenvalue problem for the operator $\tilde{A}^{k/2} : \tilde{A}^{k/2}\psi = \lambda\psi$ (see [5, 6] in the case $k = 2$). Thus, the operator $\tilde{A}$ solves also the eigenvalue problem: $\tilde{A}\psi = \lambda^{2/k}\psi$.

References

[1] Albeverio S., Bozhok R., Dudkin M., Koshmanenko V. Dense subspaces in scales of Hilbert spaces. Methods Funct. Anal. Topology 11 (2005), no. 2, 156–169.

[2] Albeverio S., Bozhok R., Koshmanenko V. The rigged Hilbert spaces approach in singular perturbation theory. Rep. Math. Phys. 58 (2006), no. 2, 227–246.

[3] Albeverio S., Brasche J.F., Koshmanenko V. Lippmann-Schwinger equation in the singular perturbation theory. Methods Funct. Anal. Topology 3 (1997), no. 1, 1–27.

[4] Albeverio S., Brasche J., Neidhardt H. On inverse spectral theory for self-adjoint extensions: mixed types of spectra. J. Funct. Anal. 154 (1998), no. 1, 130–173.

[5] Albeverio S., Dudkin M., Koshmanenko V. Rank-one singular perturbations with a dual pair of eigenvalues. Lett. Math. Phys. 63 (2003), no. 3, 219–228.

[6] Albeverio S., Dudkin M., Konstantinov A., Koshmanenko V. On the point spectrum of H_{-2}-singular perturbations. Math. Nachr. 280 (2007), no. 1-2, 20–27.

[7] Albeverio S., Gesztesy F., Høegh-Krohn R., Holden H. Solvable Models in Quantum Mechanics. Second edition. With an appendix by Pavel Exner. AMS Chelsea Publishing, Providence, RI, 2005. xiv+488 pp.

[8] Albeverio S., Hryniv R., Mykytyuk Ya. Inverse spectral problems for Sturm-Liouville operators in impedance form. J. Funct. Anal. 222 (2005), no. 1, 143–177.

[9] Albeverio S., Karwowski W., Koshmanenko V. Square powers of singularly perturbed operators. Math. Nachr. 173 (1995), 5–24.

[10] Albeverio S., Karwowski W., Koshmanenko V. On negative eigenvalues of generalized Laplace operators. Rep. Math. Phys. 48 (2001), no. 3, 359–387.

[11] Albeverio S., Konstantinov A., Koshmanenko V. The Aronszajn–Donoghue theory for rank one perturbations of the H_2-class. Integral Equations Operator Theory 50 (2004), no. 1, 1–8.

[12] Albeverio S., Konstantinov A., Koshmanenko V. Decompositions of singular continuous spectra of H_2-class rank one perturbations. Integral Equations Operator Theory 52 (2005), no. 4, 455–464.

[13] Albeverio S., Konstantinov A., Koshmanenko V. On inverse spectral theory for singularly perturbed operators: point spectrum. Inverse Problems 21 (2005), no. 6, 1871–1878.

[14] Albeverio S., Konstantinov A., Koshmanenko V. Remarks on the inverse spectral theory for singularly perturbed operators. In: Modern Analysis and Applications. The Mark Kreĭn Centenary Conference. Vol. 1: Operator Theory and Related Topics, 115–122, Oper. Theory Adv. Appl., 190, Birkhäuser Verlag, Basel, 2009.

[15] Albeverio S., Koshmanenko V. Some remarks on the Simon–Gesztezy version of rank-one perturbations, BiBoS-Preprint, Nr. 727/5/96, (1996).

[16] Albeverio S., Koshmanenko V. Singular rank one perturbations of self-adjoint operators and Kreĭn theory of self-adjoint extensions. Potential Anal. 11 (1999), no. 3, 279–287.

[17] Albeverio S., Koshmanenko V., Makarov K.A. Generalized eigenfunctions under singular perturbations. Methods Funct. Anal. Topology 5 (1999), no. 1, 13–28.

[18] Albeverio S., Koshmanenko V. On form-sum approximations of singularly perturbed positive self-adjoint operators. J. Funct. Anal. 169 (1999), no. 1, 32–51.

[19] Albeverio S., Koshmanenko V. On the problem of the right Hamiltonian under singular form-sum perturbations. Rev. Math. Phys. 12 (2000), no. 1, 1–24.

[20] Albeverio S., Koshmanenko V. On Schrödinger operators perturbed by fractal potentials. Rep. Math. Phys. 45 (2000), no. 3, 307–326.

[21] Albeverio S., Koshmanenko V., Kurasov P., Nizhnik L. On approximations of rank one H_2-perturbations. Proc. Amer. Math. Soc. 131 (2003), no. 5, 1443–1452.

[22] Albeverio S., Koshmanenko V., Kuzhel S. On a variant of abstract scattering theory in terms of quadratic forms. Rep. Math. Phys. 54 (2004), no. 3, 309–325.

[23] Albeverio S., Kurasov P. Singular Perturbations of Differential Operators. Solvable Schrödinger Type operators. London Mathematical Society Lecture Note Series, 271. Cambridge University Press, Cambridge, 2000. xiv+429 pp.

[24] Albeverio S., Kurasov P. Rank one perturbations, approximations, and self-adjoint extensions. J. Funct. Anal. 148 (1997), no. 1, 152–169.

[25] Albeverio S., Kuzhel S., Nizhnik L. Singular perturbed self-adjoint operators in scales of Hilbert spaces. Ukrain. Mat. Zh. 59 (2007), no. 6, 723–743; translation in Ukrainian Math. J. 59 (2007), no. 6, 787–810.

[26] Albeverio S., Kuzhel S., Nizhnik L.P. On the perturbation theory of selfadjoint operators. Tokyo J. Math. 31 (2008), no. 2, 273–292.

[27] Albeverio S., Nizhnik L. A Schrödinger operator with point interactions on Sobolev spaces. Lett. Math. Phys. 70 (2004), no. 3, 185–194.

[28] Alonso A., Simon B. The Birman–Kreĭn–Vishik theory of selfadjoint extensions of semibounded operators. J. Operator Theory 4 (1980), no. 2, 251–270.

[29] Adams D.R., Hedberg L.I. Function Spaces and Potential Theory. Grundlehren der Mathematischen Wissenschaften [Fundamental Principles of Mathematical Sciences], 314. Springer-Verlag, Berlin, 1996. xii+366 pp.

[30] Ando T., Nishio K. Positive selfadjoint extensions of positive symmetric operators. Tohoku Math. J. 22(2), (1970), 65–75.

[31] Ando T. Lebesgue-type decomposition of positive operators. Acta Sci. Math. (Szeged) 38 (1976), no. 3–4, 253–260.

[32] Akhiezer N.I., Glazman I.M. Theory of linear operators in Hilbert space. Vol. I. Translated from the Russian by Merlynd Nestell Frederick Ungar Publishing Co., New York 1961 xi+147 pp.

[33] Akhiezer N.I., Glazman I.M. Theory of Linear Operators in Hilbert space. Vol. II. Translated from the Russian by Merlynd Nestell, Frederick Ungar Publishing Co., New York, 1963. v+218 pp.

[34] Arlinskiĭ Yu.M. Positive spaces of boundary values and sectorial extensions of a nonnegative symmetric operator. (Russian) Ukrain. Mat. Zh. 40 (1988), no. 1, 8–14, 132; translation in Ukrainian Math. J. 40 (1988), no. 1, 5–10.

[35] Arlinskiĭ Yu.M., Kaplan V.L. Selfadjoint biextensions in rigged Hilbert spaces. (Russian) Functional analysis, No. 29 (Russian), 11–19, Ul'yanovsk. Gos. Ped. Inst., Ul'yanovsk, 1989.

[36] Arlinskii Yu.M., Tsekanovskii E.R. Some remarks on singular perturbations of self-adjoint operators. Methods Funct. Anal. Topology 9 (2003), no. 4, 287–308.

[37] Arlinskii Yu., Tsekanovskii E. Linear systems with Schrödinger operators and their transfer functions. In: Current Trends in Operator Theory and its Applications, 47–77, Oper. Theory Adv. Appl., 149, Birkhäuser, Basel, 2004.

[38] Arlinskii Yu., Tsekanovskii E.M. Kreĭn's research on semi-bounded operators, its contemporary developments, and applications. In: Modern Analysis and Applications. The Mark Kreĭn Centenary Conference. Vol. 1: Operator Theory and Related Topics, 65–112, Oper. Theory Adv. Appl., 190, Birkhäuser Verlag, Basel, 2009.

[39] Arlinskii Yu., Belyi S., Tsekanovskii E. Conservative Realizations of Herglotz–Nevanlinna Functions. Oper. Theory: Adv. Appl., 217. Birkhäuser/Springer Basel AG, Basel, 2011. xviii+528 pp.

[40] Arlinskiĭ Yury, Belyi Sergey Non-negative self-adjoint extensions in rigged Hilbert space. In: Concrete Operators, Spectral Theory, Operators in Har-

monic Analysis and Approximation, 11–41, Oper. Theory Adv. Appl., 236, Birkhäuser/Springer, Basel, 2014.

[41] Banach Stefan. Théorie des opérations linéaires. (French) Chelsea Publishing Co., New York, 1955. vii+254 pp.

[42] Berezanskii Ju.M. Expansions in Eigenfunctions of Selfadjoint Operators. Translated from the Russian by R. Bolstein, J.M. Danskin, J. Rovnyak and L. Shulman. Translations of Mathematical Monographs, Vol. 17 American Mathematical Society, Providence, R.I. 1968. ix+809 pp.

[43] Berezans'kii Yu.M. Bilinear forms and Hilbert riggings. (Russian) In: Spectral Analysis of Differential Operators, pp. 83–106, 134, Akad. Nauk Ukrain. SSR, Inst. Mat., Kiev, 1980.

[44] Berezanskii Yu.M. Selfadjoint Operators in Spaces of Functions of Infinitely Many Variables. Translated from the Russian by H.H. McFaden. Translation edited by Ben Silver. Translations of Mathematical Monographs, 63. American Mathematical Society, Providence, RI, 1986. xiv+383 pp.

[45] Berezanskii Yu.M., Kondrat'ev Yu.G. Spektral'nye Metody v Beskonechnomernom Analize. (Russian) [Spectral Methods in Infinite-dimensional analysis] "Naukova Dumka", Kiev, 1988. 680 pp.

[46] Berezansky Y.M., Kondratiev Y.G. Spectral Methods in infinite-dimensional Analysis. Vol. 1. Translated from the 1988 Russian original by P.V. Malyshev and D.V. Malyshev and revised by the authors. Mathematical Physics and Applied Mathematics, 12/1. Kluwer Academic Publishers, Dordrecht, 1995. xviii+576 pp.

[47] Berezansky Y.M., Kondratiev Y.G. Spectral Methods in Infinite-dimensional Analysis. Vol. 2. Translated from the 1988 Russian original by P.V. Malyshev and D.V. Malyshev and revised by the authors. Mathematical Physics and Applied Mathematics, 12/2. Kluwer Academic Publishers, Dordrecht, 1995. viii+432 pp.

[48] Berezansky Y.M., Sheftel Z.G., Us G.F. Functional Analysis. Vols. 1, 2. (Third edition) Institute of Mathematics NAS of Ukraine, Kyiv 2010. First edition: Functional'nyj Analiz, Vyshcha Shkola, Kiev, 1990. 600 p.; Second edition: Functional analysis. Vols. 1, 2. Translated from the 1990 Russian original. Oper. Theory Adv. App., 86. Birkhäuser Verlag, Basel, 1996. xvi+293 pp.

[49] Berezansky Y.M., Sheftel Z.G., Us G.F. Functional Aanalysis. Vol. I. Translated from the 1990 Russian original by Peter V. Malyshev. Oper. Theory Adv. App., 85. Birkhäuser Verlag, Basel, 1996. xx+423 pp.

[50] Berezansky Y.M., Sheftel Z.G., Us G.F. Functional Analysis. Vol. II. Translated from the 1990 Russian original by Peter V. Malyshev. Oper. Theory Adv. App., 86. Birkhäuser Verlag, Basel, 1996. xvi+293 pp.

[51] Berezansky Yu.M., Brasche J., Nizhnik L.P. On generalized selfadjoint operators on scales of Hilbert spaces. Methods Funct. Anal. Topology 17 (2011), no. 3, 193–198.

[52] Berezansky Yu.M., Brasche J. Generalized selfadjoint operators and their singular perturbations. Methods Funct. Anal. Topology 8 (2002), no. 4, 1–14.

[53] Berezin F.A., Faddeev L.D. Remark on the Schrödinger equation with singular potential. (Russian) Dokl. Akad. Nauk SSSR 137 (1961), 1011–1014.

[54] Birman M.S. On the theory of self-adjoint extensions of positive definite operators. (Russian) Mat. Sb. N.S. 38(80) (1956), 431–450.

[55] Birman M.S. On the spectrum of singular boundary-value problems. (Russian) Mat. Sb. (N.S.) 55 (97) 1961 125–174.

[56] Birman M.S., Solomjak M.Z. Spektral'naya Teoriya Samosopryazhennykh Operatorov v Gil'bertovom Prostranstve. (Russian) [Spectral Theory of Selfadjoint Operators in Hilbert Space] Leningrad. Univ., Leningrad, 1980. 264 pp.

[57] Bozhok R.V. On the defect of nondenseness of continuous embeddings in the scale of Hilbert spaces. (Ukrainian) Ukrain. Mat. Zh. 60 (2008), no. 5, 704–708; translation in Ukrainian Math. J. 60 (2008), no. 5, 810–815.

[58] Bozhok R.V., Koshmanenko V.D. Singular perturbations of selfadjoint operators associated with rigged Hilbert spaces. (Ukrainian) Ukrain. Mat. Zh. 57 (2005), no. 5, 622–632; translation in Ukrainian Math. J. 57 (2005), no. 5, 738–750.

[59] Bozhok R.V., Koshmanenko V.D. Parametrization of supersingular perturbations in the method of rigged Hilbert spaces. Russ. J. Math. Phys. 14 (2007), no. 4, 409–416.

[60] Brasche J.F., Exner P., Kuperin Yu.A., Seba P. Schrödinger operators with singular interactions. J. Math. Anal. Appl. 184 (1994), no. 1, 112–139.

[61] Brasche J., Nizhnik L. A generalized sum of quadratic forms. Methods Funct. Anal. Topology 8 (2002), no. 3, 13–19.

[62] Brasche J.F., Koshmanenko V., Neidhardt H. New aspects of Kreĭn's extension theory. Ukrain. Mat. Zh. 46 (1994), no. 1-2, 37–54; translation in Ukrainian Math. J. 46 (1994), no. 1-2, 34–53.

[63] Brasche J.F., Malamud M., Neidhardt H. Weyl function and spectral properties of self-adjoint extensions. Integral Equations Operator Theory 43 (2002), no. 3, 264–289.

[64] Bruk V.M. Extensions of symmetric relations. (Russian) Mat. Zametki 22 (1977), no. 6, 825–834.

[65] Cekanovskiĭ, É.R., Šmul'jan, Ju.L. Questions in the theory of the extension of unbounded operators in rigged Hilbert spaces. (Russian) Mathematical

analysis, Vol. 14 (Russian), pp. 59–100, i. (loose errata). Akad. Nauk SSSR Vsesojuz. Inst. Nauch. i Tekhn. Inform., Moscow, 1977.

[66] Cojuhari P. Estimates of the number of perturbed eigenvalues. In: Operator Theory, Operator Algebras and Related Topics, Proc. of OT 17 Conf., 97–111, Bucharest, 2000.

[67] Cojuhari P., Gheondea A. Closed embeddings of Hilbert spaces. (English summary) J. Math. Anal. Appl. 369 (2010), no. 1, 60–75.

[68] Coddington E.A., de Snoo H, Hendrik S.V. Positive selfadjoint extensions of positive symmetric subspaces. Math. Z. 159 (1978), no. 3, 203–214.

[69] Chernoff P.R., Hughes R.J. A new class of point interactions in one dimension. J. Funct. Anal. 111 (1993), no. 1, 97–117.

[70] Derkach V A., Hassi S., Malamud M.M., de Snoo H. Weyl functions and intermediate extensions of symmetric operators. Dopov. Nats. Akad. Nauk Ukr. Mat. Prirodozn. Tekh. Nauki 2001, no. 10, 33–39.

[71] Derkach V., Hassi S., Malamud M., de Snoo H. Boundary relations and their Weyl families. Trans. Amer. Math. Soc. 358 (2006), no. 12, 5351–5400.

[72] Derkach V.A., Hassi S., Malamud M.M., de Snoo H.S.V. Generalized resolvents of symmetric operators and admissibility. Methods Funct. Anal. Topology 6 (2000), no. 3, 24–55.

[73] Derkach V.A., Malamud M.M. Generalized resolvents and the boundary value problems for Hermitian operators with gaps. J. Funct. Anal. 95 (1991), no. 1, 1–95.

[74] Derkach V.A., Malamud M.M. The extension theory of Hermitian operators and the moment problem. Analysis. 3. J. Math. Sci. 73 (1995), no. 2, 141–242.

[75] Dunford, Nelson, Schwartz, Jacob T. Linear operators. Part II. Spectral theory. Selfadjoint Operators in Hilbert Space. With the assistance of William G. Bade and Robert G. Bartle. Reprint of the 1963 original. Wiley Classics Library. A Wiley-Interscience Publication. John Wiley and Sons, Inc., New York, 1988.

[76] Dudkin M.E., Koshmanenko, V.D. On the point spectrum of selfadjoint operators that arise under singular perturbations of finite rank. (Ukrainian) Ukrain. Mat. Zh. 55 (2003), no. 9, 1269–1276; translation in Ukrainian Math. J. 55 (2003), no. 9, 1532–1541.

[77] Dudkin N.E., Koshmanenko V.D. Commutativity properties of singularly perturbed operators. (Russian) Teoret. Mat. Fiz. 102 (1995), no. 2, 183–197; translation in Theoret. Math. Phys. 102 (1995), no. 2, 133–143.

[78] Eyvazov E.H. Higher order differential operators with finite number of δ-interactions in multivariate case. Journal of Mathematical Physics, Analysis, Geometry. 6 (2010), no. 3, 266–276, 337–340.

[79] Exner P., Kondej S. Strong-coupling asymptotic expansion for Schrödinger operators with a singular interaction supported by a curve in R^3. Rev. Math. Phys. 16 (2004), no. 5, 559–582.

[80] Exner P., Kondej S. Hiatus perturbation for a singular Schrödinger operator with an interaction supported by a curve in R^3. J. Math. Phys. 49 (2008), no. 3, 032111, 19 pp.

[81] Exner P., Neidhardt H., Zagrebnov V.A. Potential approximations to δ': an inverse Klauder phenomenon with norm-resolvent convergence. Comm. Math. Phys. 224 (2001), no. 3, 593–612.

[82] Faris W.G. Self-adjoint Operators. Lecture Notes in Mathematics, Vol. 433. Springer-Verlag, Berlin-New York, 1975. vii+115 pp.

[83] Feshchenko I. Private communication, 2012.

[84] Fedik M.M. An operator that is conjugate to a related sesquilinear form. (Ukrainian) Visnik L'viv. Derzh. Univ. Ser. Mekh.-Mat. No. 28 (1987), 43–46,

[85] Fragela A.K. Perturbation of a polyharmonic operator with delta-like potentials. (Russian) Mat. Sb. (N.S.) 130(172) (1986), no. 3, 386–393, 431–432.

[86] Friedrichs K. Spektraltheorie halbbeschränkter Operatoren und Anwendung auf die Spektralzerlegung von Differentialoperatoren. (German) Math. Ann. 109 (1934), no. 1, 465–487.

[87] Gel'fand I.M., Vilenkin N.Ja. Obobshchennye Funktsii, Vyp. 4. Nekotorye Primeneniya Garmonicheskogo Analiza. Osnashchennye Gil'bertovy Prostranstva. (Russian) [Generalized Functions, No. 4. Some Applications of Harmonic Analysis. Rigged Hilbert spaces] Gosudarstv. Izdat. Fiz.-Mat. Lit., Moscow] 1961 472 pp.

[88] Gel'fand I.M., Shilov G.E. Generalized Functions. Vol. 2. Spaces of Fundamental and Generalized Functions. Translated from the Russian by Morris D. Friedman, Amiel Feinstein and Christian P. Peltzer. Academic Press [Harcourt Brace Jovanovich, Publishers], New York-London, 1968 [1977]. x+261 pp.

[89] Gesztesy F., Simon B. Rank-one perturbations at infinite coupling. J. Funct. Anal. 128 (1995), no. 1, 245–252.

[90] Golovaty Yu.D., Hryniv R.O. On norm resolvent convergence of Schrödinger operators with δ'-like potentials. J. Phys. A 43 (2010), no. 15, 155204, 14 pp.

[91] Gorbacuk M.L., Gorbacuk V.I. The scattering problem for first-order differential equations with operator coefficients. (Russian) Ukrain. Mat. Zh. 30 (1978), no. 4, 452–461.

[92] Gorbachuk V.I., Gorbachuk M.L. Boundary Value Problems for Operator Differential Equations. Translated and revised from the 1984 Russian origi-

nal. Mathematics and its Applications (Soviet Series), 48. Kluwer Academic Publishers Group, Dordrecht, 1991. xii+347 pp.

[93] Gorbacuk M.L., Mihailec V.A. Semibounded selfadjoint extensions of symmetric operators. (Russian) Dokl. Akad. Nauk SSSR 226 (1976), no. 4, 765–767.

[94] Golovaty Yuriy D., Man'ko Stepan S. Corrigendum: "Solvable models for the Schrödinger operators with δ'-like potentials". Ukr. Mat. Visn. 8 (2011), no. 3, 458–459.

[95] Hassi S., de Snoo H. On rank one perturbations of selfadjoint operators. Integral Equations Operator Theory 29 (1997), no. 3, 288–300.

[96] Hassi S., Sebestyen Z., de Snoo H. Lebesgue type decompositions for nonnegative forms. J. Funct. Anal. 257 (2009), no. 12, 3858–3894.

[97] Hryniv R.O., Mykytyuk Ya.V. 1-D Schrödinger operators with periodic singular potentials. Methods Funct. Anal. Topology 7 (2001), no. 4, 31–42.

[98] Karataeva T.V., Koshmanenko V.D. Generalized sum of operators. (Russian) Mat. Zametki 66 (1999), no. 5, 671–681; translation in Math. Notes 66 (1999), no. 5–6, (2000), 556–564

[99] Karwowski W., Koshmanenko V.D. Additive regularization of singular bilinear forms. (Russian) Ukrain. Mat. Zh. 42 (1990), no. 9, 1199–1204; translation in Ukrainian Math. J. 42 (1990), no. 9, 1066–1069.

[100] Karwowski W., Koshmanenko V. On the definition of singular bilinear forms and singular linear operators. Ukrain. Mat. Zh. 45 (1993), no. 8, 1084–1089; translation in Ukrainian Math. J. 45 (1993), no. 8, (1994), 1208–1214.

[101] Karwowski W., Koshmanenko V.D. Regular restrictions of singular bilinear forms. (Russian) Funktsion. Anal. Prilozhen. 29 (1995), no. 2, 79–81; translation in Funct. Anal. Appl. 29 (1995), no. 2, 136–137.

[102] Karwowski W., Koshmanenko V. Singular quadratic forms: regularization by restriction. J. Funct. Anal. 143 (1997), no. 1, 205–220.

[103] Karwowski W., Koshmanenko V. Schrödinger operator perturbed by dynamics of lower dimension. In: Differential Equations and Mathematical Physics (Birmingham, AL, 1999), 249–257, AMS/IP Stud. Adv. Math., 16, Amer. Math. Soc., Providence, RI, 2000.

[104] Karwowski W., Koshmanenko V., Ôta, S. Schrödinger operator perturbed by operators related to null sets. Positivity 2 (1998), no. 1, 77–99.

[105] Karwowski W., Koshmanenko V. The generalized Laplace operator in $L_2(\mathbb{R}^n)$. In: Stochastic Processes, Physics and Geometry: New Interplays, II (Leipzig, 1999), 385–393, CMS Conf. Proc., 29, Amer. Math. Soc., Providence, RI, 2000.

[106] Karwowski W., Kondej S. The Laplace operator, null set perturbations and boundary conditions. In: Operator Methods in Ordinary and Partial Dif-

ferential Equations (Stockholm, 2000), 233–244, Oper. Theory Adv. Appl., 132, Birkhäuser, Basel, 2002.

[107] Kato Tosio. Perturbation Theory for Linear Operators. Die Grundlehren der mathematischen Wissenschaften, Band 132 Springer-Verlag New York, Inc., New York, 1966. xix+592 pp.

[108] Kiselev A., Simon B. Rank one perturbations with infinitesimal coupling. J. Funct. Anal. 130 (1995), no. 2, 345–356.

[109] Kochubeĭ, A.N. Elliptic operators with boundary conditions on a subset of measure zero. (Russian) Funktsion. Anal. Prilozhen. 16 (1982), no. 2, 74–75.

[110] Kochubei A.N. On extensions of nondnsely defined symmetric operators. Sibirsk. Matem. Zh. 18.2 (1977). 314–320.

[111] Kondej S. On the eigenvalue problem for self-adjoint operators with singular perturbations. Math. Nachr. 244 (2002), 150–169.

[112] Kondej S. Hamiltonian with delta type interaction: perturbation by dynamics. J. Phys. Stud. 15 (2011), no. 1, 1006, 4 pp.

[113] Košhmanenko V.D. An operator representation for nonclosable quadratic forms, and the scattering problem. (Russian) Dokl. Akad. Nauk SSSR 245 (1979), no. 2, 295–298.

[114] Košhmanenko V.D. Formulation of the scattering problem with singular perturbation in rigged spaces. (Russian) In: Operators of Mathematical Physics and Infinite-dimensional Analysis (Russian), pp. 58–72, 142, Akad. Nauk Ukrain. SSR, Inst. Mat., Kiev, 1979.

[115] Košhmanenko V.D. Spectral decompositions in the problem of scattering with singular perturbation. (Russian) Dokl. Akad. Nauk SSSR 252 (1980), no. 3, 531–535.

[116] Koshmanenko V.D. Singular bilinear forms and selfadjoint extensions of symmetric operators. (Russian) In: Spectral Analysis of Differential Operators, pp. 37–48, 133, Akad. Nauk Ukrain. SSR, Inst. Mat., Kiev, 1980.

[117] Koshmanenko V.D. Closable extensions of bilinear forms with exit to a new space. (Russian) Mat. Zametki 30 (1981), no. 6, 857–864, 959.

[118] Koshmanenko V.D. Classification of Singularly Perturbed Operators. Preprint, Institute of Mathematics, Kyiv, NAS Ukraine, (1982) no. 82.34, 47 p.

[119] Koshmanenko V.D. Singular perturbations defined by forms, Application of Self-adjoint Extensions in Quantum Physics Eds. P. Exner, P. Šeba, Lecture Notes in Phys., **324**, Springer (1987), 55–66.

[120] Koshmanenko V.D. Uniqueness of a singularly perturbed operator. (Russian) Dokl. Akad. Nauk SSSR 300 (1988), no. 4, 786–789; translation in Soviet Math. Dokl. 37 (1988), no. 3, 718–721

[121] Koshmanenko V.D. Perturbations of selfadjoint operators by singular bilinear forms. (Russian) Ukrain. Mat. Zh. 41 (1989), no. 1, 3–19, 134; translation in Ukrainian Math. J. 41 (1989), no. 1, 1–14

[122] Koshmanenko V.D. δ-potential perturbations of an n-dimensional Laplace operator. (Russian) Translated in Selecta Math. Soviet. 9 (1990), no. 4, 355–364. The spectral theory of operator-differential equations (Russian), 70–79, v, Akad. Nauk Ukrain. SSR, Inst. Mat., Kiev, 1986.

[123] Koshmanenko Y.D. A contribution to the theory of rank-one singular perturbations of selfadjoint operators. Ukrain. Mat. Zh. 43 (1991), no. 11, 1559–1566; translation in Ukrainian Math. J. 43 (1991), no. 11, 1450–1457 (1992)

[124] Koshmanenko V.D. Singular operators and forms in the scale of Hilbert spaces. In: Methods of Functional Analysis in Problems of Mathematical Physics, 73–87, Acad. Sci. Ukraine, Inst. Math., Kiev, 1992.

[125] Koshmanenko V.D. Dense subspace in A-scale of the Hilbert space, Preprint ITP UWr 835/93, Wroclaw, 22 p. (1993).

[126] Koshmanenko V. Singularly perturbed operators. In: Mathematical Results in Quantum Mechanics (Blossin, 1993), 347–351, Oper. Theory Adv. Appl., 70, Birkhäuser, Basel, 1994.

[127] Koshmanenko V.D. Singular Wick monomials in an orthogonally extended Fock space [translation of Boundary Value Problems for Differential Equations (Russian), 78–82, iv, Akad. Nauk Ukrain. SSR, Inst. Mat., Kiev, 1988. Selected translations. Selecta Math. 13 (1994), no. 1, 17–25.

[128] Koshmanenko V.D. When the n-power of symmetric operator has dense domain, Proceedings of Seventh Crimean Autumn Math. School-Symposium, Simferopol, Crimea, Ukraine, **5**, 20–22 (1996).

[129] Koshmanenko V.D. Form-sum approximations in singular perturbation theory. In: Proceedings of Seventh Crimean Autumn Math. School-Symposium, Simferopol, Crimea, Ukraine, **7**, 2–5 (1996).

[130] Koshmanenko V. Singularly perturbed operators of type $-\Delta + \lambda\delta$. In: Algebraic and Geometric Methods in Mathematical Physics (Kaciveli, 1993), 433–437, Math. Phys. Stud., 19, Kluwer Acad. Publ., Dordrecht, 1996.

[131] Koshmanenko V.D., Ôta Sh. On the characteristic properties of singular operators. (Ukrainian) Ukrain. Mat. Zh. 48 (1996), no. 11, 1484–1493; translation in Ukrainian Math. J. 48 (1996), no. 11, (1997), 1677–1687.

[132] Koshmanenko V.D., Samoilenko O.V. Singular perturbations of finite rank. Point spectrum. (Ukrainian) Ukrain. Mat. Zh. 49 (1997), no. 9, 1186–1194; translation in Ukrainian Math. J. 49 (1997), no. 9, (1998), 1335–1344.

[133] Koshmanenko V.D. Singular perturbations with an infinite coupling constant. (Russian) Funkt. Anal. Prilozhen. 33 (1999), no. 2, 81–84; translation in Funct. Anal. Appl. 33 (1999), no. 2, 148–150.

[134] Koshmanenko V.D. Singulyarnye Bilineinye Formy v Teorii Vozmushchenii Samosopryazhennykh Operatorov. (Russian) [Singular Bilinear Forms in the Theory of Perturbations of Selfadjoint Operators] "Naukova Dumka", Kiev, 1993. 176 pp.

[135] Koshmanenko V. Singular Quadratic Forms in Perturbation Theory. Translated from the 1993 Russian original. Mathematics and its Applications, 474. Kluwer Academic Publishers, Dordrecht, 1999. viii+308 pp.

[136] Koshmanenko V.D. On negative eigenvalue problem under singular perturbations of self-adjoint operators. In: Intern. Conference "Spectral Analysis of Second Order Differential and Difference Operators", Banach Center, Warsaw, p. 11, (2000)

[137] Koshmanenko V. Singular operator as a parameter of self-adjoint extensions. Operator Theory and Related Topics, Vol. II (Odessa, 1997), 205–223, Oper. Theory Adv. Appl., 118, Birkhäuser, Basel, 2000.

[138] Koshmanenko V.D. Regularized approximations of singular perturbations from the $\mathcal{H}_{-2}$-class. (Ukrainian) Ukrain. Mat. Zh. 52 (2000), no. 5, 626–637; translation in Ukrainian Math. J. 52 (2000), no. 5, 715–728 (2001)

[139] Koshmanenko V. Towards the spectral analysis of Schrödinger operator with fractal perturbation. In: Partial Differential Equations and Spectral Theory (Clausthal, 2000), 169–178, Oper. Theory Adv. Appl., 126, Birkhäuser, Basel, 2001.

[140] Koshmanenko V. A variant of inverse negative eigenvalues problem in singular perturbation theory. Methods Funct. Anal. Topology 8 (2002), no. 1, 49–69.

[141] Koshmanenko V. Distribution of negative eigenvalues for singular perturbations and associated Jacobi matrices. In: Intern. Conf. "Operator Theory and Applications in Mathematical Physics". Bedlewo, p. 15 (2002).

[142] Koshmanenko V.D., Tugaĭ G.V. On the structure of the resolvent of a singularly perturbed operator that solves the eigenvalue problem. (Ukrainian) Ukrain. Mat. Zh. 56 (2004), no. 9, 1292–1297 (2005); translation in Ukrainian Math. J. 56 (2004), no. 9, 1538–1545.

[143] Koshmanenko V. Construction of singular perturbations by the method of rigged Hilbert spaces. J. Phys. A 38 (2005), no. 22, 4999–5009.

[144] Koshmanenko V.D., Tugaĭ G.V. Jacobi matrices associated with the inverse eigenvalue problem in the theory of singular perturbations of selfadjoint operators. (Ukrainian) Ukrain. Mat. Zh. 58 (2006), no. 12, 1651–1662; translation in Ukrainian Math. J. 58 (2006), no. 12, 1876–1890.

[145] Kostenko A.S., Malamud M.M. On the one-dimensional Schrödinger operator with δ-interactions. (Russian) Funkts. Anal. Prilozhen. 44 (2010), no. 2, 87–91; translation in Funct. Anal. Appl. 44 (2010), no. 2, 151–155.

[146] Kostenko A.S., Malamud M.M. Schrödinger operators with local point interactions on a discrete set. J. Differential Equations 249 (2010), no. 2, 253–304.

[147] Krasnosel'skii M.A. On self-adjoint extensions of Hermitian operators. (Russian) Ukrain. Mat. Zurnal 1, (1949). no. 1, 21–38.

[148] Kreĭn M. The theory of self-adjoint extensions of semi-bounded Hermitian transformations and its applications. I. (Russian) Rec. Math. [Mat. Sbornik] N.S. 20(62), (1947). 431–495.

[149] Kreĭn M G., Yavryan V.A. Spectral shift functions that arise in perturbations of a positive operator. (Russian) J. Operator Theory 6 (1981), no. 1, 155–191.

[150] Kurasov P. Distribution theory for discontinuous test functions and differential operators with generalized coefficients. J. Math. Anal. Appl. 201 (1996), no. 1, 297–323.

[151] Kurasov P., Pavlov Yu.V. On field theory methods in singular perturbations theorey. Lett. Math. Phys. 64 (2003), no. 2, 171–184.

[152] Kuzhel A.V., Kuzhel S.A. Regular Extensions of Hermitian Operators. Translated from the Russian. VSP, Utrecht, 1998. xii+273 pp.

[153] Lax P.D. Symmetrizable linear transformations. Comm. Pure Appl. Math., 7, no 4, (1954), 633–647.

[154] Lyantse V.E., Storozh O.G. Metody teorii neogranichennykh operatorov. (Russian) [Methods of the theory of unbounded operators] "Naukova Dumka", Kiev, 1983. 211 pp.

[155] Lyantse V.E., Fedik M.N. Properties of operators generated by sesquilinear forms. (Russian) Dokl. Akad. Nauk Ukrain. SSR Ser. A (1987), no. 2, 27–29, 87.

[156] Malamud M.M., Neidhardt H. On the unitary equivalence of absolutely continuous parts of self-adjoint extensions. J. Funct. Anal. 260 (2011), no. 3, 613–638.

[157] Maz'ja Vladimir G. Sobolev spaces. Translated from the Russian by T.O. Shaposhnikova. Springer Series in Soviet Mathematics. Springer-Verlag, Berlin, 1985. xix+486 pp.

[158] Mikhailets V.A. Spectral properties of the one-dimensional Schrödinger operator with point intersections. In: Proceedings of the XXVII Symposium on Mathematical Physics (Torun, 1994). Rep. Math. Phys. 36 (1995), no. 2-3, 495–500.

[159] Moren K. Metody Gil'bertova Prostranstva. (Russian) [Hilbert space Methods] Translated from the Polish by V.E. Ljance. Edited by Yu.M. Berezanskii and E.A. Gorin Izdat. "Mir", Moscow 1965 570 pp.

[160] Neĭman-z M.I., Shkalikov A.A. Schrödinger operators with singular potentials from spaces of multipliers. (Russian) Mat. Zametki 66 (1999), no. 5, 723–733; translation in Math. Notes 66 (1999), no. 5-6, (2000), 599–607.

[161] von Neumann J. Allgemeine Eigenwertstheorie Hermitescher Funktinaloperatoren,. Math. Ann., 102 (1929) 49-131.

[162] Nizhnik L.P. On rank one singular perturbations of selfadjoint operators. Methods Funct. Anal. Topology 7 (2001), no. 3, 54–66.

[163] Nizhnik L.P. On point interaction in quantum mechanics. (Russian) Ukrain. Mat. Zh. 49 (1997), no. 11, 1557–1560; translation in Ukrainian Math. J. 49 (1997), no. 11, 1752–1755 (1998)

[164] Nizhnik L.P. A one-dimensional Schrödinger operator with point interactions in Sobolev spaces. (Russian) Funkt. Anal. Prilozhen. 40 (2006), no. 2, 74–79; translation in Funct. Anal. Appl. 40 (2006), no. 2, 143–147

[165] Ôta S. Erratum: "On a singular part of an unbounded operator" Z. Anal. Anwendungen 7 (1988), no. 5, 417.

[166] Plesner A.I. Spectral Theory of Linear Operators. Vols. I, II. Translated from the Russian by Merlynd K. Nestell and Alan G. Gibbs, Frederick Ungar Publishing Co., New York 1969 Vol. I: xii+235 pp. Vol. II: viii+271 pp.

[167] Posilicano A. A Kreĭn-like formula for singular perturbations of self-adjoint operators and applications. J. Funct. Anal. 183 (2001), no. 1, 109–147.

[168] Riesz F., Sz.-Nagy B. Functional Analysis. Translated from the second French edition by Leo F. Boron. Reprint of the 1955 original. Dover Books on Advanced Mathematics. Dover Publications, Inc., New York, 1990. xii+504 pp.

[169] Reed M., Simon B. Methods of Modern Mathematical Physics. I. Functional Analysis. Second edition. Academic Press, Inc. [Harcourt Brace Jovanovich, Publishers], New York, 1980. xv+400 pp.

[170] Reed M., Simon B. Methods of Modern Mathematical Physics. II. Fourier Analysis, Self-adjointness. Academic Press [Harcourt Brace Jovanovich, Publishers], New York-London, 1975. xv+361 pp.

[171] Reed M., Simon B. Methods of Modern Mathematical Physics. III. Scattering Theory. Academic Press [Harcourt Brace Jovanovich, Publishers], New York-London, 1979. xv+463 pp.

[172] Reed M., Simon B. Methods of modern mathematical physics. IV. Analysis of Operators. Academic Press [Harcourt Brace Jovanovich, Publishers], New York-London, 1978. xv+396 pp.

[173] Richtmyer R.D. Principles of Advanced Mathematical Physics. Vol. I. Texts and Monographs in Physics. Springer-Verlag, New York-Heidelberg, 1978. xv+422 pp.

[174] Rofe-Beketov F.S. Selfadjoint extensions of differential operators in a space of vector-valued functions. (Russian) Dokl. Akad. Nauk SSSR 184 1969 1034–1037.

[175] Savchuk A.M., Shkalikov A.A. Sturm-Liouville operators with singular potentials. (Russian) Mat. Zametki 66 (1999), no. 6, 897–912; translation in Math. Notes 66 (1999), no. 5-6, 741–753 (2000)

[176] Savchuk A.M., Shkalikov A.A. Inverse problems for the Sturm-Liouville operator with potentials in Sobolev spaces: uniform stability. (Russian) Funkt.

Anal. Prilozhen. 44 (2010), no. 4, 34–53; translation in Funct. Anal. Appl. 44 (2010), no. 4, 270–285

[177] Simon B. A canonical decomposition for quadratic forms with applications to monotone convergence theorems. J. Funct. Anal. 28 (1978), no. 3, 377–385.

[178] Stone M.H. Linear Transformations in Hilbert Space. Reprint of the 1932 original. American Mathematical Society Colloquium Publications, 15. American Mathematical Society, Providence, RI, 1990. viii+622 pp.

[179] Straus A.V. Extensions of a semi-bounded operator. (Russian) Dokl. Akad. Nauk SSSR 211 (1973), 543–546.

[180] Triebel H. Fractals and Spectra. Related to Fourier Analysis and Function Spaces. Modern Birkhäuser Classics. Birkhäuser Verlag, Basel, 2011. viii+271 pp.

[181] Višhik, M.I. On general boundary problems for elliptic differential equations. (Russian) Trudy Moskov. Mat. Obsc. 1, (1952). 187–246.

[182] Vladimirov V.S. Generalized Functions in Mathematical Physics. Moscow, Mir Publishers (1979). xii+362 pp.

[183] Vladimirov V.S. Equations of Mathematical Physics. (2nd ed.), Moscow, Mir Publishers (1983). 464 pp.

[184] Zolotaryuk A.V. Point interactions of the dipole type defined through a three-parametric power regularization. J. Phys. A 43 (2010), no. 10, 105302, 21 pp.

Subject Index

Notation Index

Zeitfracht Medien GmbH
Ferdinand-Jühlke-Straße 7
99095 Erfurt, Deutschland
produktsicherheit@kolibri360.de